# Principles of Turbomachinery

Second edition

R.K. TURTON
Senior Lecturer in Mechanical Engineering
Loughborough University of Technology

Springer

First edition 1984

Second edition 1995

ISBN 0-412-60210-5



A catalogue record for this book is available from the British Library.

Library of Congress Catalog Card Number: 94-72652

**Turton: Principles of Turbomachinery, Second edition**



First Indian Reprint 2010
Second Indian Reprint 2013

ISBN 978-81-8489-604-6



This edition is published by Springer (India) Private Limited, (a part of Springer Science+Business Media), Registered Office: 7th Floor, Vijaya Building, 17 Barakhamba Road, New Delhi – 110 001, India.

Printed in India by Rashtriya Printers, Delhi.

# Contents

# Preface to the second edition

The objectives outlined in the preface to the first edition have remained unchanged in preparing this edition as they have continued to be the basis of my teaching programme. This edition is therefore not radically different from the first, which to my pleasure and relief was well received by those who obtained and used the book.

I have taken the opportunity to correct errors that occurred, have improved some diagrams and added others, and brought all the material on cavitation together into Chapter 3: I hope that this gives a more connected account of this very important topic. I have added some updated material in places, have added some references, and hope that by this means the reader can pursue some topics in more depth after reading this introduction.

The worked examples that were included in the text have been retained, and extra exercises have been added where students have commented on the need for further clarification. A major change has been the addition of sets of problems for solution by the reader. These are given at the end of all chapters but four, five and ten. These are based in most cases on the questions set over the years in the Finals in the course on Turbomachinery at Loughborough University of Technology, and I am grateful for the permission granted by the University authorities to use them. While the problems are placed at the end of each chapter, the solutions are collected together at the end of the book. It is hoped that readers will attempt the problems first and then turn to the end for help.

I hope that this edition is free from error and ambiguity, and as an earnest seeker after truth will be grateful for comments and suggestions.

I must acknowledge the invaluable help of Mrs Janet Redman for her translation of my sketches and of Mrs Gail Kirton who typed the new chapters. Finally, my thanks to my dear wife who has been patient and helpful as always.

# Preface to the first edition

This text outlines the fluid and thermodynamic principles that apply to all classes of turbomachines, and the material has been presented in a unified way. The approach has been used with successive groups of final year mechanical engineering students, who have helped with the development of the ideas outlined. As with these students, the reader is assumed to have a basic understanding of fluid mechanics and thermodynamics. However, the early chapters combine the relevant material with some new concepts, and provide basic reading references.

Two related objectives have defined the scope of the treatment. The first is to provide a general treatment of the common forms of turbomachine, covering basic fluid dynamics and thermodynamics of flow through passages and over surfaces, with a brief derivation of the fundamental governing equations. The second objective is to apply this material to the various machines in enough detail to allow the major design and performance factors to be appreciated. Both objectives have been met by grouping the machines by flow path rather than by application, thus allowing an appreciation of points of similarity or difference in approach. No attempt has been made to cover detailed points of design or stressing, though the cited references and the body of information from which they have been taken give this sort of information.

The first four chapters introduce the fundamental relations, and the succeeding chapters deal with applications to the various flow paths. The last chapter covers the effects of cavitation, solids suspensions, gas content and pumped storage systems, and includes a short discussion of the control of output. These topics have been included to highlight the difficulties encountered when the machine is not dealing with a clean Newtonian fluid, or in systems where problems are posed that can only be solved by compromise. Chapter 5 discusses all the conventional centrifugal machines, covering in a uniform manner the problems faced with liquids and gases: since high pressure rise machines have a number of stages, the ways in which fluid is guided from stage to stage are introduced. Thrust load problems are described and the common solutions adopted are outlined. The discussion of axial machines has been divided between two chapters, as the technologies of pumps, fans and water turbines are similar but differ from those used in compressible machines. Radial flow turbines form the subject matter of Chapter 8, and the common designs in use in industry and in turbochargers are discussed.

Worked examples have been included in all chapters but the last. They are intended to provide illustration of the main points of the text, and to give a feel for both the shape of the velocity triangles and the sizes of the velocity vectors that normally apply. They are of necessity simplified, and must not be regarded as representing current practice in all respects. No problems for student solution have been provided. Teachers normally prefer to devise their own material, and may obtain copies of examination questions set by other institutions if they wish.

As a matter of course the SI system of units has been used throughout, except in some diagrams. To assist the reader, a list of symbols used in the early chapters, together with a statement of the conventional dimensions used, follows the Preface. As far as possible the British Standard on symbols has been followed but, where current and hallowed practice dictates the use of certain symbols, these have been used; it is hoped that where the same symbol appears to have different meanings the context makes the usage clear.

The material presented forms the core of a lecture course of about 46 hours, and the author hopes that in the inevitable distillation no ambiguities have occurred. He will be grateful for comments and suggestions, as he is still an earnest 'seeker after truth'.

Finally, it is necessary to offer some words of thanks, especially to Mrs Redman, who ensured that the diagrams were clear, to Mrs Smith and Mrs McKnight, who helped with the typing, and finally to my dear wife, who was so patient and who ensured that the proof-reading was done properly.

# Symbols used: their meaning and dimensions

| Symbol | Meaning | Dimensions |
|---|---|---|
| $a$ | acoustic velocity | $m\,s^{-1}$ |
| $b$ | passage height | m |
| $C_L$ | lift coefficient (Table 4.1) | |
| $C_D$ | drag coefficient (Table 4.1) | |
| $C_p$ | pressure rise coefficient (equation 4.15) | |
| $C_p$ | specific heat at constant pressure | $kJ\,kg^{-1}\,K^{-1}$ |
| $C_v$ | specific heat at constant volume | $kJ\,kg^{-1}\,K^{-1}$ |
| $D$ | diameter | m |
| $D$ | drag force on an aerofoil | N |
| $F_a$ | force acting in the axial direction on a foil or blade | N |
| $F_t$ | force acting in the tangential direction on a foil or blade | N |
| $g$ | acceleration due to gravity | $m\,s^{-2}$ |
| $gH$ | specific energy | $J\,kg^{-1}$ |
| $h$ | specific enthalpy | $J\,kg^{-1}$ |
| $H$ | head | m of liquid |
| $K$ | lattice coefficient (equation 4.11) | |
| $k$ | an alternative to $\gamma$ $(= C_p/C_v)$ | |
| $k_s$ | dimensionless specific speed | |
| $L$ | lift force on an aerofoil | N |
| $M$ | pitching moment acting on a foil | N m |
| $M_n$ | Mach number $(= V/a)$ | |
| $\dot{m}$ | mass flow rate | $kg\,s^{-1}$ |
| N | rotational speed | $rev\,min^{-1}$ |
| NPSE | net positive suction energy | $kJ\,kg^{-1}$ |
| $NPSE_a$ | net positive suction energy available | $kJ\,kg^{-1}$ |
| $NPSE_R$ | net positive suction energy required | $kJ\,kg^{-1}$ |
| NPSH | net positive suction head | m of liquid |
| $N_s$ | specific speed | |
| $o$ | opening or throat in a turbine cascade | m |

| | | |
|---|---|---|
| $p$ | pressure | $\mathrm{N\,m^{-2}}$ |
| $P_0$ | stagnation pressure | $\mathrm{N\,m^{-2}}$ |
| $p_v$ | vapour pressure | $\mathrm{N\,m^{-2}}$ |
| $P$ | power | $\mathrm{J\,s^{-1} = W}$ |
| $Q$ | volumetric flow rate | $\mathrm{m^3\,s^{-1}}$ |
| $R$ | reaction (Section 1.3) | |
| $R$ | specific gas constant | $\mathrm{kJ\,kg^{-1}\,K^{-1}}$ |
| $R_e$ | Reynolds number | |
| $R_{em}$ | model Reynolds number | |
| $S$ | suction specific speed | |
| $t$ | blade thickness | m |
| $t$ | blade passage minimum width or throat | m |
| $T$ | temperature (absolute) | K |
| $T_0$ | stagnation temperature (absolute) | K |
| $T$ | torque | N m |
| $u$ | peripheral velocity | $\mathrm{m\,s^{-1}}$ |
| $V$ | absolute velocity | $\mathrm{m\,s^{-1}}$ |
| $V_a$ | axial component of absolute velocity | $\mathrm{m\,s^{-1}}$ |
| $V_n$ | normal component of absolute velocity | $\mathrm{m\,s^{-1}}$ |
| $V_{is}$ or ($V_{isen}$) | isentropic velocity (equation 1.34) | $\mathrm{m\,s^{-1}}$ |
| $V_R$ | radial component of absolute velocity | $\mathrm{m\,s^{-1}}$ |
| $V_u$ | peripheral component of absolute velocity | $\mathrm{m\,s^{-1}}$ |
| $W$ | relative velocity | $\mathrm{m\,s^{-1}}$ |
| $W_u$ | peripheral component of relative velocity | $\mathrm{m\,s^{-1}}$ |
| $Y_N$, $Y_R$ | loss coefficients (equation 4.27) | |
| $Z$ | blade number or position | |
| $\alpha$ | angle made by absolute velocity | degrees |
| $\beta$ | angle made by relative velocity | degrees |
| $\gamma$ | ratio of specific heats | |
| $\gamma$ | stagger angle | degrees |
| $\delta$ | deviation angle | degrees |
| $\varepsilon$ | fluid deflection | degrees |
| $\zeta$ | loss coefficient (equation 4.13) | |
| $\eta$ | efficiency | |
| $\eta_{SS}$ | static to static efficiency | |
| $\eta_{TS}$ | total to static efficiency | |
| $\eta_{TT}$ | total to total efficiency | |
| $\theta$ | camber angle | degrees |
| $\kappa$ | elastic modulus | $\mathrm{kg\,m^{-1}\,s^{-2}}$ |
| $\mu$ | absolute viscosity | $\mathrm{kg\,m^{-1}\,s^{-1}}$ |
| $\nu$ | kinematic viscosity | $\mathrm{m^2\,s^{-1}}$ |
| $\xi$ | Markov's loss coefficient (equation 4.26) | |
| $\rho$ | density | $\mathrm{kg\,m^{-3}}$ |

| | | |
|---|---|---|
| $\sigma$ | Thoma's cavitation parameter | |
| $\sigma$ | velocity ratio (equation 4.29) | |
| $\phi$ | flow coefficient ($V_a/u$) | |
| $\psi$ | specific energy coefficient | |
| $\psi$ | $\eta_{TT}/2\sigma_{is}^2$ (equation 4.30) | |
| $\Omega$ | Howell's work done factor | |
| $\omega$ | angular velocity | $rad\,s^{-1}$ |

Subscripts 1, 2 etc. indicate the point of reference.

For a complete definition of blade terminology please refer also to Fig. 4.2 and Table 4.1.

# 1 Fundamental principles

## 1.1 Introduction

An important class of fluid machine has, as its characteristic, the transfer of energy between a continuous stream of fluid and an element rotating about a fixed axis. Such a machine is classed as a turbomachine: fans, pumps, compressors and turbines come into this group. Discussion is limited in this book to those machines where the fluid is at all times totally enclosed by the machine elements, so that it is controlled by passage walls. This restriction excludes the Pelton turbine and wind turbines.

The machines will be categorized by flow path and by function, as indicated by the simple line diagrams in Fig. 1.1 of the typical machines to be covered. The ideal performance laws are introduced first: the discussion centres on the Euler equation and its applications, it being assumed that basic fluid mechanics and the principles of vector diagrams are understood. The incompressible cases are treated first, and then attention is paid to the problems posed by compressible considerations. Shock wave theory and basic gas dynamics are also taken to be understood by the reader, who is referred to basic texts like those by Shapiro (1953) and Rogers and Mayhew (1967).

## 1.2 Euler equation

An outward flow radial machine is illustrated in Fig. 1.2. Fluid approaches along the suction pipe, is picked up and operated upon by the rotor and is discharged into the casing at a higher level of energy. The rotor has imparted both a velocity and a radial position change to the fluid, which together result in momentum changes and resultant forces on the rotor. The resulting axial and radial forces on the rotating system are treated later: present concern centres on the changes experienced by the fluid.

In the pump in Fig. 1.2 a typical stream surface is examined which

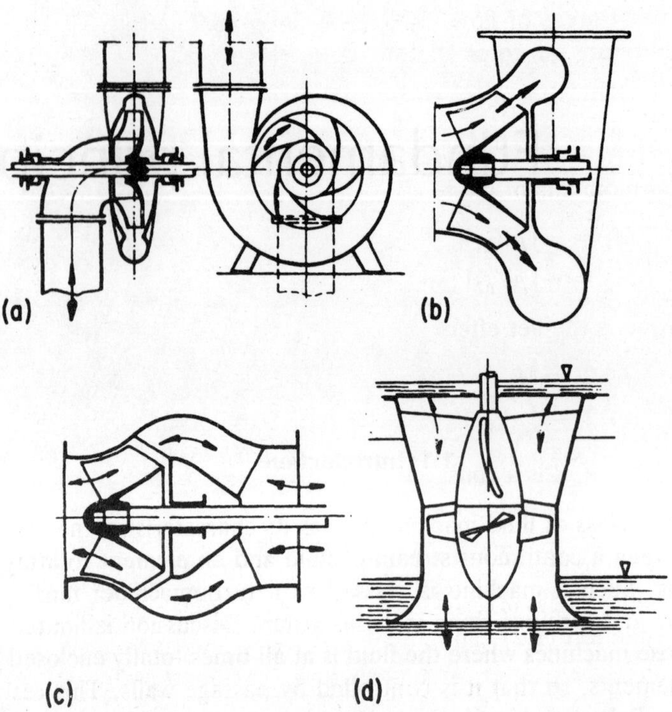


**Figure 1.1** Typical flow paths in machines: (a) centrifugal or centripetal; (b) mixed flow; (c) bulb or bowl; (d) axial.

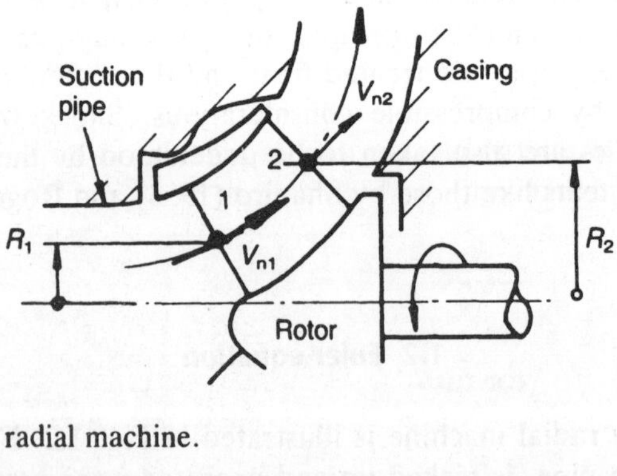


**Figure 1.2** Typical radial machine.

intersects the inlet edge at 1 and the outlet edge at 2. Since momentum changes in the tangential direction give rise to a torque and thus to work, moment of momentum equations for elemental areas of flow at the points of entry and exit will be written down. The normal fluid velocities are $V_{n1}$ and

$V_{n2}$. If elemental areas of flow $da_1$ and $da_2$ are examined, the moments of momentum entering the rotor at 1 and 2 are given by

$$dM_1 = (\rho V_{n1} da_1) V_{u1} R_1$$
$$dM_2 = (\rho V_{n2} da_2) V_{u2} R_2$$

Thus the total moments of momentum are

$$M_1 = \int \rho V_{n1} V_{u1} R_1 da_1 \quad \text{entering plane (1)}$$
$$M_2 = -\int \rho V_{n2} V_{u2} R_2 da_2 \quad \text{leaving plane (2)}$$

The fluid torque is the net effect given by

$$T = M_1 + M_2$$
$$T = \int \rho V_{n1} V_{u1} R_1 da_1 - \int \rho V_{n2} V_{u2} R_2 da_2 \qquad (1.1)$$

It is assumed that $V_u R$ is a constant across each surface, and it is noted that $\int \rho V_n da$ is the mass flow rate $\dot{m}$. Then equation (1.1) becomes

$$T = \dot{m}(V_{u1} R_1 - V_{u2} R_2) \qquad (1.2)$$

The rate of doing work is $\omega T$, and since $\omega R$ is the rotor peripheral velocity $u$ at radius $R$, equation (1.2) can be transformed to give work done per unit mass:

$$gH = u_1 V_{u1} - u_2 V_{u2} \qquad (1.3)$$

This is one form of the **Euler equation.**

To distinguish this $gH$ the suffix E (Euler) will be used, and the two forms of the Euler equation used are:

$$\text{for pumps} \qquad gH_E = u_2 V_{u2} - u_1 V_{u1} \qquad (1.4)$$

$$\text{for turbines} \qquad gH_E = u_1 V_{u1} - u_2 V_{u2} \qquad (1.5)$$

If $gH$ is the specific energy change experienced by the fluid and $\eta_h$ is the hydraulic efficiency, then

$$\text{for pumps} \qquad \eta_h = \frac{gH}{gH_E} \qquad (1.6)$$

$$\text{for turbines} \qquad \eta_h = \frac{gH_E}{gH} \qquad (1.7)$$

## 1.3 Reaction

This concept is much used in axial flow machines as a measure of the relative proportions of energy transfer obtained by static and dynamic pressure change. It is often known as the degree of reaction, or more simply as reaction. The conventional definition is that reaction is given by,

$$R = \frac{\text{energy change due to, or resulting from, static pressure change in the rotor}}{\text{total energy change for a stage}}$$

or, in simple enthalpy terms,

$$R = \frac{\text{static enthalpy change in rotor}}{\text{stage static enthalpy change}} \tag{1.8}$$

## 1.4 Application to a centrifugal machine

A simple centrifugal pump is illustrated in Fig. 1.3. Liquid passes into the rotor from the suction pipe, is acted upon by the rotor whose channels are wholly in the radial plane, and passes out into the volute casing which collects the flow and passes it into the discharge pipe.

The velocity triangles of Fig. 1.4 assume that the fluid enters and leaves the impeller at blade angles $\beta_1$ and $\beta_2$, and that the heights $V_{R1}$ and $V_{R2}$ are obtained from relations like $V_R = Q/\pi Db$. Applying the Euler equation,

$$gH_E = u_2V_{u2} - u_1V_{u1} \tag{1.9}$$

Also it can readily be shown from the triangles that

$$gH_E = \tfrac{1}{2}[(V_2^2 - V_1^2) + (u_2^2 - u_1^2) + (W_1^2 - W_2^2)] \tag{1.10}$$

On the right-hand side (RHS) of equation (1.10), the first bracket is the change in fluid absolute kinetic energy, the second is effectively the energy change due to the impeller rotation, and the third is the change in relative

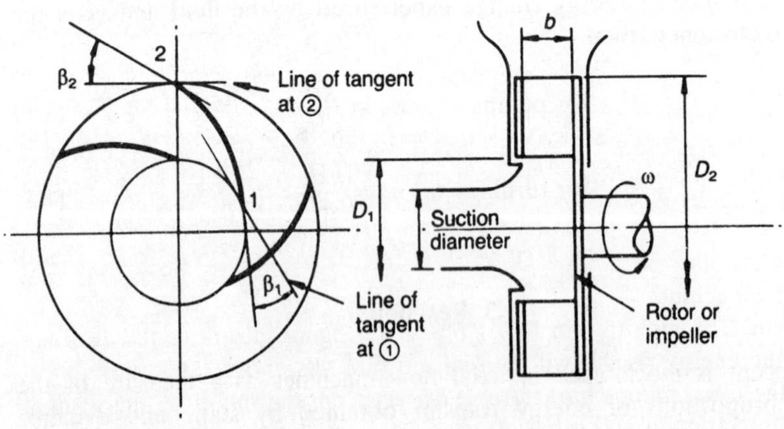


**Figure 1.3** Simple centrifugal pump.

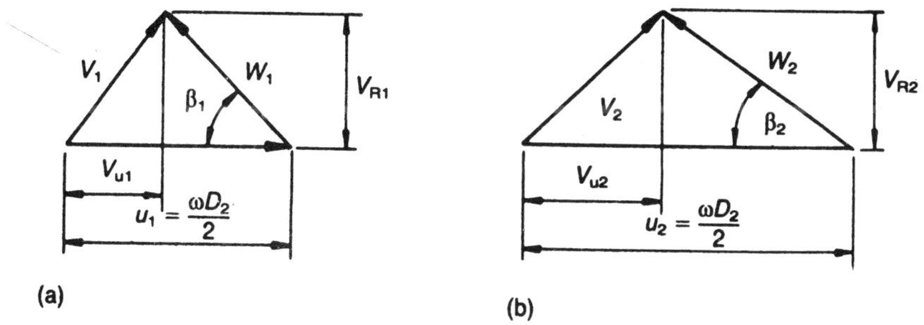


**Figure 1.4** (a) Inlet and (b) outlet velocity triangles for the pump in Fig. 1.3.

kinetic energy. Equation (1.10) is thus a statement that the total energy change is the sum of velocity energy change (the first bracket) and the static equivalent energy change (the sum of the second and third brackets).

Consider now the ideal case where $V_{u1} = 0$ (called the zero inlet whirl case). The inlet triangle is now right-angled and, depending on the blading layout, the outlet triangle can take one of the three forms shown in Fig. 1.5. The outlet triangles are based on the same peripheral and radial velocities, and they demonstrate how the absolute velocity $V_2$ and its peripheral component increase with $\beta_2$. This increase, if the fluid is compressible, could lead to high outlet Mach numbers, and in pumps would lead to casing pressure recovery problems, so that the forward curved geometry tends only to be used in some high performance fans.

It is instructive also to study the effect of $\beta_2$ on the impeller energy change. So, considering zero inlet whirl, equation (1.4) becomes

$$\begin{aligned} gH_E &= u_2V_{u2} \\ &= u_2(u_2 - V_{R2}\cot\beta_2) \end{aligned} \qquad (1.11)$$

Since $V_{R2} = Q/A_2$ ($A_2 = \pi D_2 b$ in Fig. 1.3),

$$gH_E = \left(1 - \frac{Q}{A_2u_2}\cot\beta_2\right)u_2^2 \qquad (1.12)$$

or

$$gH_E = u_2^2 - k_2Qu_2 \qquad (1.13)$$

for a given machine.

Equation 1.13 is plotted in Fig. 1.6 and this figure illustrates the effect of $\beta_2$ upon the energy rise. It will be realized that this only relates to the ideal case, as the usual pump characteristic departs considerably from the straight line owing to friction and other effects.

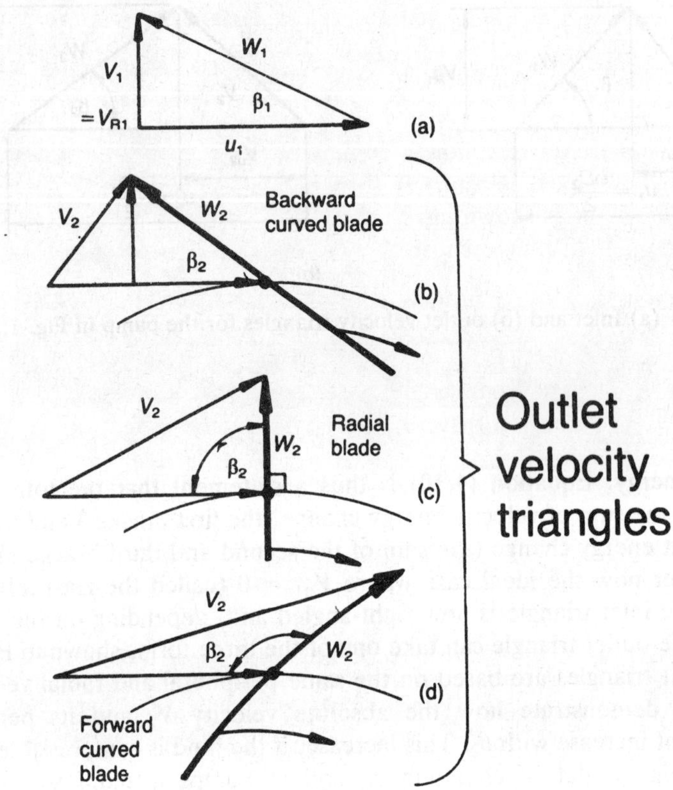


**Figure 1.5** Effect of outlet angle on the outlet triangles for a centrifugal pump.

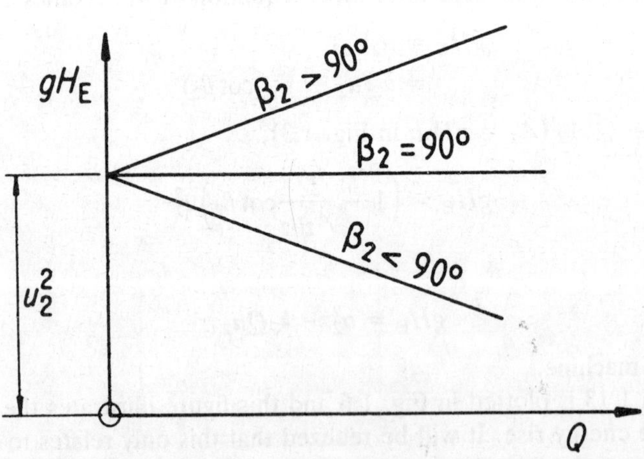


**Figure 1.6** Influence of outlet angle on the ideal centrifugal pump characteristic.

## 1.5 Application to axial pumps and turbines

### 1.5.1 Axial pump or fan

In these idealized cases, flow is assumed to be passing through the machine parallel to the machine centre of rotation at all radii; the annulus dimensions remain constant as shown in Fig. 1.7.

If one plane of flow is chosen, say at radius $R$, the peripheral velocity $u$ is

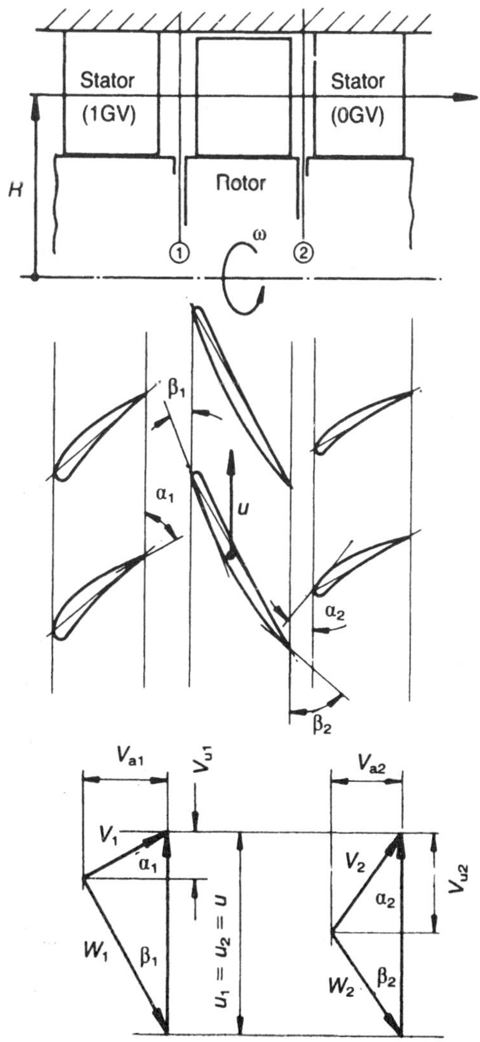


**Figure 1.7** Axial flow pump or compressor stage and the ideal velocity triangles (GV = guide vane).

$\omega R$ and is constant through the machine. This, and the assumption that $V_a$ is constant, allows the triangles in Fig. 1.7 to be drawn.

The Euler equation, equation (1.4), reduces to

$$gH_E = u(V_{u2} - V_{u1}) = u\Delta V_u \tag{1.14}$$

or, using the extended form of equation (1.10),

$$gH_E = \tfrac{1}{2}[(V_2^2 - V_1^2) + (W_1^2 - W_2^2)] \tag{1.15}$$

The reaction, equation (1.8), can be expressed as

$$R = 1 - \frac{1}{2}\left(\frac{V_{u1} + V_{u2}}{u}\right) \tag{1.16}$$

A convenient way of drawing the velocity triangles is to draw them together on a common base (Fig. 1.8) or on a common height, which is not so usual. The common base will be used throughout, and allows $\Delta V_u$ and changes in velocity size and direction to be seen at a glance.

To illustrate how the shape of the velocity triangles is related to the degree of reaction, Fig. 1.9 has been drawn for three cases (a) axial inlet velocity, (b) axial outlet velocity and (c) 50% reaction. Using equation (1.16), the reaction in case (a) is between 0.5 and 1, depending on the size of $V_{u2}$. In case (b) $R$ is always greater than 1. The 'symmetrical' case (c) (50% reaction) yields $R$ as 0.5, and has been much used in compressors since $V_1 = W_2$ and $V_2 = W_1$. The triangles are symmetrical on a common base, making them easy to draw and understand.

### 1.5.2 Axial turbine stage

An axial turbine stage consists of a stator row, usually called a nozzle ring, directing fluid into the rotor row. Again the assumption is that fluid is

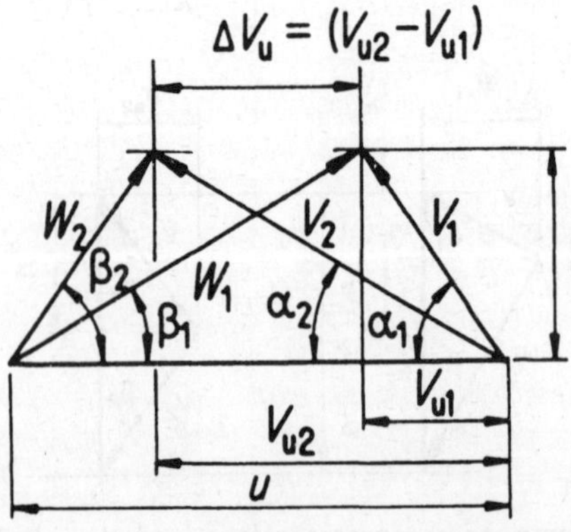


**Figure 1.8** Axial velocity triangles based on a common base for an axial pump stage.

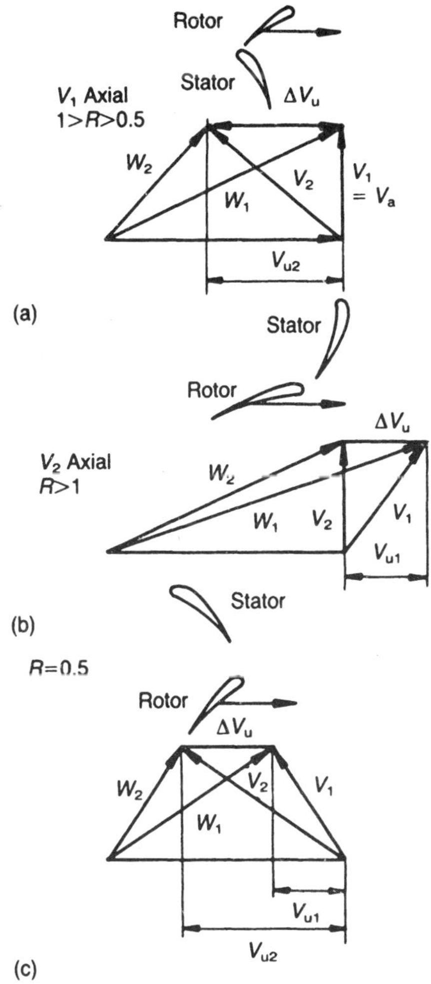


**Figure 1.9** Effect of reaction choice on the velocity triangles for an axial compressor stage.

moving parallel to the axis of rotation, and equations (1.5) and (1.10) become

$$gH_E = u(V_{u1} - V_{u2}) \tag{1.17}$$

or

$$gH_E = \tfrac{1}{2}[(V_1^2 - V_2^2) + (W_2^2 - W_1^2)] \tag{1.18}$$

Figure 1.10 illustrates blade layout, velocity diagrams and the common base velocity triangles, which may be considered typical.

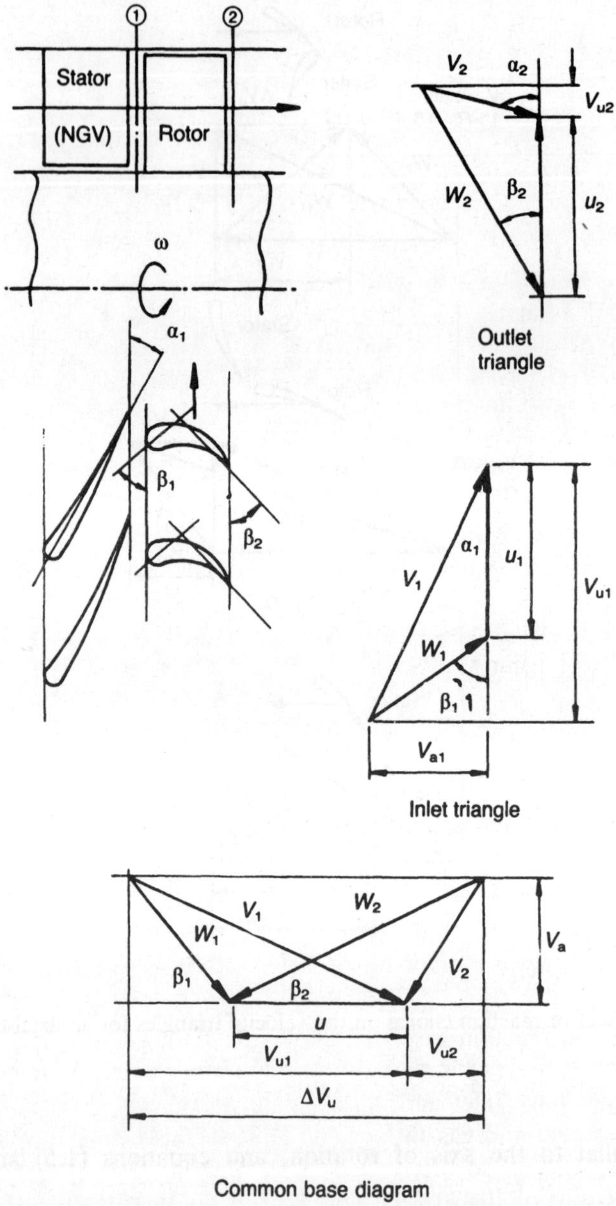


**Figure 1.10** Velocity triangles for an axial turbine stage (NGV = nozzle guide vane).

For an axial machine where the axial velocity remains constant through the stage it can be shown that the reaction is given by the equation

$$R = \frac{W_{u2} - W_{u1}}{2u} \tag{1.19}$$

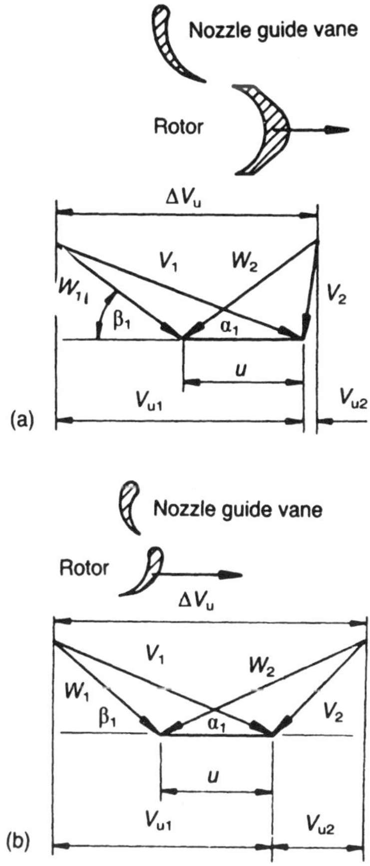


**Figure 1.11** Effect of reaction on an axial turbine stage velocity triangle: (a) zero reaction, $R = 0$ ($\beta_1 = \beta_2$, $W_1 = W_2$); (b) 50% reaction, $R = 0.5$ ($W_1 = V_2$, $W_2 = V_1$).

Two common reaction cases are sketched in Fig. 1.11, (a) zero and (b) 50% reaction. Case (a) is the so-called 'impulse' layout, where the rotor blade passages have inlet and outlet angles the same, and are of constant area. Many steam and gas turbine designs use this layout. Case (b) is the 50% reaction layout, so often used in compressors. Many turbine designs use low reactions of 10–20% design, as then the 'leaving loss' due to $V_2$ is minimized.

## 1.6 Alternative operating modes

Before going on to discuss cavitation and thermodynamic limitations on machine performance, a comment will be made on the possible modes in

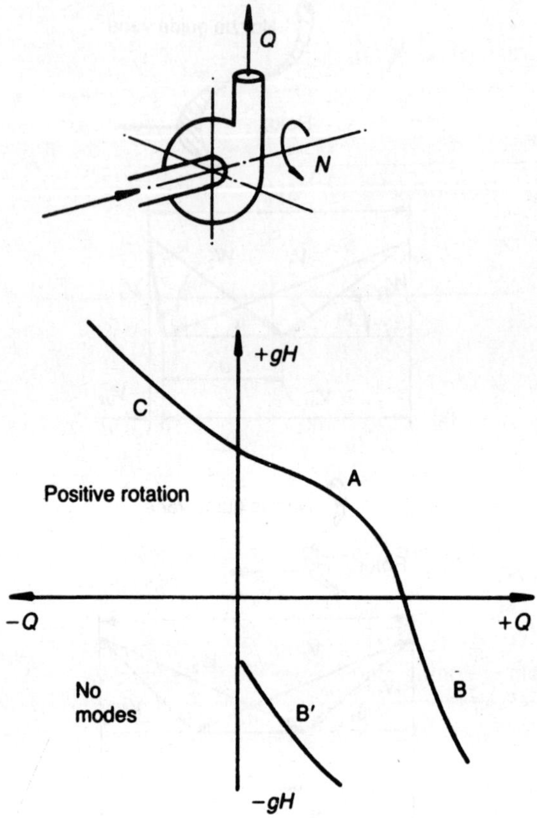


**Figure 1.12** Possible operating modes for a radial machine rotating in the positive (normal) direction.

which a machine may be used. For the centrifugal pump sketched in Fig. 1.3, two directions of rotation may be applied: the normal for pumping, identified as positive, and the reverse. Possible characteristics are sketched in Figs 1.12 and 1.13. These possible modes are briefly discussed in Table 1.1.

## 1.7 Compressible flow theory

### 1.7.1 General application to a machine

The preceding sections assumed that the density was constant throughout the machine concerned. If the density does change considerably it is necessary to relate enthalpy changes to the equations already discussed.

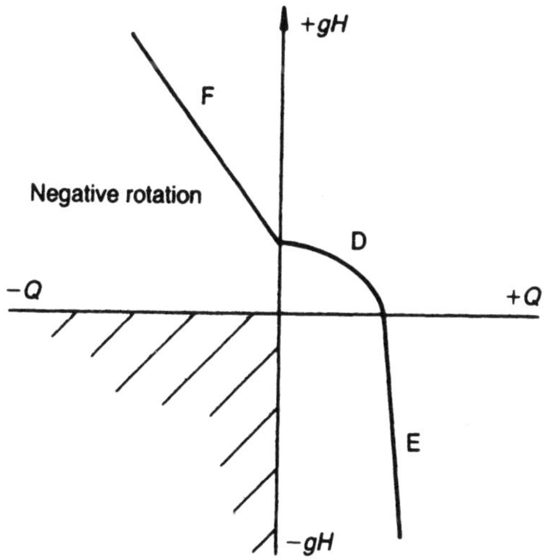


**Figure 1.13** Possible operating modes for a radial machine rotating in the negative (reverse) direction.

**Table 1.1** Possible operating modes for a radial machine

| Mode | Flow | Change in $gH$ | Rotation | Comment |
|---|---|---|---|---|
| A | + | + | + | Normal pumping mode |
| B | + | − | + | Energy dissipation, outward flow turbine, line B′ in Fig. 1.12 |
| C | − | + | + | Energy dissipation, rotor resisting back flow |
| D | + | + | − | Pump being driven wrong way round |
| E | + | − | − | Energy dissipation |
| F | − | + | − | Turbine operation |

Suppose that a compressor is considered to be the control volume (Fig. 1.14) and that the fluid system moves through the control volume in time $dt$ as sketched. If the heat added in the time interval is $dQ$ and the work extracted is $dW$, then application of the first law of thermodynamics to the system at time $t$ and time $t + dt$ in Fig. 1.14 yields the equation (following Shapiro, 1953)

$$dQ - dW = (E_B - E_A) + (me_2 - me_1) + (p_2mv_2 - p_1mv_1) \quad (1.20)$$

On the RHS of equation (1.20), the first bracket is the change in internal energy of the fluid in the control volume, the second is the difference

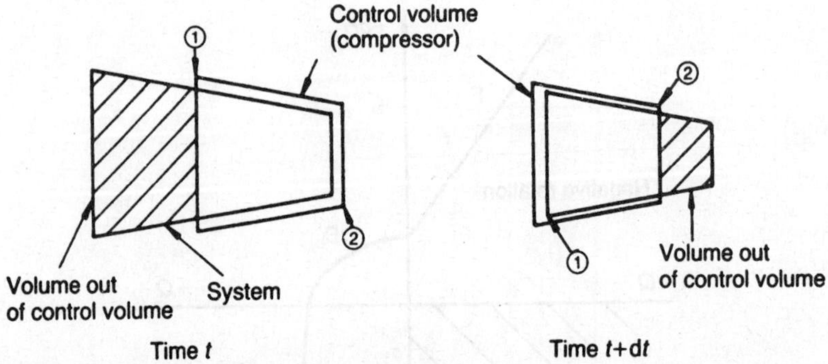


**Figure 1.14** Compressible flow theory example.

between the internal energy at exit from and inlet to the control volume, and the third is the net work done by the system on the surrounding fluid as $m$ passes through the control volume. If flow is steady, the first bracket is zero and the second and third brackets can be rewritten in terms of enthalpy, velocity and potential energy as follows:

$$dQ - dW = m[(h_2 - h_1) + (V_2^2 - V_1^2)/2 + g(Z_2 - Z_1)] \qquad (1.21)$$

This can be expressed in terms of work rate and mass flow rate. The suppression of heat exchange results, if stagnation enthalpy is used, in the equation

$$\dot{W} = \dot{m}(h_{02} - h_{01}) \qquad (1.22)$$

or

$$\dot{W} = \dot{m}C_p(T_{02} - T_{01}) \qquad (1.23)$$

In a turbomachine,

$$\dot{W} = \dot{m}(gH) = (u_2V_{u2} - u_1V_{u1})\dot{m}$$

Thus

$$C_p(T_{02} - T_{01}) = (u_2V_{u2} - u_1V_{u1}) \qquad (1.24)$$

or

$$(h_{02} - h_{01}) = (u_2V_{u2} - u_1V_{u1}) \qquad (1.25)$$

Thus stagnation enthalpy change relates to the velocity triangles. The way this relates to the compression and expansion processes will now be discussed.

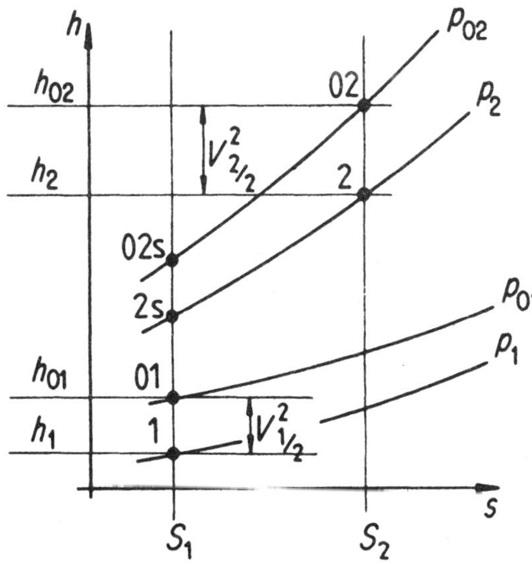


**Figure 1.15** *h–s* diagram for compression between two planes in a machine.

1.7.2 Compression process

Figure 1.15 illustrates the compression from state 1 to state 2; both static and stagnation conditions are depicted. Actual compression is shown as accompanied by an increase in entropy from $S_1$ to $S_2$. The isentropic compression is shown as increasing stagnation enthalpy from point 01 to point 02s. Isentropic efficiency statements may be written as total to total

$$\eta_{TT} = \frac{h_{02s} - h_{01}}{h_{02} - h_{01}} \tag{1.26}$$

or as static to static

$$\eta_{SS} = \frac{h_{2s} - h_1}{h_2 - h_1} \tag{1.27}$$

The choice of efficiency depends on the system in which the compressor is placed, $\eta_{TT}$ being appropriate for situations where stagnation conditions at inlet and outlet are proper indexes of performance. The equations apply to the whole process of compression but, because of the divergence of the constant pressure lines (Fig. 1.16), the sum of the stage isentropic rises exceeds the overall rise. Thus, if $\eta_p$ is the small stage efficiency and $\eta_c$ the overall efficiency,

$$\eta_p > \eta_c \tag{1.28}$$

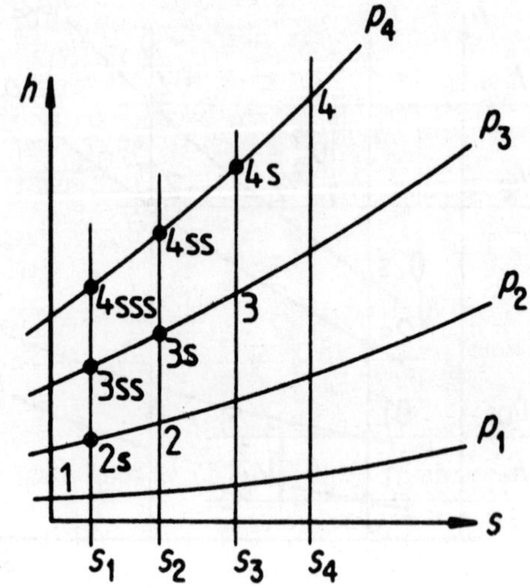


**Figure 1.16** Preheat effect in a multistage compressor.

The preheat factor is given by

$$\frac{\eta_p}{\eta_c} = \frac{\Sigma \text{ stage rise}}{\text{overall rise}} \tag{1.29}$$

The small stage or polytropic efficiency is given by

$$\eta_p = \frac{dh_{is}}{dh} = \frac{v dp}{C_p dT} = \frac{RT dp}{C_p p dT}$$

Thus

$$\frac{dT}{T} = \frac{(\gamma - 1)}{\gamma n_p} \frac{dp}{p}$$

since

$$C_p = \left(\frac{\gamma}{\gamma - 1}\right) R$$

When integrated for the compressor this yields, if $\eta_p$ is the same for all stages.

$$\frac{T_2}{T_1} = \left(\frac{p_2}{p_1}\right)^{(\gamma-1)/\eta_p \gamma} \tag{1.30}$$

If $\eta_p = 1$ for the ideal compression process,

$$\frac{T_{2s}}{T_1} = \left(\frac{p_2}{p_1}\right)^{(\gamma-1)/\gamma}$$

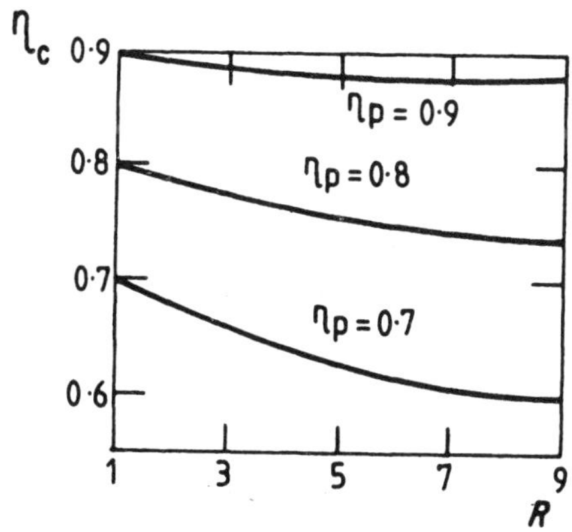


**Figure 1.17** Plot of the variation of polytropic and overall efficiencies with pressure ratio in compression.

Thus

$$\eta_c = \frac{(p_2/p_1)^{(\gamma-1)/\gamma} - 1}{(p_2/p_1)^{(\gamma-1)/\gamma\eta_p} - 1} \tag{1.31}$$

Figure 1.17 is a plot of $\eta_c$ against the pressure ratio at various $\eta_p$.

Figure 1.18 illustrates a simple centrifugal compressor. Figure 1.19 is the $h$–$s$ diagram relating to the compression flow path, which has been constructed on the basis that $h_{02} = h_{03} = h_{04}$. From this figure,

$$\eta_{TT} = \frac{h_{04sss} - h_{01}}{h_{04} - h_{01}} \tag{1.32}$$

$$\eta_{TS} = \frac{h_{4sss} - h_1}{h_4 - h_1} \tag{1.33}$$

### 1.7.3 Expansion process

In compressible machines some energy transformation takes place in a nozzle before the fluid passes through the turbine rotor. If heat losses in a nozzle (Fig. 1.20) are neglected, the stagnation enthalpy remains constant. The isentropic heat drop is $h_{01} - h_{2s} = \Delta h_{isen}$. If this could be utilized, the outlet velocity would be

$$V_{is} = \sqrt{(2\Delta h_{isen})} \tag{1.34}$$

Since the actual $\Delta h = h_{01} - h_2$, a commonly used nozzle efficiency may be written as

$$\eta_{nozzle} = \frac{h_{01} - h_2}{\Delta h_{isen}} \tag{1.35}$$

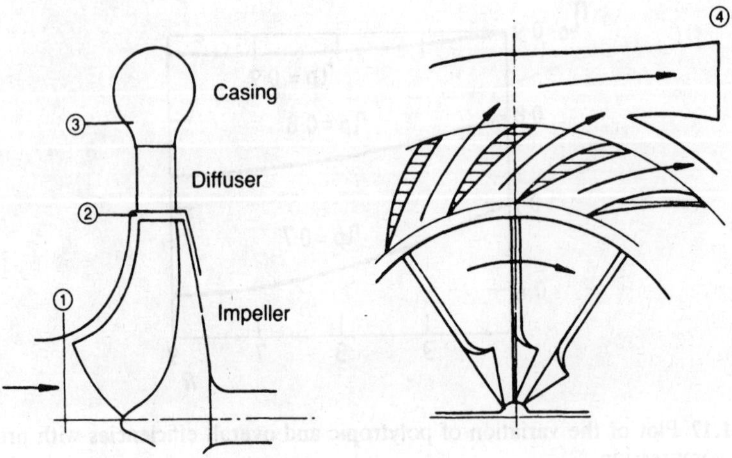


**Figure 1.18** Centrifugal compressor.

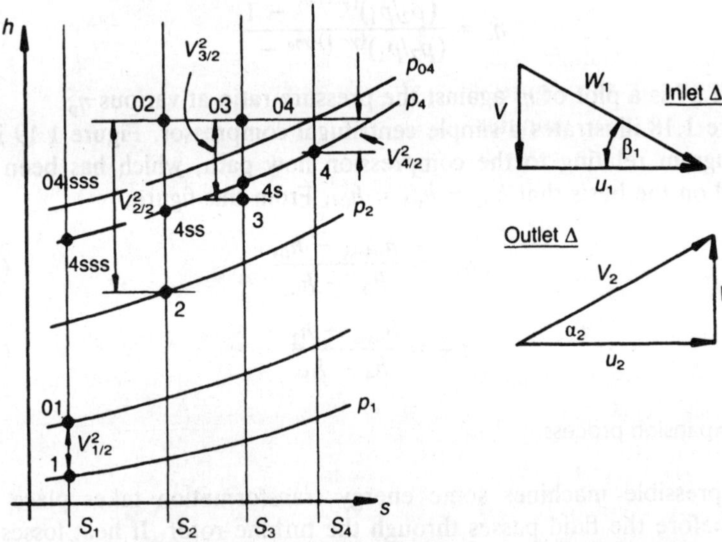


**Figure 1.19** *h–s* diagram and velocity triangles for a centrifugal compressor.

Two other definitions are quoted by Kearton (1958):

$$\eta_n = \frac{h_1 - h_2}{h_1 - h_{2s}} \tag{1.36}$$

and

$$\eta_n = \frac{(h_{01} - h_2) - \eta_{con}(h_{01} - h_1)}{h_1 - h_{2s}} \tag{1.37}$$

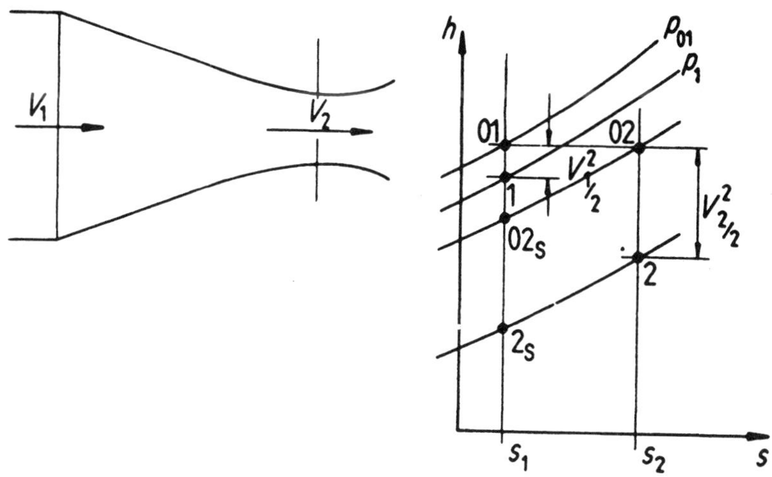


**Figure 1.20** Simple expansion process.

where $\eta_{con}$ is the efficiency with which the inlet kinetic energy is converted.

If a complete stage is considered, the $h$–$s$ diagram in Fig. 1.21 results. Also shown in this figure are the rotor inlet and velocity triangles, and a sketch of a typical annulus. The $h$–$s$ diagram for the nozzle is based on $h_{01} = h_{02}$, and the rotor $h$–$s$ is based on $h_{02\text{relative}} = h_{03\text{relative}}$, since the velocity vectors $W_2$ and $W_3$ allow total enthalpy relative to the rotor blades to be calculated. It can be shown by using equation (1.25) that for an axial stage

$$h_{02R} - h_{03R} = \frac{u_2^2}{2} - \frac{u_3^2}{2} = 0$$

as assumed. Using the $h$–$s$ diagram,

$$\eta_{TT} = \frac{h_{01} - h_{03}}{h_{01} - h_{03ss}} \tag{1.38}$$

$$\eta_{TS} = \frac{h_{01} - h_{03}}{h_{01} - h_{3ss}} \tag{1.39}$$

In general, $\eta_{TS}$ is a more realistic statement for machines in systems where the outlet velocity energy is not utilized, and $\eta_{TT}$ for applications like the pass-out steam turbine where the energy is utilized.

Another efficiency much used in basic steam turbine theory is the so-called diagram efficiency:

$$\eta_D = \frac{\text{work from velocity triangles}}{\text{energy available to rotor blades}} \tag{1.40}$$

The $h$–$s$ diagram in Fig. 1.21 is drawn for a reaction machine layout, but it is necessary to comment upon the distinction between 'impulse' and 'zero

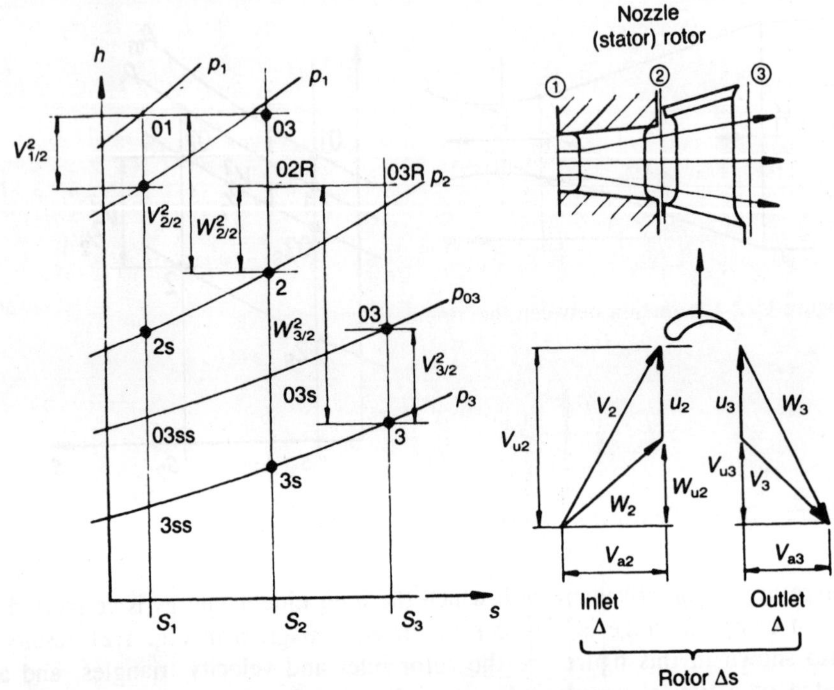


**Figure 1.21** *h–s* diagram and velocity triangles for an axial turbine stage.

reaction' systems. By zero reaction it is understood that the whole of the stage enthalpy drop takes place over the nozzles and none in the rotor, so that the rotor *h–s* diagram results in the shape of Fig. 1.22(a). If the true impulse concept is applied, $p_2 = p_3$. The result is Fig. 1.22(b), and an enthalpy increase from $h_2$ to $h_3$ follows, so the impulse concept is strictly one of negative reaction.

Figure 1.23 illustrates the 'reheat' effect which follows from the divergence of the pressure lines. The reheat factor is given by

$$R_H = \frac{\Sigma \Delta h_{is}}{h_{is}} = \frac{\eta_T}{\eta_p} \tag{1.41}$$

where $\eta_T$ is turbine efficiency and $\eta_p$ is 'small stage' efficiency.

Figure 1.24 is a plot of $\eta_T$ against pressure ratio for various $\eta_p$, obtained by an argument similar to that for the compressor.

Figure 1.25 illustrates a radial inflow turbine, and Fig. 1.26 shows the *h–s* diagram and the ideal velocity triangles. For the turbine,

$$\eta_{TT} = \frac{h_{01} - h_{03}}{h_{01} - h_{03ss}} \tag{1.42}$$

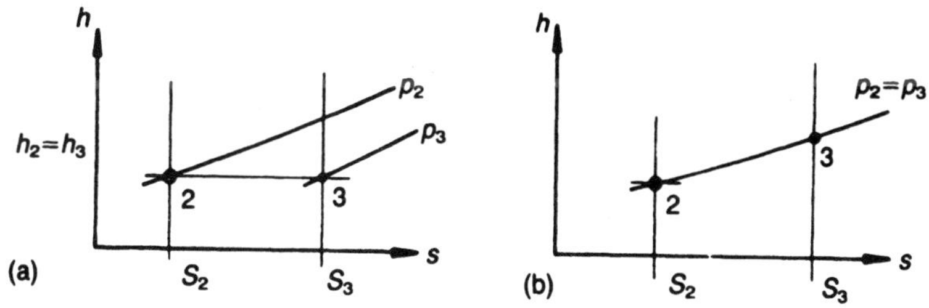


**Figure 1.22** Distinction between the 'impulse' and 'zero reaction' concepts.

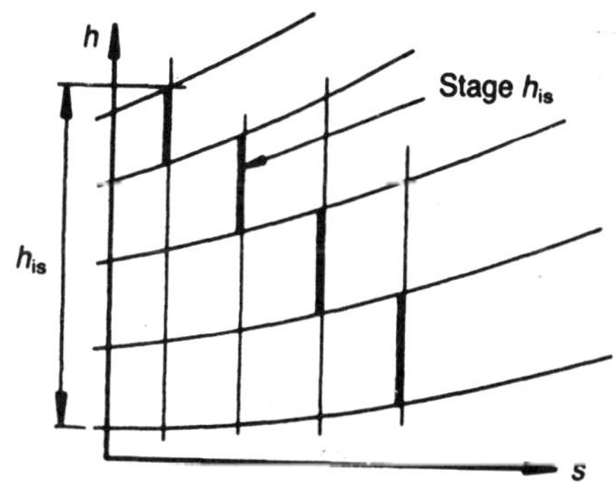


**Figure 1.23** Reheat factor.

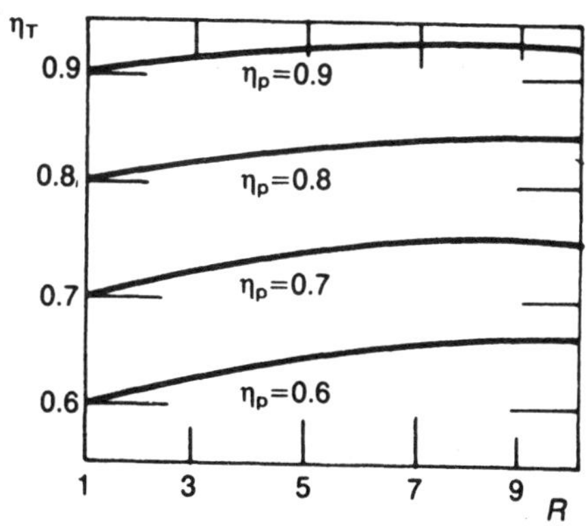


**Figure 1.24** Effect on overall efficiency and polytropic efficiency of expansion ratio.

$$\eta_{TS} = \frac{h_{01} - h_{03}}{h_{01} - h_{3ss}} \tag{1.43}$$

It will be noted that it was assumed that $h_{01} = h_{02}$, but that since peripheral velocities $u_2$ and $u_3$ are dissimilar, the rotor portion of the diagram takes the shape shown.

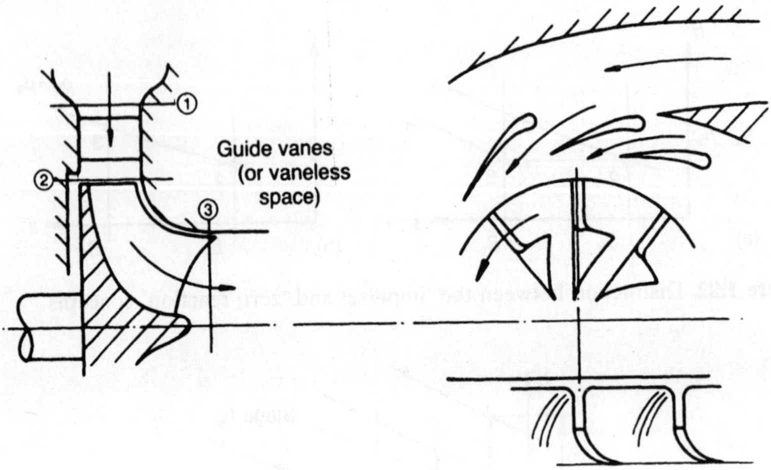


**Figure 1.25** Radial inflow turbine.

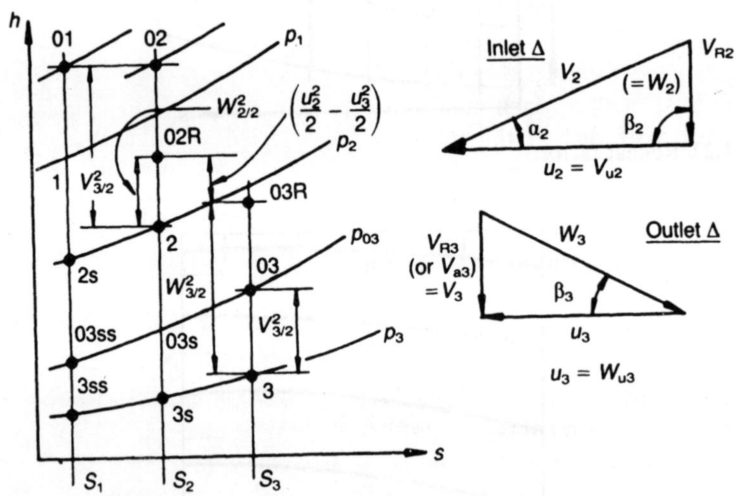


**Figure 1.26** $h$–$s$ diagram and velocity diagrams for a radial inflow turbine.

## 1.8 Shock wave effects

The reader is referred to Shapiro (1953) and Ferri (1949) for shock wave theory, as the object here is simply to comment that when shock waves form they create fluid dynamic difficulties.

In compressors, the conditions at inlet to a row typically cause shocks to form, and these have an effect on the flow into the passages in addition to

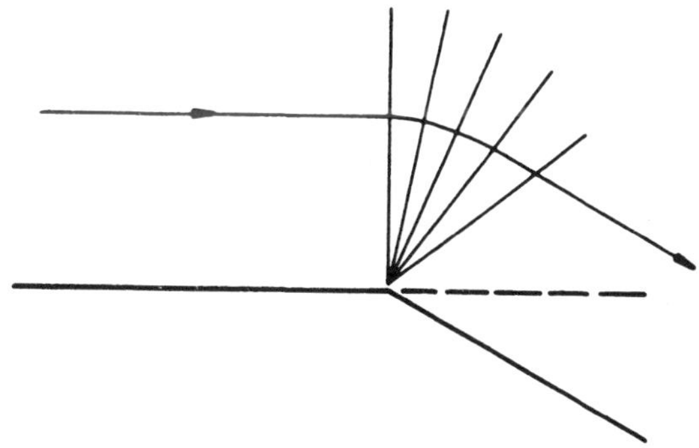

**Figure 1.27** Prandtl–Meyer shock-wave effect.

the flow instabilities caused by the wakes of the preceding blade rows in a multistage machine.

In turbines, flow near the trailing edges of nozzles is of particular importance. If expansion ratios are selected that give rise to shock conditions, the flow is deflected by a Prandtl–Meyer expansion (Fig. 1.27), the misdirection caused being several degrees. Horlock (1966), for example, shows that a deflection of 12° resulted from a nozzle with an expansion ratio over the critical, with a Mach number of unity in the throat expanding to a back pressure half that at the throat. Clearly, a substantial correction is needed in design calculations.

## 1.9 Illustrative examples

### 1.9.1 Radial outflow machine (pump)

The pump sketched in Fig. 1.28 is driven at 1470 rev min$^{-1}$ and delivers 100 l s$^{-1}$ with a specific energy change of 400 J kg$^{-1}$. Sketch the inlet and outlet triangles, assuming a hydraulic efficiency and zero inlet whirl.

The shapes of the triangles are shown in Fig. 1.28. The calculations are as follows:

$$u_1 = \frac{1470 \times \pi \times 0.2}{30 \times 2} = 15.39\,\mathrm{m\,s^{-1}}$$

$$u_2 = 28.48\,\mathrm{m\,s^{-1}}$$

The radial normal velocities $V_{R1}$ and $V_{R2}$ are found using the flow rate:

$$V_{R1} = 0.1/\pi \times 0.2 \times 0.03 = 5.31\,\mathrm{m\,s^{-1}}$$

$$V_{R2} = 0.1/\pi \times 0.37 \times 0.03 = 2.87\,\mathrm{m\,s^{-1}}$$

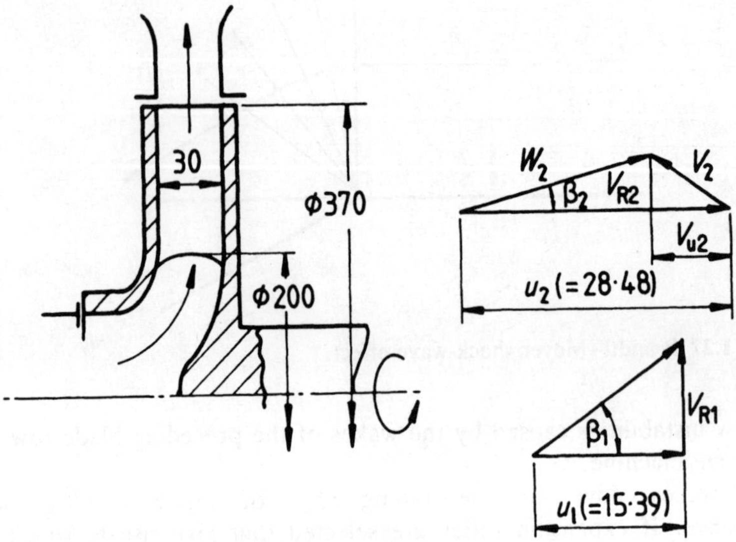


**Figure 1.28** Radial outflow machine.

The Euler specific energy rise is given by

$$\frac{400}{0.85} = 28.48V_{u2}$$

Thus

$$V_{u2} = 16.52\,\mathrm{m\,s^{-1}}$$

From the inlet triangle,

$$\beta_1 = \tan^{-1} 5.31/15.39$$
$$= 19.04°$$

Similarly, from the outlet triangle,

$$\beta_2 = \tan^{-1} 2.87/(28.48 - 16.5) = 13.47°$$

1.9.2 Axial pump and turbine

The axial machine sketched in Fig. 1.29 is driven at 45 rad s$^{-1}$. If the energy change is 120 J kg$^{-1}$, sketch the velocity triangles for both pumping and turbining modes of operation assuming $V_a = 12\,\mathrm{m\,s^{-1}}$. Ignore efficiency, and assume zero inlet whirl for the pump and zero outlet whirl for the turbine.

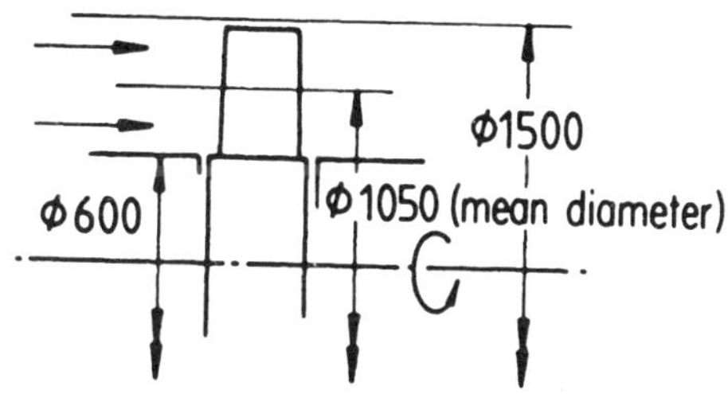


**Figure 1.29** Axial machine.

Work on the mean diameter.

$$u_{\text{mean}} = 23.63\,\text{m}\,\text{s}^{-1}$$

For the pump,

$$120 = 23.63 V_{u2}$$

Therefore

$$V_{u2} = 5.08\,\text{m}\,\text{s}^{-1}$$

The velocity triangles thus take the shapes shown in Fig. 1.30 with

$$\beta_1 = 26.9^\circ$$
$$\beta_2 = 32.9^\circ$$

For a turbine, the triangles follow as shown in Fig. 1.31 with

$$\beta_1 = 32.9^\circ$$
$$\beta_2 = 26.9^\circ$$

### 1.9.3 Compressible flow problem

A simple air turbine of the axial type has a nozzle angle of 20° referred to the peripheral direction. The nozzle efficiency is 90%, and the temperature drop over the nozzle is 125 K. Construct the velocity triangles if the rotor outlet angle is 30°, and suggest the air power available. Assume a rotor tangential velocity of $250\,\text{m}\,\text{s}^{-1}$, no flow losses through the rotor, a flow rate of $4\,\text{kg}\,\text{s}^{-1}$, and zero outlet whirl. (Assume also that $C_p$ = $1.005\,\text{kJ}\,\text{kg}^{-1}\,\text{K}^{-1}$.)

Using equation (1.34), and introducing the nozzle efficiency from equation (1.35), the nozzle outlet velocity is given by

$$V_1 = \sqrt{(2 \times 1.005 \times 10^3 \times 125 \times 0.9)} = 475.53\,\text{m}\,\text{s}^{-1}$$

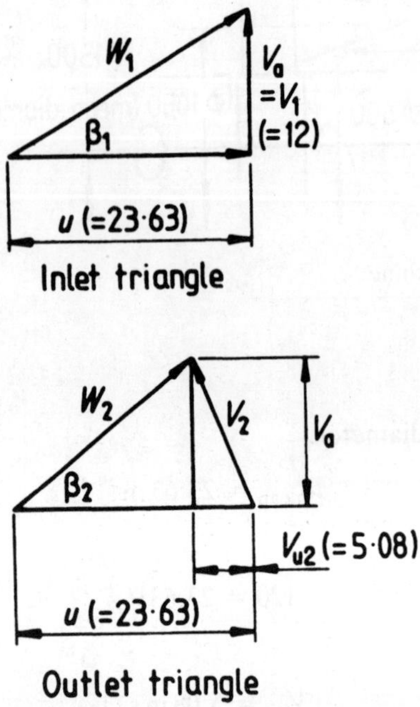


**Figure 1.30** Velocity triangles.

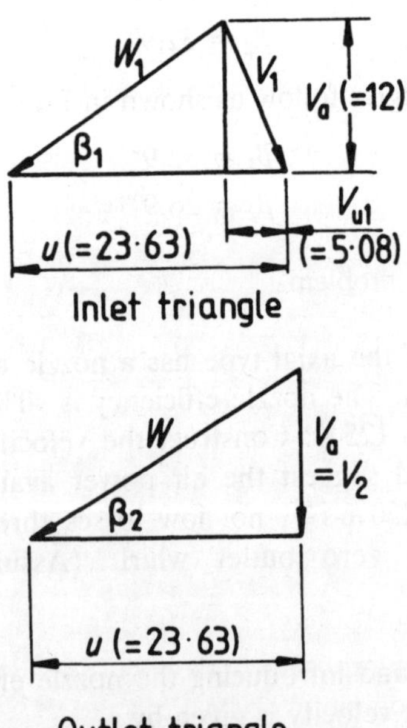


**Figure 1.31** Velocity triangles for a turbine.

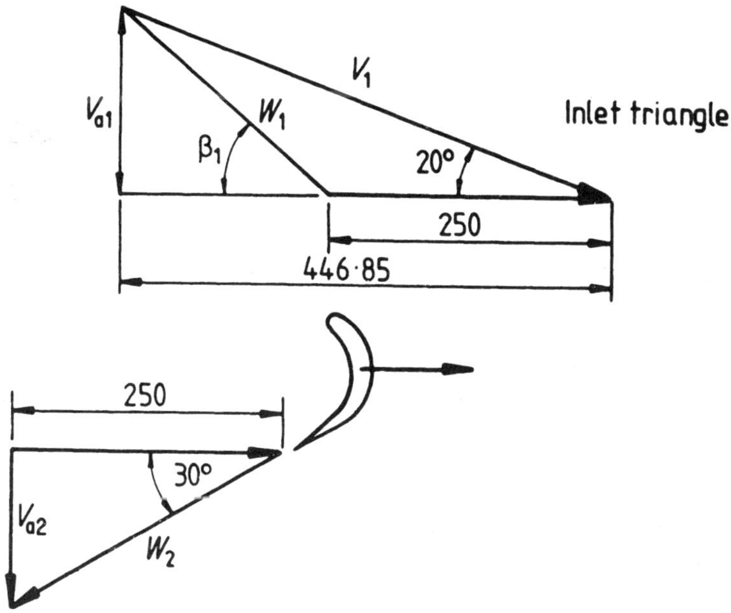


**Figure 1.32** Velocity triangles for compressible flow problem.

In the inlet velocity triangle (Fig. 1.32) it is now possible to calculate $V_{u1}$, and hence

$$\beta_1 = 59.23°$$
$$W_1 = 189.29 \, \mathrm{m\,s^{-1}}$$

With zero outlet whirl the outlet velocity triangle shape is as shown in Fig. 1.32 and

$$W_2 = 288.68 \, \mathrm{m\,s^{-1}}$$

The air power is given by

$$muV_u = 4 \times 250 \times 446.85$$
$$= 4.468 \times 10^5 W$$

## 1.10 Exercises

1.1 A radial outflow pump has an impeller with an outside diameter of 305 mm, an inside diameter of 75 mm, and passage height of 49 mm. If the blade inlet angle is 45° to the tangent and the outlet angle is 30° to the tangent, find for a rotational speed of 1500 revolutions per minute the flow rate, pressure rise, torque and power needed, ignoring losses and assuming zero inlet whirl. Assume water is pumped.

1.2 An axial flow pump has a rotor tip diameter 1.5 m and hub diameter 0.75 m. If the rotational speed in 45 $\text{rad s}^{-1}$, the axial velocity through the rotor is 10 $\text{m s}^{-1}$, sketch the mean diameter velocity triangles and estimate the Euler specific energy rise. Assume ideal flow, zero inlet whirl, the outlet blade angle referred to the tangential direction is 25° and that water is the fluid pumped.

1.3 If the machine in 1.2 is used as a fan passing air of density 1.2 $\text{kg m}^{-3}$ suggest, basing calculations on the mean section, the volume rate of flow and the ideal pressure rise.

1.4 An axial flow fan has a tip diameter of 2 m, a hub diameter of 0.8 m, and rotates at 1450 revolutions per minute. For the condition of zero inlet whirl estimate the velocity diagrams at the tip section, if the inlet absolute velocity is 55 $\text{m s}^{-1}$ the air has a density of 1.2 $\text{kg m}^{-3}$ and losses are ignored. Estimate also the fluid power, if $\Delta p$ is 5 $\text{kN m}^{-2}$.

1.5 A small water turbine is of the axial type, and is installed in a river barrage where the level drop is 8 m. The rotor tip diameter is 1.5 m, the hub diameter is 0.62 m, the rotational speed is 45 $\text{rad s}^{-1}$, the axial velocity through the rotor is 10 $\text{m s}^{-1}$, the hydraulic efficiency is 92% and the machine mechanical losses plus electrical losses reduce the turbine output by 5%. Estimate the mean radius blade angles assuming ideal flow conditions, and the power output in MW.

# 2 Principles and practice of scaling laws

## 2.1 Introduction

This section covers the fundamentals of similarity, scaling, and the problems involved when models are used to predict full size machine performance. The fundamental hydrodynamic principles leading to Euler's equation are introduced and applied to incompressible and compressible machines, and the problems associated with their use are discussed.

Three machine flow paths are possible – radial, mixed and axial flow, as indicated in Fig. 1.1 – and the flow direction and rotation is related to the machine action, either as energy input or as extraction.

## 2.2 Performance laws

In a simple approach, a turbomachine can be considered as a black box. Figure 2.1 describes a functional unit. A shaft transmitting torque is the external work output/input $P$, flow is passing through at a volume rate $Q$, the fluid is experiencing an energy change $gH$, and the rotor speed is $\omega$. How the energy $gH$ is achieved, and its relation to $Q$ and $P$, were discussed in Chapter 1. The quantities listed are all measurable externally, and can be used to indicate machine performance in a way relevant for the user.

If a pumping machine is considered, the input power $P$ is a function of flow rate $Q$, specific energy rise $gH$, machine size (characteristic dimension $D$), rotational speed $\omega$, and liquid properties $\mu$, $\rho$ and $\kappa$ (modulus). Using the principles of dimensional analysis, sets of non-dimensional groups may be obtained, one such being

$$\underset{(1)}{\frac{P}{\rho\omega^3D^5}} = f\left[\underset{(2)}{\frac{Q}{\omega D^3}}, \underset{(3)}{\frac{gH}{\omega^2D^2}}, \underset{(4)}{\frac{\rho\omega D^2}{\mu}}, \underset{(5)}{\frac{\rho\omega^2D^2}{\kappa}}\right] \tag{2.1}$$

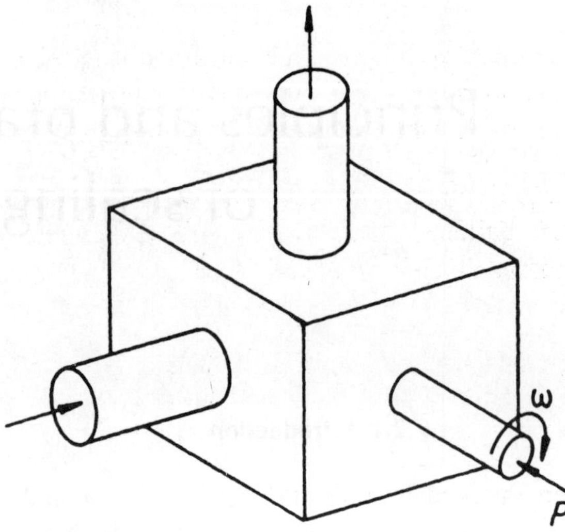


**Figure 2.1** Machine 'black box'.

The following may be noted for equation (2.1).

- Grouping 1 is a power coefficient.
- Grouping 2 is often known as a flow coefficient $\phi$ (or, since $Q \propto VD^2$, and $\omega D \propto u$, $\phi = V/u$, which is a velocity coefficient).
- Since $\omega^2 D^2 \propto u^2$, grouping 3 can be written as $gH/u^2$, and without $g$ is known as $\psi$, the head coefficient.
- Grouping 4 is a Reynolds number based on a typical machine dimension.
- Grouping 5 is a form of Mach number irrelevant in pumps, fans and water turbines, but not in other turbomachines.

One use of the first three groupings is to present the typical characteristics, shown in Fig. 2.2, in non-dimensional form as in Fig. 2.3.

They may also be used to predict the probable dynamically similar performance of the same pump running at different speeds, or of a pump in the same 'family', using the 'scaling' laws

$$\frac{P}{\rho\omega^3 D^5} = \text{constant}$$

$$\frac{Q}{\omega D^3} = \text{constant} \tag{2.2}$$

$$\frac{gH}{\omega^2 D^2} = \text{constant}$$

Figure 2.4 illustrates the point.

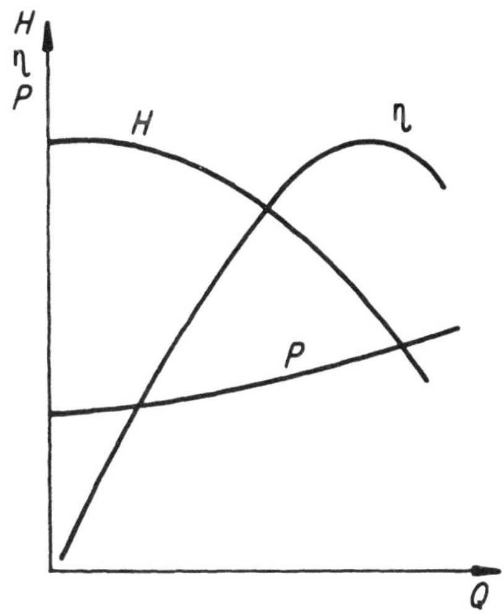


**Figure 2.2** Constant speed characteristics of a pump.

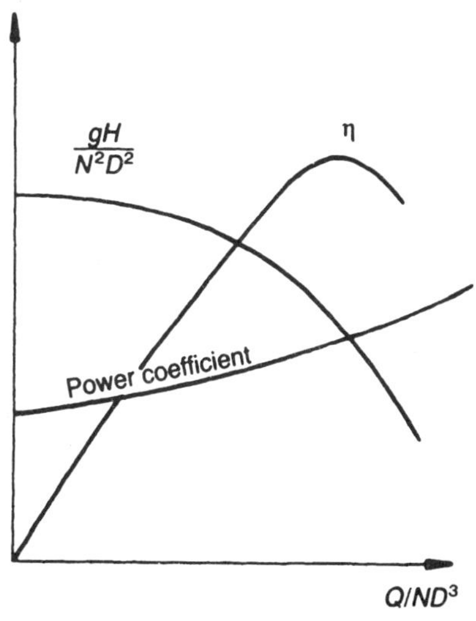


**Figure 2.3** Non-dimensional presentation of the complete performance characteristics of a pump.

Equation (2.1) can thus be seen as a basic performance law, which should be valid for all fluids, compressible and incompressible, providing account is taken of changes in fluid properties. For compressible fluids it is conventional to express quantities in terms of inlet conditions, and instead of

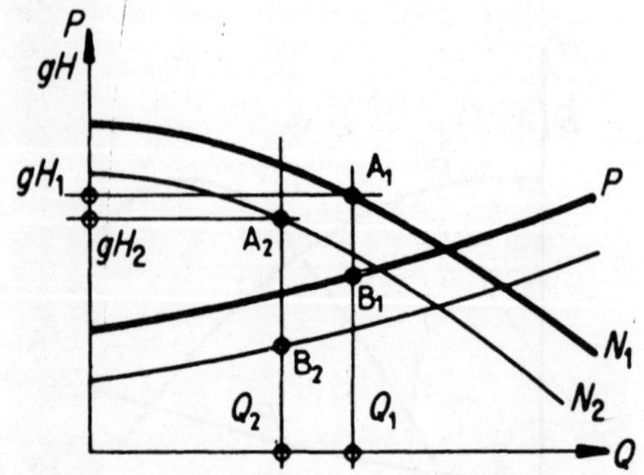


**Figure 2.4** Application of scaling laws to pump performance translation.

equation (2.1) there is an alternative form, much used in the gas turbine industry in particular:

$$\frac{p_{02}}{p_{01}} = f\left[\frac{\omega D}{\sqrt{RT_{01}}}, \frac{\dot{m}\sqrt{(RT_{01})}}{D^2 p_{01}}, R_e, M_n \ldots\right] \tag{2.3}$$

Conventionally the gas constant and $D$ are not quoted explicitly (the gas and size being specified separately), so equation (2.3) is often written as:

$$\frac{p_{02}}{p_{01}} = f\left[\frac{\omega}{\sqrt{T_{01}}}, \frac{m\sqrt{T_{01}}}{p_{01}}, R_e, M_n \ldots\right] \tag{2.4}$$

and performance plots are presented in Fig. 2.5 for compressors and turbines.

## 2.3 Concept of specific speed

The classical view of the problem of characterizing the performance of a pump without involving its dimensions directly is discussed very clearly by, among others, Addison (1955). He defines a pump as being of standardized 'size' when it delivers energy at the rate of one horsepower when generating a head of one foot. This imaginary wheel has a speed termed the specific speed. He shows that the specific speed $N_s$ is given by

$$N_s = K\frac{N\sqrt{Q}}{H^{3/4}} \tag{2.5}$$

where $K$ contains the fluid density and the horsepower equivalent, $N$ is in revolutions per minute (rpm), $Q$ in gallons per minute (gpm), and $H$ in feet

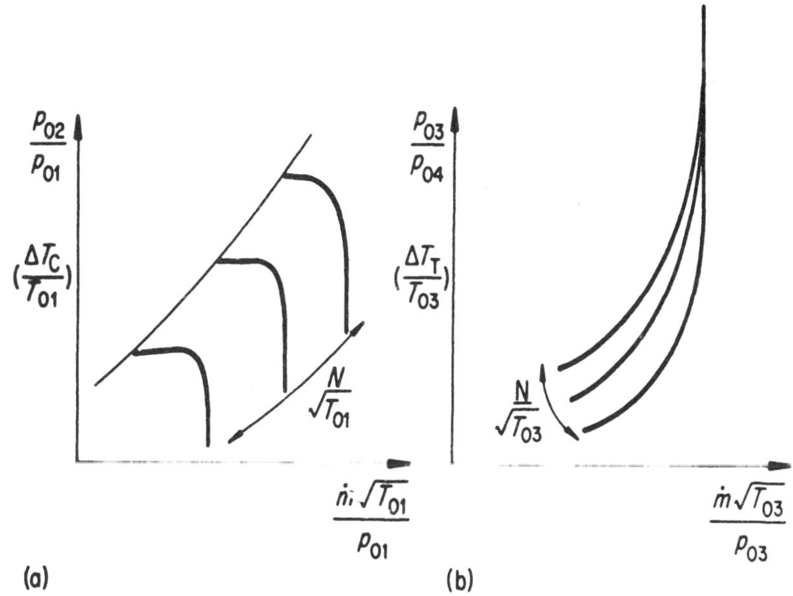


**Figure 2.5** Non-dimensional plots of compressor and turbine performance; (a) compressor; (b) turbine.

of fluid. It is customary to suppress $K$, and for many years the accepted form of specific speed has been

$$N_s = \frac{N\sqrt{Q}}{H^{3/4}} \tag{2.6}$$

Though called specific speed, the dimensions of $N_s$ vary with the units involved. $N$ is usually in rpm, but $Q$ can be in imperial or US gpm, litres per second or cubic metres per second, and $H$ in feet or metres. Caution therefore is needed when $N_s$ is used.

The equivalent expression relating to hydraulic turbines takes the form

$$N_s = \frac{N\sqrt{P}}{H^{5/4}} \tag{2.7}$$

Here $P$ may be in kilowatts or horsepower, so that care is needed in reading literature. Neither of equations (2.6) or (2.7) is non-dimensional, and care has to be taken to determine the dimensions used when using them. This fact, and the introduction of the SI system, has given rise to the so-called characteristic number $k_s$:

$$k_s = \frac{\omega\sqrt{Q}}{(gH)^{3/4}} \tag{2.8}$$

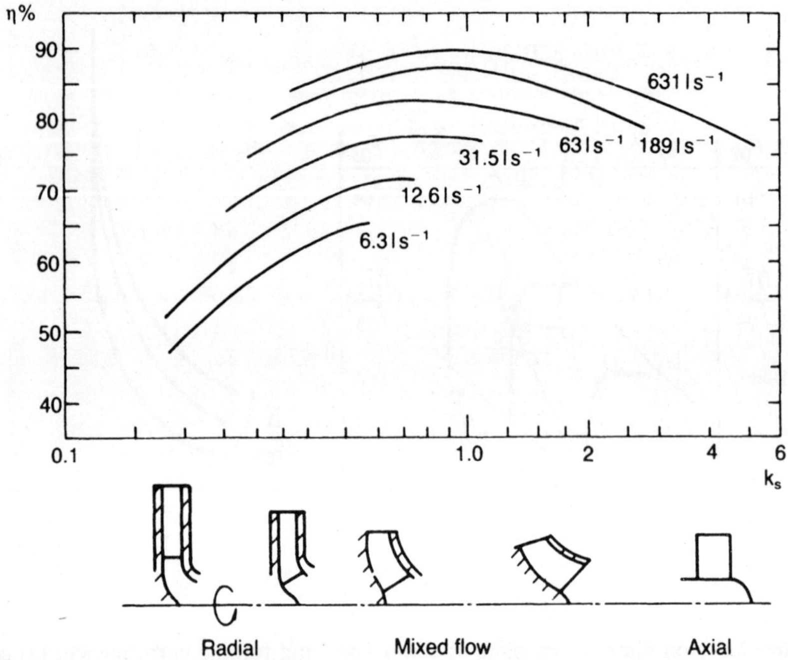


**Figure 2.6** Variation of overall efficiency with non-dimensional specific speed (after the well known Worthington plot).

where $\omega$ is in radians per second, $Q$ in cubic metres per second and $gH$ is the specific energy change. This gives a non-dimensional result, as does the $\Omega$ number (where $Q$ and $gH$ are in the foot, pound, second system) due to Csanady (1964).

Figure 2.6 indicates how this concept may be used as an index to flow path shape and type of machine in a range of applications. This figure is a well known chart for pumps. It illustrates how flow path changes as the specific speed increases. A low number means that the flow rate is low relative to energy change, and a radial impeller is needed; a large value suggests a high flow relative to energy change, and an axial rotor is required. It can readily be seen that specific speed could be a basis for presenting dimensions, as discussed by Balje (1981), for example.

## 2.4 Scale effects in incompressible units

### 2.4.1 Hydraulic machines

The use of models for predicting full size water turbine performance is well established, but the same technique for pumps is relatively recent, having

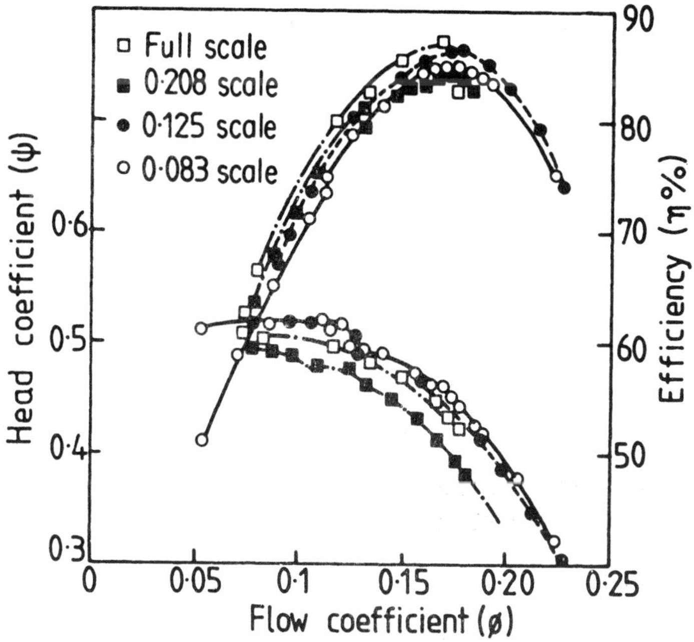


**Figure 2.7** Non-dimensional presentation of model and full size performance for the Eggborough pump (adapted from Nixon and Cairney, 1972a).

been made necessary by the increasing size of cooling water pumps for thermal power stations. Significant differences in efficiency between model and full size occur, suggesting departures from strict dynamic similarity. It is argued by both turbine and pump authorities that losses differ, and that work/capacity curves differ too. All are agreed that exact mathematical treatment is not possible, and attempt empirical formulation from experimental data. The references cited, which show wide variations, may be consulted as an extension of this section. The discussion here is based on the reports by Nixon (1965), and Nixon and coauthors (1972a,b), which are admirable examples of a logical approach.

As these authors suggest, departures from the scaling laws are due to:

- geometrical dissimilarities due to tolerances
- clearance variations
- surface finish
- hydrodynamic effects
- testing errors
- installation effects.

These were examined for the prototype and four models, three of which were in aluminium and the other in fibreglass, but being one-eighth scale for comparison with the aluminium surface produced. Figure 2.7 illustrates the variations observed from the prototype machine when the three aluminium models were tested.

Geometrical dissimilarities and their effects were examined. For example blade shape, as determined by blade angle, was compared. A variation of $\pm 0.5°$ about a mean value which was $1.5°$ less than design was found for the full size machine, and in the models the scatter was of the same order, apart from the glass-fibre impeller in which it was much worse. It must be commented that much higher variations are quite common in commercial cast impellers. Since the small aluminium models were shell mouldings the area variations were small, but the larger one was floor moulded, giving at least a 10% variation in area from passage to passage. Again, larger variations have been observed, particularly in cast impellers. The volute throat is a most important area, and correction is needed if variations occur.

Clearances tend to be of the same order in models as on the prototype, so leakage losses tend to be larger, and model surface roughness cannot be supersmooth in scale without large cost. This is therefore a significant problem because it affects boundary layers, as discussed in the following section.

Turning to hydrodynamic problems, Nixon and Cairney (1972), Osterwalder (1978), and Osterwalder and Ettig (1977), suggest the following relation:

$$1 - \eta_Y = \underset{\substack{\text{total}\\ \text{loss}}}{\delta_T} = \underset{\substack{\text{mechanical}\\ \text{loss}}}{\delta_M} + \underset{\substack{\text{leakage}\\ \text{loss}}}{\delta_L} + \underset{\substack{\text{disc}\\ \text{friction}}}{\delta_D} + \underset{\substack{\text{skin}\\ \text{friction}}}{\delta_F} + \underset{\substack{\text{inertia}\\ \text{loss}}}{\delta_I} \tag{2.9}$$

Here, $\delta_M$ and $\delta_I$ are unaffected by the Reynolds number and $\delta_I$ is usually assumed to remain the same. $\delta_M$ is considered to vary as speed, in contrast to the other hydrodynamic losses which tend to follow an $N^3$ law, and reduces with reducing speed at a lesser rate, thus being proportionally more important at low speeds. Nixon and Cairney (1972) present a method of finding $\delta_M$, and suggest that prediction from low speed tests be limited to differential head readings.

The estimation of disc friction loss has been a subject for argument, as the classical work was done on plain thin discs rotating in a close fitting closed casing. Nixon used work by Necce and Daily (1960) and Watabe (1958) for 'smooth' and 'rough' discs, and showed an error from measured data of about 10%. Sutton (1968) studied this problem, particularly the effect of leakage flow through wear rings and its relation to disc friction. Osterwalder (1978) commented that there is little current data of general applicability, but Kurokawa and Toyokura (1976) and Wilson and Goulburn (1976) extended the database.

The same situation is found in attempting to correlate $\delta_F$. Both Nixon and Osterwalder suggest the applicability of Nikuradse and Colebrook data with

**Table 2.1** A selection of model scale formulae (as quoted for example by Nixon (1965). Subscript m denotes 'model'

Moody (1942):

$$\frac{1-\eta}{1-\eta_m} = \left(\frac{D_m}{D}\right)^n \qquad 0 \leqslant n \leqslant 0.26$$

Moody:

$$\frac{1-\eta}{1-\eta_m} = \left(\frac{H_m}{H}\right)^{0.01}\left(\frac{D_m}{D}\right)^{0.25}$$

Anderson:

$$\frac{1-\eta}{1-\eta_m} = \frac{0.94 - Q^{-0.32}}{0.94 - Q_m^{-0.32}}$$

Pfleiderer:

$$\frac{1-\eta}{1-\eta_m} = \left(\frac{R_{em}}{R_e}\right)^{0.01}\left(\frac{D_m}{D}\right)^{0.25} \qquad \text{valid between } 1/12 < R_{em}/R_e < 20$$

Hutton:

$$\frac{1-\eta}{1-\eta_m} = 0.3 + 0.7\left(\frac{R_{em}}{R_e}\right)^{0.2}$$

Ackeret:

$$\frac{1-\eta}{1-\eta_m} = 0.5\left[1 + \left(\frac{R_{em}}{R_e}\right)^{0.2}\right]$$

a limiting Reynolds number criterion for 'transition'. Nixon proposes a relatively simple approach which is applicable and gives reasonable accuracy. Osterwalder surveys the published material and does not propose a general correlation, but indicates that a computer-based in-house study is relevant.

Nixon (1965), illustrating the deficiencies of a number of formulae in predicting pump efficiency from a model (some of which are in Table 2.1), demonstrated that there was little correlation. Anderson (1977) preferred a 'design' approach using a diagram based on data from many pumps and turbines to estimate probable efficiency.

### 2.4.2 Fans and blowers

The prediction of fan performance from models was studied at the National Engineering Laboratory (NEL), UK, resulting in a report by Dalgleish and Whitaker (1971) in which work on three small fans was used to predict the performance of fans at a scale of 1.5:1 and at a scale of about 2.5:1. The report underlined Nixon's comments on tolerancing and proposed a formula for $\eta_A$ similar to the pump equations just discussed:

$$\frac{(1-\eta_A)_p}{(1-\eta_A)_m} = 0.3 + 0.7\left(\frac{R_{em}}{R_{ep}}\right)^{0.2} \tag{2.10}$$

where $\eta_A$ is the air efficiency and $R_e = \rho\omega D^2/\mu$ for a range $0.8 \times 10^6 < R_e < 6.5 \times 10^6$. They comment that clearances are important, and that the best efficiency moved to higher flow coefficients than suggested by scaling as size increased. It is of interest to note that improvement in surface roughness does not give better efficiency except for small high-speed fans.

## 2.5 Scale effects in compressible machines

Equation (2.1) was quoted as applying to compressible machines, but limits of application apply, as for incompressible machines. If single-stage machines are considered, the effect of compressibility may be neglected for low Mach numbers (below about 0.5); the divergence caused increases with Mach number. Work with compressors using refrigerant and other 'heavy' gases indicates that the effect of the adiabatic exponent may be neglected for few stages. However, as the overall pressure ratio and hence the number of stages increase, density change particularly is important. As an example, if a pressure ratio of 8:1 is chosen, with air ($k = 1.4$) the density ratio is 4.41:1; if $k$ is 1.05 this becomes 7.23:1. This is clearly important in a multistage compressor where the stages are designed assuming a constant flow coefficient for air, but the machine is used for a heavier gas ($k = 1.05$) even though use of the gas reduces power demand for dynamically similar conditions since acoustic velocity goes down. The only technique proved satisfactory is to blend gases to give the right $k$, as Csanady (1964) showed.

The scaling problems in pumps, discussed in the previous section, occur in compressors. For example, Henssler and Bhinder (1977) studied the influence of size on a family of small turbocharger compressors; these had 60 mm, 82 mm and 94 mm impeller diameters, and other dimensions were scaled, but surface finishes had the same surface roughness. They show how performance varied with flow coefficient and Reynolds number based on peripheral speed. The authors comment that these changes are not predicted by the similarity laws, but do not attempt to suggest the correlation that different laws apply to different families of machines.

A contribution by Miller (1977) surveyed earlier correlations of scaling predictions for axial compressors. His approach was based on the need to conserve rig power consumption by testing at reduced pressure levels and dealing with the associated problems of correcting efficiency pressure ratio and flow to normal operating level, and was also concerned with scaling model tests to full size performance. Miller examined a number of Reynolds number correction approaches and concluded that although careful testing would yield effective prediction for one compressor design, this could not be applied to another, and that pressure level effects appear to be more pronounced than scale effects.

Others have contributed to the discussion as part of their study; for example, McKenzie (1980) shows a scatter in his predictions. No completely satisfactory general prediction appears to be available at the moment, although individual companies and groups use their own approaches and satisfy their own needs.

## 2.6 Illustrative examples

### 2.6.1 Similarity laws applied to a water turbine

The turbines in a river barrage hydroelectric plant are designed to give 55 MW each when the level difference is 25 m and they are running at 94.7 rev min$^{-1}$. The designed overall efficiency is 93%, and the runner diameter is 6 m. A model with a runner diameter of 300 mm is to be tested under the same level difference. Suggest the probable rotational speed, flow rate, efficiency and power produced when the model is operating in dynamically similar conditions.

The full size flow rate is first calculated:

$$\frac{55 \times 10^6}{0.93} = 25g \times 10^3 \times Q$$

Therefore

$$Q = 241\,\text{m}^3\,\text{s}^{-1}$$

Applying the scaling laws, equations (2.2),

$$\frac{Q}{ND^3} = \text{constant}$$

$$\frac{gH}{N^2D^2} = \text{constant}$$

and substituting the known data, it is found that

$$Q_{\text{model}} = 0.602\,\text{m}\,\text{s}^{-1}$$
$$N_{\text{model}} = 1894\,\text{rev}\,\text{min}^{-1}$$

The model efficiency must now be found using one of the equations in Table 2.1. The well known turbine equation due to Hutton will be used:

$$\frac{1-\eta}{1-\eta_{\text{model}}} = 0.3 + 0.7\left(\frac{1894 \times 0.3^2}{94.7 \times 6^2}\right)^{0.2}$$

from which the model efficiency is found to be

$$\eta_{\text{model}} = 89.77\%$$

The model power developed, assuming the same mechanical efficiency, is therefore

$$P_{\text{model}} = 25g \times 10^3 \times 0.602 \times 0.8977$$
$$= 132.54\,\text{kW}$$

### 2.6.2 Compressor performance prediction problem

A compressor for hydrogen duty is to deliver $18\,\text{kg}\,\text{s}^{-1}$ while increasing the pressure from $1.01 \times 10^5\,\text{N}\,\text{m}^{-2}$ to $16.5\,\text{N}\,\text{m}^{-2}$. The inlet temperature is expected to be 300 K, and the rotational speed $2900\,\text{rev}\,\text{min}^{-1}$.

For development purposes a half-scale machine is to be tested using air as the medium, with inlet conditions $10^5\,\text{N}\,\text{m}^{-2}$ and 288 K. Suggest the model mass flow rate, delivery pressure and rotational speed for dynamical similarity.

The full size pressure ratio is 16.5/1.01 = 16.34 : 1. The model outlet pressure will be $16.34 \times 10^5\,\text{N}\,\text{m}^{-2}$. Following equation (2.3), for similarity,

$$\frac{\dot{m}\sqrt{(RT_{01})}}{p_{01}D^2} = \text{constant}$$

and

$$\frac{ND}{\sqrt{(RT_{01})}} = \text{constant}$$

For hydrogen, $R = 4.124\,\text{kJ}\,\text{kg}^{-1}\,\text{K}^{-1}$; for air, $R = 0.287\,\text{kJ}\,\text{kg}^{-1}\,\text{K}^{-1}$. Thus

$$\frac{18\sqrt{(4.124 \times 10^3 \times 300)}}{1.01 \times 10^5 \times D^2} = \frac{\dot{m}\sqrt{(0.287 \times 10^3 \times 288)}}{10^5 \times (0.5D)^2}$$
$$\dot{m} = 17.24\,\text{kg}\,\text{s}^{-1}$$

and

$$\frac{2900D}{\sqrt{(4.124 \times 10^3 \times 300)}} = \frac{N \times 0.5D}{\sqrt{(0.287 \times 10^3 \times 288)}}$$
$$N = 1499.15\,\text{rev}\,\text{min}^{-1}$$

## 2.7 Exercises

2.1 A centrifugal pump rotates at $185\,\text{rad}\,\text{s}^{-1}$ and at best efficiency has a pressure rise of $4.5 \times 10^5\,\text{N}\,\text{m}^{-2}$ when pumping water at the rate of

$0.28\,m^3\,s^{-1}$. Predict the corresponding best efficiency flow rate and pressure rise when rotating at 80% of the design speed. If the efficiency is 85% in both cases estimate the input power required.

2.2 If a pump geometrically similar to that in Exercise 2.1 is increased in diameter to give the same pressure rise of $4.5 \times 10^5\,N\,m^{-2}$ when rotating at $148\,rad\,s^{-1}$ suggest the scale ratio, and the flow rate and power input if the efficiency is 85%.

2.3 A multistage pump is to lift water at a rate of $0.03\,m^3\,s^{-1}$ from a mine 820 m deep. If the motor speed is 2900 rpm, estimate the number of stages needed if the specific speed in 20 (rpm $m^3\,s^{-1}$.m). (Remember that the specific speed applies to one stage only in a multistage machine.)

2.4 The specified duty for a hydraulic turbine is 37.5 MW at 90 rpm under a head of 18 m with an efficiency of 93%. The factory test bay can give a head of 6 m and absorb 45 kW. Suggest the necessary scale ratio for a model, its rotational speed and the flow rate. Assume that the model efficiency is 93%.

2.5 A one-third scale model of a ventilating fan is to be constructed. The machine to be modelled is to run at 1450 rpm, with a duty flow rate of $5\,m^3\,s^{-1}$ and pressure rise of $450\,N\,m^{-2}$ and overall efficiency of 78% when the intake conditions are 5 °C and $0.989 \times 10^5\,N\,m^{-2}$. The model fan will draw air in at 20 °C and $10^5\,N\,m^{-2}$. Predict the model performance and rotational speed when performing under dynamically similar conditions.

2.6 A water turbine is to develop 100 MW when running at 93.7 rpm under a head of 20 m with an efficiency of 92%. The runner diameter is to be 7 m. If a water model is to be tested and the maximum test head and flow available are 5 m and $0.9\,m^3\,s^{-1}$ estimate the scale ratio, the model runner diameter and rotational speed and the power to be absorbed. Discuss the possibilities of using air models to study such a machine.

2.7 A new design of compressor is to be tested. It is to deliver helium with a pressure ratio of 12:1 when rotating at 3000 rpm with a mass flow rate of $12\,000\,kg\,h^{-1}$. The inlet conditions are 10 bar and 300 K. For in-house testing air is to be used with intake conditions of 1 bar and 300 K. Suggest the pressure ratio, mass flow rate, and rotational speed to give dynamically similar conditions and discuss the validity of your approach and the use of air to model other gases.

2.8 A gas turbine tested with air passed 15 kg s$^{-1}$ when the expansion ratio was 1.6:1, the rotational speed was 6000 rpm, and the inlet static conditions were $10 \times 10^4$ N m$^{-2}$ and 288 K. The same machine when used for its designed purpose was supplied with combustion products at the inlet conditions of $1.4 \times 10^6$ N m$^{-2}$ and 660 K. If the expansion ratio was also 1.6:1, estimate for dynamically similar conditions the rotational speed and mass flow rate. (For combustion products R = 2291 J kg$^{-1}$ K$^{-1}$.)

# 3 Cavitation

## 3.1 Introduction

Cavities form in a flowing liquid in an area where the local pressure approaches vapour pressure; the entrained and dissolved gas form bubbles which grow and then collapse. In a venturi section of a pipe, Fig. 3.1, bubbles form near the throat and are swept downstream to an area where the pressure is higher and where they then collapse. There are a number of contributory factors: liquids absorb gas into solution (Henry's law) and the capacity to retain gas in solutions is a function of the local pressure, thus gas is liberated as pressure falls. Gas and vapour nuclei act as centres of growth and liberated gas helps bubble growth. (Henry's law suggests that water will absorb air through a free surface at atmospheric pressure at the rate of 2% of the water volume; hydrocarbon fluids have a greater capacity to absorb gas.) Bubble growth and collapse is rapid, as Figs 3.2 and 3.3 indicate. It has been observed that bubbles can burst and reform several times, indicating that the process is unstable. This gives rise to flow instability and in machines the result is noise, vibration and surface damage.

Since cavities form where the pressure is low, cavitation is a problem in the suction zone of pumps and in the area of discharge from a water turbine. Figure 3.4 illustrates the effect on the characteristics of a pump of reduction in the suction pressure leading to cavitation, and Fig. 3.5 illustrates the effect of cavitation on the behaviour of a turbine.

In a centrifugal pump, Fig. 3.6, the rotational effect of the rotor is increasingly imposed on the liquid as it approaches the impeller blades. This results in an increasing absolute velocity with a reduction in pressure which continues until the leading edges of the impeller blades are reached. Usually, the pressure then increases on the pressure side of the blades but reduces further on the suction surface with bubbles forming if the vapour pressure is approached. Flow in the blade passages is affected by the formation and collapse of the cavities and this gives rise to a reduction in performance, as sketched in Fig. 3.4, and also to noise and vibration with, in addition, damage to surfaces (section 3.3).

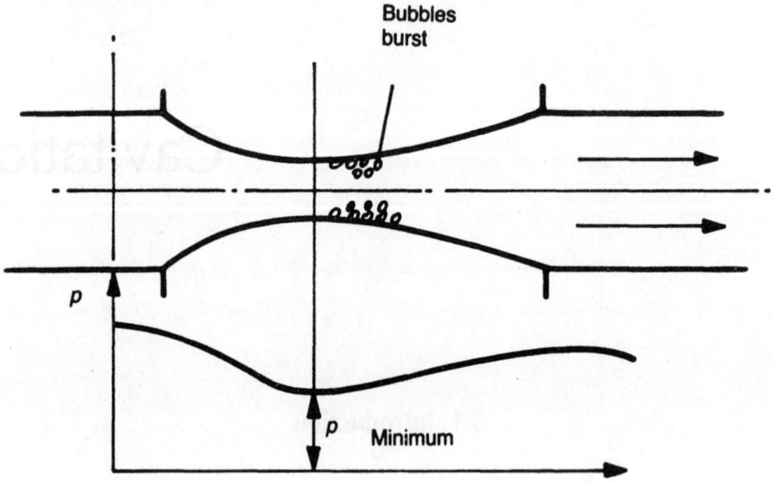


**Figure 3.1** Cavitating flow through a venturi nozzle.

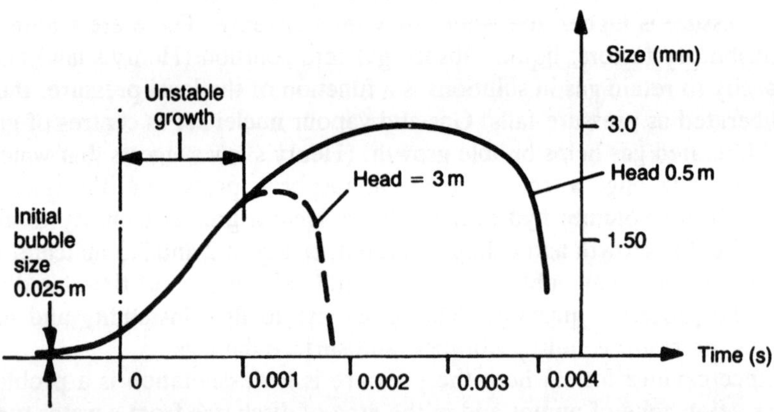


**Figure 3.2** The life history of a bubble when growing and bursting in regions at two different pressure heads.

## 3.2 Net positive suction energy (NPSE) or head (NPSH)

Hydraulic machinery engineers have used an expression for a long time which compares the local pressure with the vapour pressure and which states the local pressure margin above the vapour pressure level. In the era before the SI system it was always known as the net positive suction head (NPSH) and must now be known as the net positive suction energy (NPSE). Two definitions are used, that which relates to the pressure presented to the

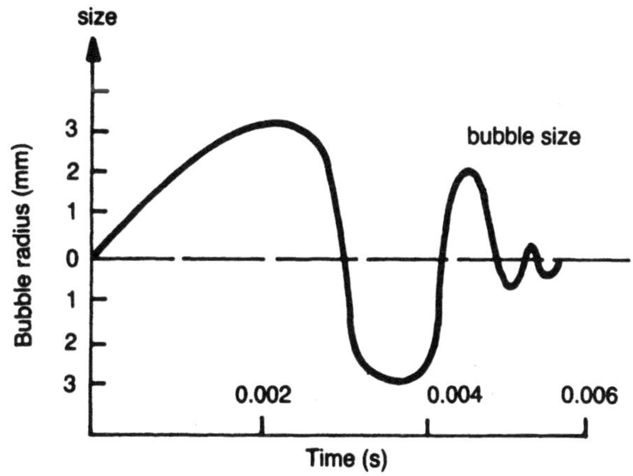


**Figure 3.3** The life history of growth and collapse for a single cavity over 6 microseconds (based on data in Knapp and Daily, 1970).

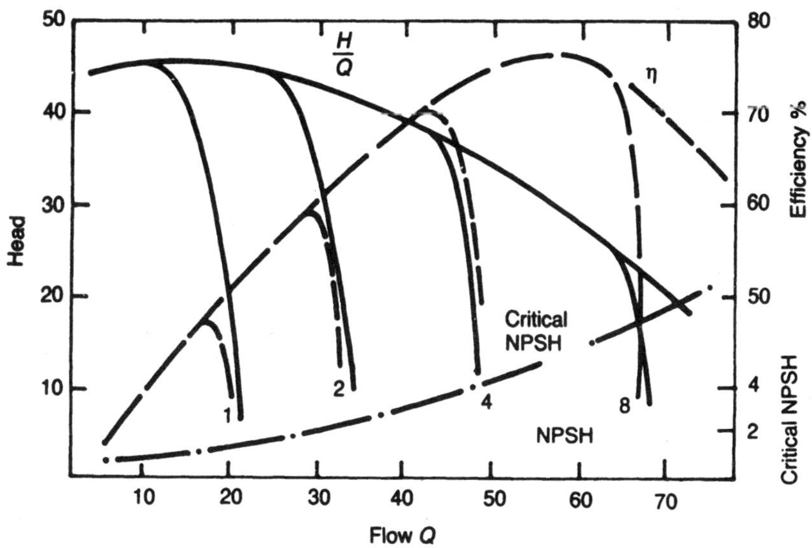


**Figure 3.4** The effect of net positive suction head on a pump characteristic.

machine by the system ($NPSE_a$) and that generated by the dynamic action of the machine ($NPSE_R$).

### 3.2.1 NPSE available ($NPSE_a$) or $NPSH_a$

There are two main suction systems applied to pumps: the drowned system and the suction lift system (Fig. 3.7). The lowest pressure $P_s$ in the suction

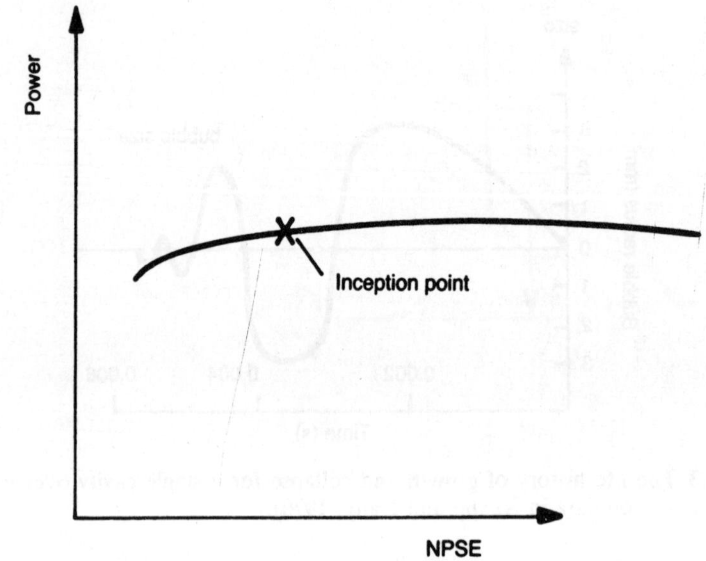


**Figure 3.5** The effect of cavitation inception on a water turbine.

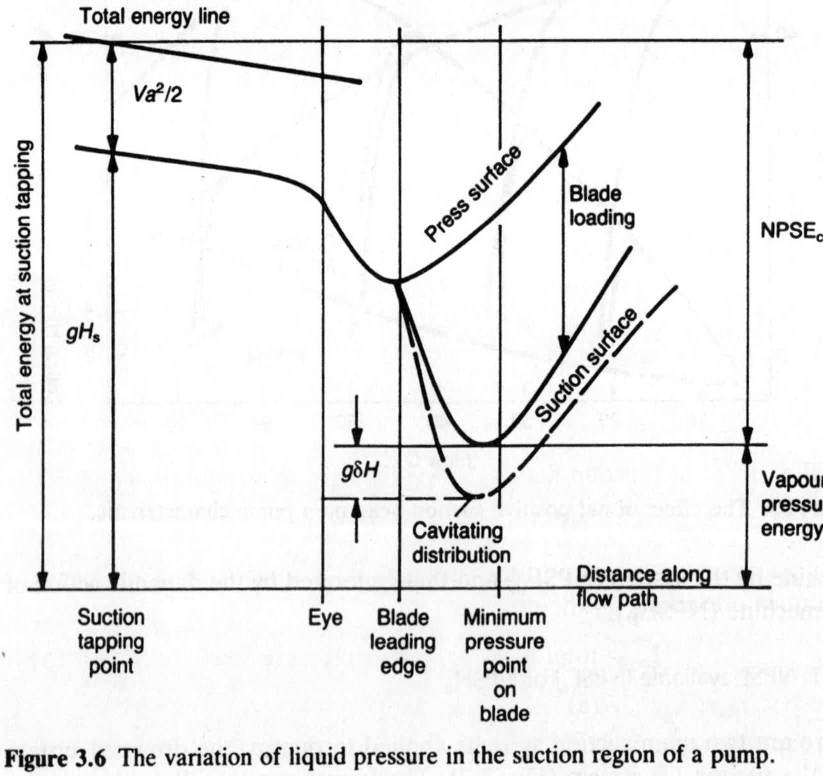


**Figure 3.6** The variation of liquid pressure in the suction region of a pump.

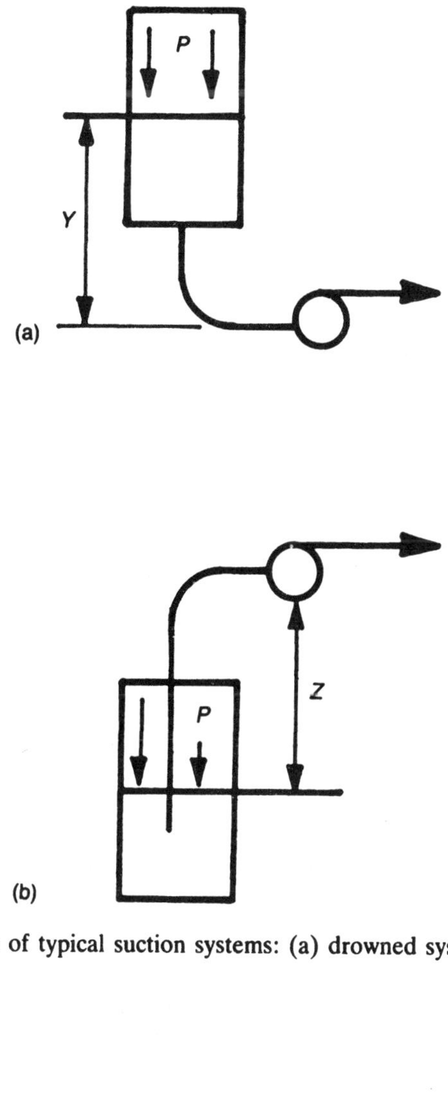


**Figure 3.7** Sketches of typical suction systems: (a) drowned system; (b) suction lift system.

line will occur at the suction flange. $NPSE_a$ is then defined by the equation

$$NPSE_A = P_s/\rho - P_v/\rho \quad (3.1a)$$

where $P_s$ is the total pressure of the fluid and $P_v$ the vapour pressure at the temperature prevailing. In the older system

$$NPSH_A = \text{total head} - \text{vapour pressure head} \quad (3.1b)$$

For the system in Fig. 3.7(a)

$$NPSE_a = P/\rho + gH - \text{flow losses in the suction line} - P_v/\rho \quad (3.2)$$

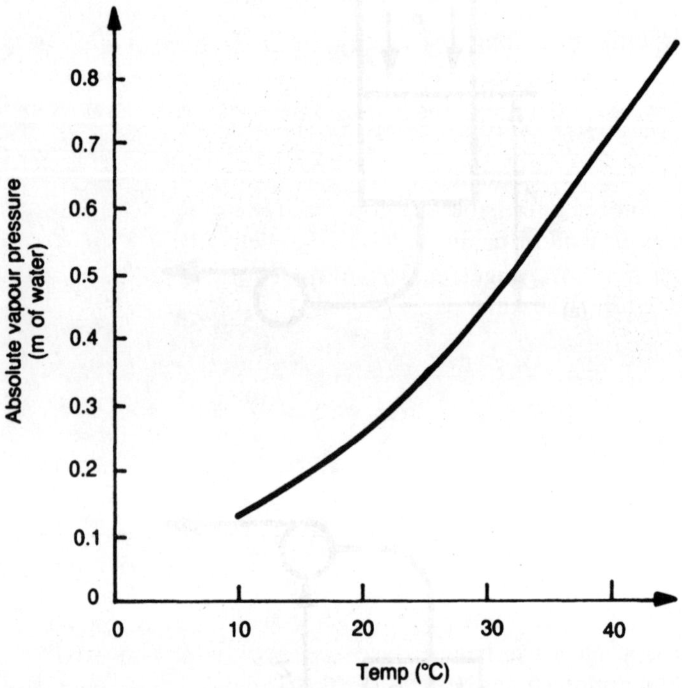


**Figure 3.8** Variation of vapour pressure of water with temperature.

and, for the system in Fig. 3.7(b),

$$\text{NPSE}_a = P/\rho - gH - \text{flow losses in the suction line} - P_v/\rho \quad (3.3)$$

where $P$ is absolute pressure.

The flow losses must be calculated for the flow rate that the pump is to deliver. An example of a calculation for the system in Fig. 3.7(b) now follows.

A water pump has its centreline 5 m above the free surface in a sump which is at atmospheric pressure. The water temperature is 20 °C. If the flow losses are calculated to be 10 J kg$^{-1}$ estimate the NPSE$_a$. From data sheets and Fig. 3.8, $P_v$ is 2.354 × 10$^3$. Thus,

$$\text{NPSE}_a = \frac{10^5}{10^3} - 5g - 10 - \frac{2.354}{10^3} = 38.596\,\text{J kg}^{-1}\ (\text{NPSH}_a = 3.93\,\text{m})$$

As an example of an NPSE$_a$ calculation for a drowned suction, consider a tank that has a free surface 2.3 m above the pump centreline, containing *n*-butane which is held at its boiling point of 37.8 °C. Friction losses are computed to be 5 J kg$^{-1}$ at the flow condition under study – estimate NPSE$_a$

if the vapour pressure for $n$-butane at 37.8 °C is 3.59 bar, and the relative density is 0.56.

Since the liquid is 'boiling' the free surface is at 3.59 bar, therefore

$$\text{NPSE}_a = \frac{3.59 \times 10^5}{0.56 \times 10^3} + 2.5g - 5 - \frac{3.59 \times 10^5}{0.56 \times 10^3} = 19.525\,\text{J kg}^{-1}$$

or, $\text{NPSH}_a$ is 1.99 m of butane.

If the suction pressure is obtained from a pressure gauge mounted in the suction flange instead of being estimated from the system data, it has to be remembered that the gauge reads static pressure, so the total pressure is obtained by adding the kinetic term,

$$\frac{p}{\rho} = \frac{p_{\text{gauge}}}{\rho} + \frac{V^2}{2} \tag{3.4}$$

### 3.2.2 NPSE required ($\text{NPSE}_R$)

This is the statement of the margin above vapour pressure that the pump can cope with by its dynamic action. It is clearly a function of the impeller geometry and is very difficult to estimate with any accuracy. The classic text by Pfleiderer (1961) contains equations that will given an estimate, but in general the approach has been to use empirical data to give $\text{NPSE}_R$. The most common approach has been to use the Thoma number, first developed for water turbines:

$$\sigma = \frac{\text{NPSE}_R}{\text{Machine } gH} \tag{3.5a}$$

or

$$\sigma = \text{NPSH}_R/\text{machine head rise} \tag{3.5b}$$

Figures 3.9 and 3.10 are plots of $\sigma$ against characteristic number $k_s$, based on well known diagrams to be found in the classic texts. This approach gives conservative values of $\text{NPSE}_R$, and yields values which are based on **design** flow and energy rise.

### 3.2.3 Critical or limiting NPSE

The two statements of NPSE are based on different criteria, so their relation to one another must be discussed. If a pump is considered, $\text{NPSE}_a$ and $\text{NPSE}_R$ can be plotted against flow rate (Fig. 3.11). $\text{NPSE}_a$ will fall with increasing flow rate since the flow losses vary as (flow rate)$^2$; $\text{NPSE}_R$ varies in the manner shown. The point at which the curves cross is called the critical flow, as at higher flow rates the pump cannot sustain the suction pressure presented to it, and at lower rates the pump is supplied with

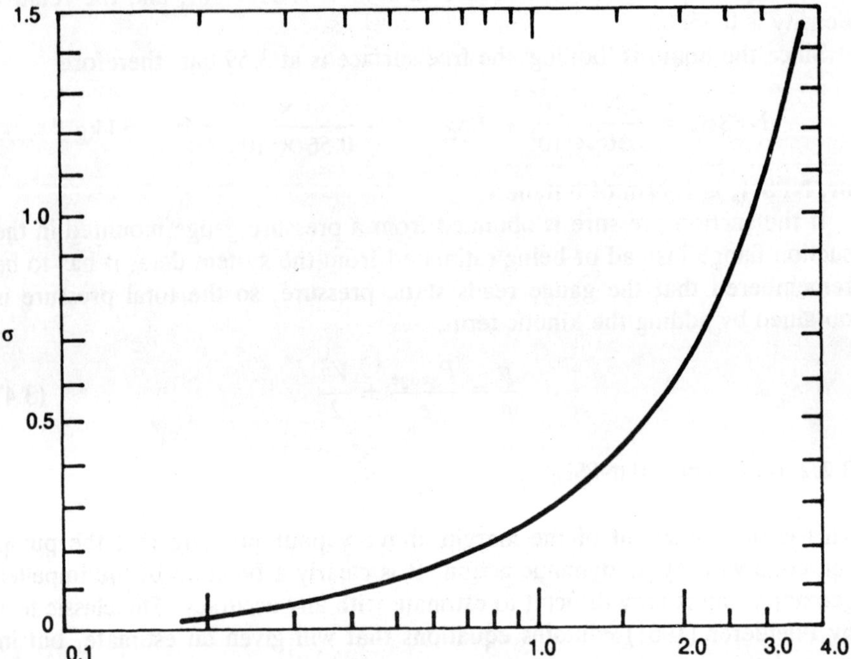


**Figure 3.9** Variation of Thoma's cavitation parameter with characteristic number for pumps.

much larger suction pressures than it needs and cavitation should not be a problem. It can be argued that at flow rates higher than the critical value the pump will cavitate badly, and at lower flow rates it should behave satisfactorily. While this is a general rule, studies over the last 15 years have supported the view put forward by Bush *et al.* (1976) that at low flow rates recirculation in the suction line and between impeller and casing can give rise to cavitation-induced damage and instability (section 3.4).

In general, pumps are tested for their cavitation performance by reducing the NPSE whilst maintaining the pump speed and flow constant. The resulting curve (Fig. 3.12) shows a level energy rise until at low suction pressure it falls away. The point at which the performance is stated to be unacceptable is when the energy rise falls by $x$% (usually 3% for standard pumps as laid down by the American Hydraulic Institute). The corresponding NPSE is called the critical value.

Hydraulic turbines, when model tested, are usually tested at the design full size head drop and the point at which cavitation is observed noted. In turbines the effect is not so marked as cavitation is downstream and air is injected into the draft tube to reduce pulsations. Damage is, of course,

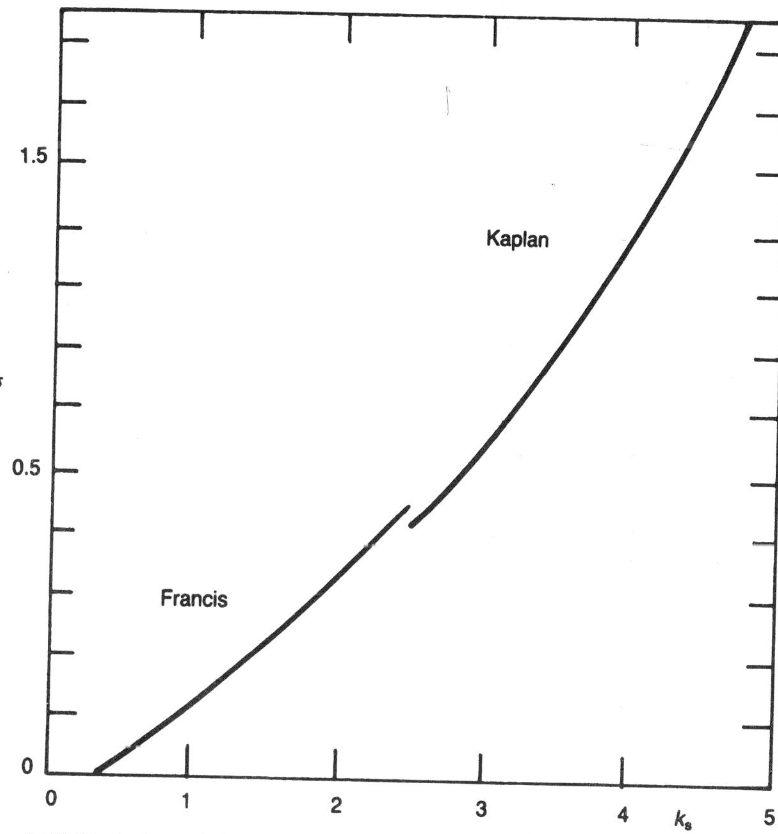


**Figure 3.10** Variation of Thoma's cavitation parameter with characteristic number for Francis and Kaplan water turbines.

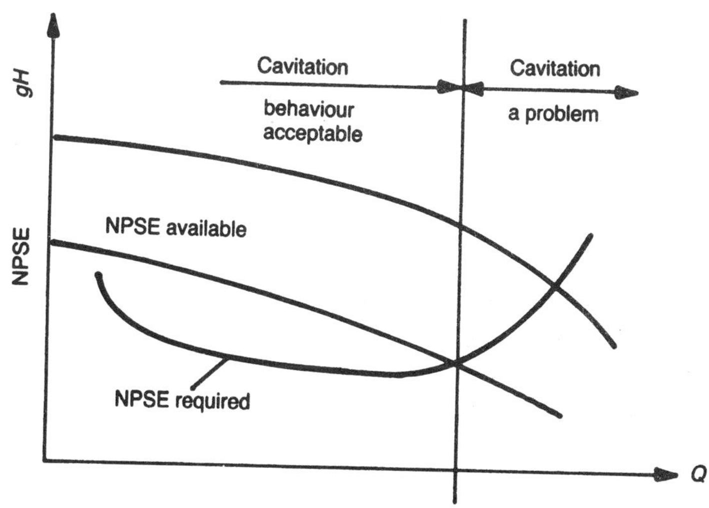


**Figure 3.11** NPSE available and NPSE required curves presented on a base of flow rate to illustrate the critical flow rate where the curves cross.

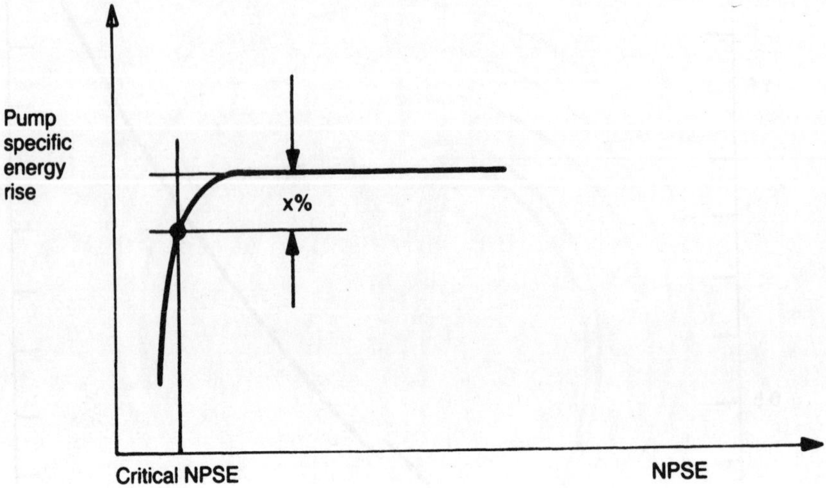


**Figure 3.12** The usual way of testing and presenting pump cavitation behaviour. (The test is performed of constant speed and constant 'design' flow.)

important in all machines so effort is expended to reduce cavitation where possible.

As an example of $NPSE_R$ calculation consider a pump with a design duty of $35\,m^3\,h^{-1}$ when delivering against a specific energy rise of $409\,J\,kg^{-1}$ and rotating at 3000 rpm. The system NPSE is computed to be $30\,J\,kg^{-1}$. Comment on the risk of cavitation.

$$k_s = \frac{\dfrac{3000\pi}{30}\sqrt{\left(\dfrac{35}{3600}\right)}}{490^{3/4}} = 0.297$$

from Fig. 3.9 $\sigma = 0.05$, thus

$$NPSE_r = 0.05 \times 490 = 24.5\,J\,kg^{-1}$$

or, $NPSH_R = 2.5\,m$ of liquid.

$NPSE_a$ is greater than $NPSE_r$ by a small margin. This indicates that if cold water is the liquid that there may be little difficulty, if the suction system does not ingest air. If the fluid is volatile and is bubbly, there will not be enough margin and some thinking is needed about the levels of tank and pump and the suction system.

## 3.3 Cavitation damage

Surface damage is a serious problem in pumps and turbines. Many studies have been made over the years of the causes and of the ways in which

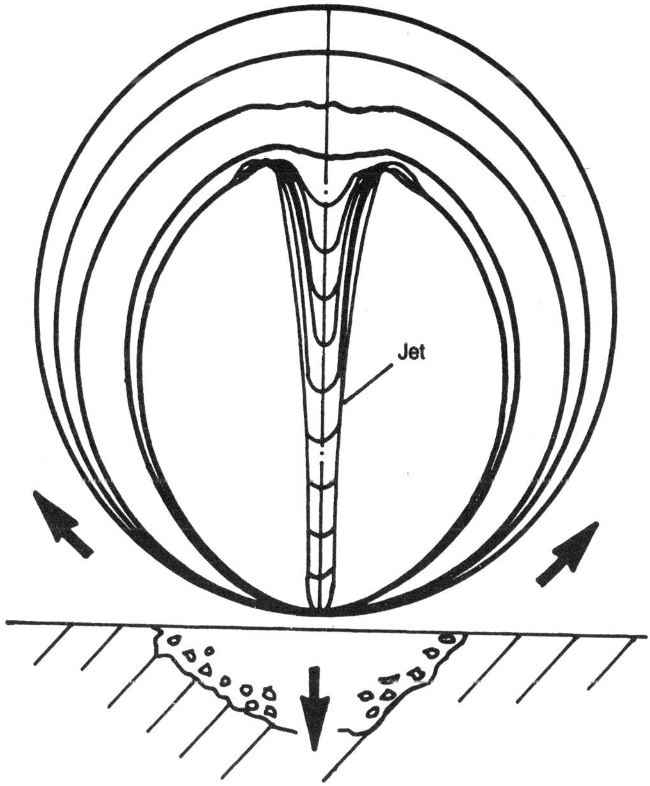


**Figure 3.13** The jet collapse model for damage due to cavitation as illustrated by Lush (1987).

damage can be reduced. Lush (1987a,b) summarized findings about the way the bubble collapse gives rise to very high local pressures on the surfaces of the machine channels. The existence of a microjet is proposed which, as the cavity collapses, bursts across the void of the bubble to hit the opposite surface at very high velocity (Fig. 3.13). The impact results in very high stresses which are equal to or higher than the ultimate strength of the material in many cases. There is a debate about the mechanism that leads to surface damage, but there appears to be agreement that the material loss is due to a combination of mechanical, chemical and electrolytic actions. A suggested mechanism, based on the computed stresses being of the order of 2000 bar in some cases, is that work hardening with attendant temperature rise occurs in the material at the surface. This can lead to small changes in chemical composition and a spongy subsurface that leads to cracking of the hardened surface with subsequent penetration of the material. Eventually, the molecular bonding breaks down and erosion and corrosion occurs.

**Table 3.1** Relative losses of material under comparable conditions (obtained in a venturi test device using water at 20 °C)

| Material | Relative volumetric loss |
|---|---|
| Stellite | 1 |
| Cast stainless steel: 12.88% Cr, 0.17% Ni, 0.43% Mn, 0.38% Si | 7 |
| Stainless steel 18:8 Cr:Ni | 5 |
| Monel | 16 |
| 0.33% carbon steel | 37 |
| 14% Cr stainless steel (forged or drawn) | 98 |
| Manganese bronze | 118 |
| Gun metal | 230 |
| Cast iron (as cast without skin) | 374 |
| Typical cast aluminium alloy | 1176 |

Typical damage is described by Tillner *et al.* (1990) and Pearsall (1978) who show ways that damage can be produced in the laboratory for test and development purposes. The extent of the damage suffered depends on the fluid, the materials and the hydrodynamic system, and it has been found that even with advanced material loss the machine has developed the duty required and damage has only been found during routine maintenance. In pumps repair is usually by replacement, but in hydroelectric plant it is a routine procedure to deposit metal in damaged areas and then to return the surface to the high finish required. Table 3.1 summarizes the resistance of common materials used in hydraulic machines.

## 3.4 Hydrodynamic effects

The effect of bubble formation and collapse have been discussed briefly in the introduction to this chapter. The effect of reducing suction pressure was illustrated in Fig. 3.4, with the flow range being progressively restricted as suction pressure falls. Pearsall (1973) discussed this and proposed ways of reducing the effects by increasing the suction diameter. While this is a valid approach care is needed to ensure that the suction diameter does not get too large as then recirculation can be set up at flow rates close to the design flow and cause instability; reference may be made to Bush *et al.* (1976) and to Grist (1986). The studies by Pearsall and his team at NEL (1978) demonstrated a close relation between noise and damage and they proposed a noise criterion for cavitation prediction. Wolde *et al.* (1974) described the problems to be solved in noise studies and demonstrated the validity of the technique.

The effects on water turbines have been discussed by Knapp and Daily (1970) and by Klein (1974) who with others at the same conference illustrated the effect of cavitation on the performance envelope.

## 3.5 Thermodynamic effects on pump cavitation

It is well known that the cavitation performance of a pump varies with the fluid condition. For example, when pumping water the required NPSE is highest when passing cold water and decreases as the temperature rises. A similar effect is noted when other liquids are being pumped. The change is too large to be explained by the so-called Reynolds number effects. An empirical approach to this problem is known as the $B$-factor or $\beta$-factor method outlined by Knapp and Daily (1970) and Stahl and Stepannof (1956); the technique correlates suction hydraulic behaviour with vapour volume in the cavitating region.

When examining bubbles in cold water it has always been assumed with some justification that all energy terms involving the vapour in the cavities are negligible compared with those of the surrounding liquid. When the temperature increases such an assumption may not be applied since the latent heat required to supply vapour to the cavity can no longer be ignored, and neither can the energy exchange during the expansion and contraction of the bubble. If they are neglected and the NPSE calculated using cold data the resulting value will be very conservative. As a consequence, boiler feed pumps in the 1950s began to be uneconomic in size as both pressures and temperatures rose with turbine steam conditions; eventually designers developed techniques to design for lower cavitation numbers than those for cold duties.

The heat required by vaporization must come from the liquid surrounding the cavity, thus causing a drop in temperature and vapour pressure in the immediate vicinity of the bubble. This has the effect of reducing the bubble size from that which would apply in cold liquid, thus reducing the effect on the flow. This reasoning has led to the approach of Stahl and Stepannof (1956). Figure 3.14 shows the conventional NPSE variation curves for a cold and a hot liquid. The two critical points where the 3% head drop applies are B and C. Point B is known for the cold test, and the NPSE reduction can be found to estimate point C.

The volume ratio $B$ or $\beta$ is defined by Stahl and Stepannof as

$$B = \frac{\text{volume of vapour}}{\text{volume of liquid}} \tag{3.6}$$

they showed that using the Claperyron–Clausius equation

$$B = \Delta(\text{NPSE})C_pT\left(\frac{v_v}{v_l h_{fg}}\right)^2 \tag{3.7}$$

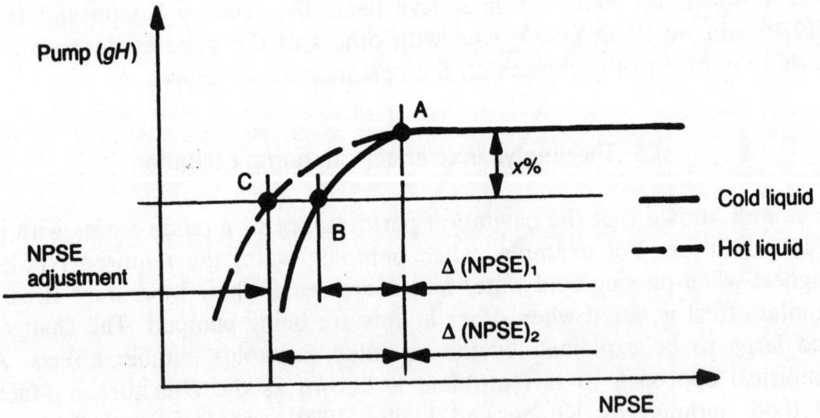


**Figure 3.14** The variation on NPSE required of liquid temperature.

or

$$B' = \frac{B}{\Delta \text{NPSE}} = C_p T \left( \frac{v_v}{v_1 h_{fg}} \right)^2 \qquad (3.8)$$

Figure 3.15, taken from Stahl and Stepannof (1956), plots $B'$ for a number of fluids based on refinery pumps of the double suction design and a 3% fall in $gH$. Also shown are lines of NPSE adjustment in the relation

$$\sigma_{\text{corrected}} = \sigma \frac{-\text{NPSE adjustment}}{gH_{\text{pump}}} \qquad (3.9)$$

The method is based on the assumptions that the cavities are uniformly distributed across the flow cross-section and that there is the 3% drop criterion. There are many other approaches but these indicate that there is a considerable difference in opinion (see for example Hutton and Furness, 1974).

### 3.6 Inducer/pump combinations

In some applications the suction pressure is too low for the pump and it cavitates. Where it is not possible to change the system design there is a solution, an inducer, which is an axial flow stage that is placed just before the impeller (Fig. 3.16). The device is a high specific-speed machine and is required to give a small pressure rise that raises the pressure at inlet to the main impeller to a level that it can sustain without cavitation. It has long passages so that bubbles that form burst before leaving the inducer, so that any damage is confined to the inducer blades. The blade shapes are

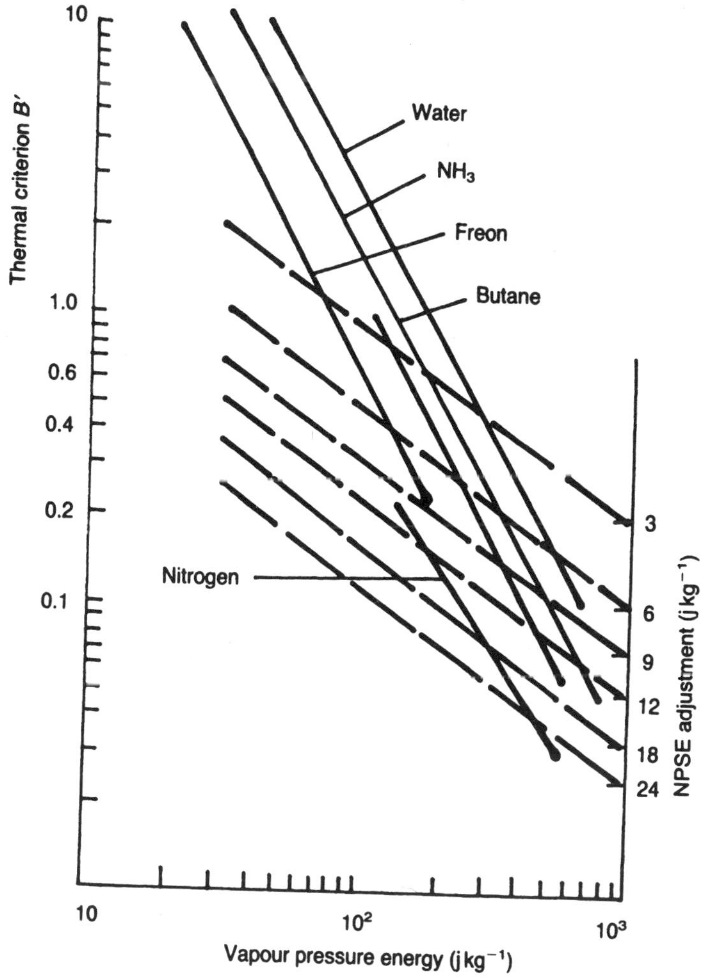


**Figure 3.15** Plot of thermodynamic adjustment factors (from Stahl and Stepannof, 1956).

archimedean spirals with a very acute inlet angle referred to the tangential direction.

The improvement in NPSE performance that an inducer can produce is illustrated in Fig. 3.17. Work has shown that the improvement is only in the flow range close to the design point, and away from that zone, because the incidence angles are higher, the impeller inducer combination can give worse NPSE values than the impeller alone. Contributions by Pearsall (1978), Susada *et al.* (1974), Lakshminarayana (1982), Turton (1984), Tillner *et al.* (1990) and Oschner (1988), may be consulted for more information on both the design and behaviour of inducers when applied to pumps.

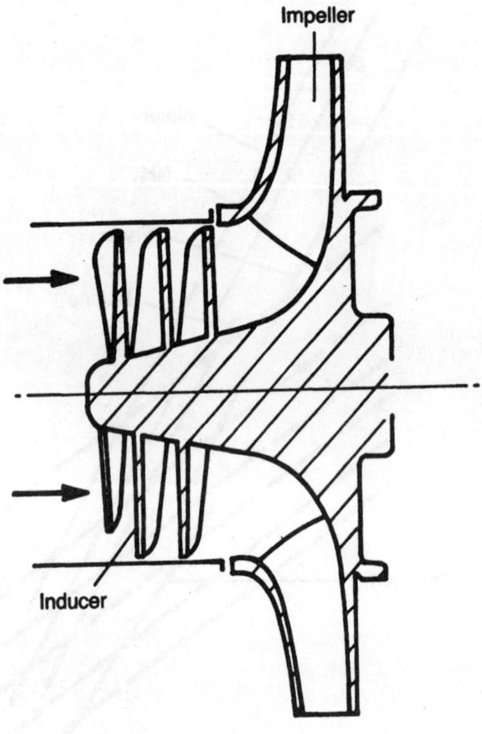


**Figure 3.16** Sketch of a pump/inducer design.

## 3.7 Concluding comments

The undesirable effects of cavitation, flow instability, noise, and surface damage have been described, and basic guidelines for avoidance have been introduced. For detailed discussion more advanced material may be found in the references cited and in the textbooks listed.

## 3.8 Exercises

3.1 A pump is sited 5.5 m above a water sump, and the friction energy loss at design flow is computed to be $8.5\,\mathrm{kg}^{-1}$. If the water temperature is 25 °C compute the $NPSE_A$ at sea level and at a level where the atmospheric pressure is $0.85 \times 10^5\,\mathrm{N\,m}^{-2}$.

3.2 A water pump is supplied by a tank with a free surface 4 m above the pump centreline. At the design flow the suction system loss is $50\,\mathrm{J\,kg}^{-1}$.

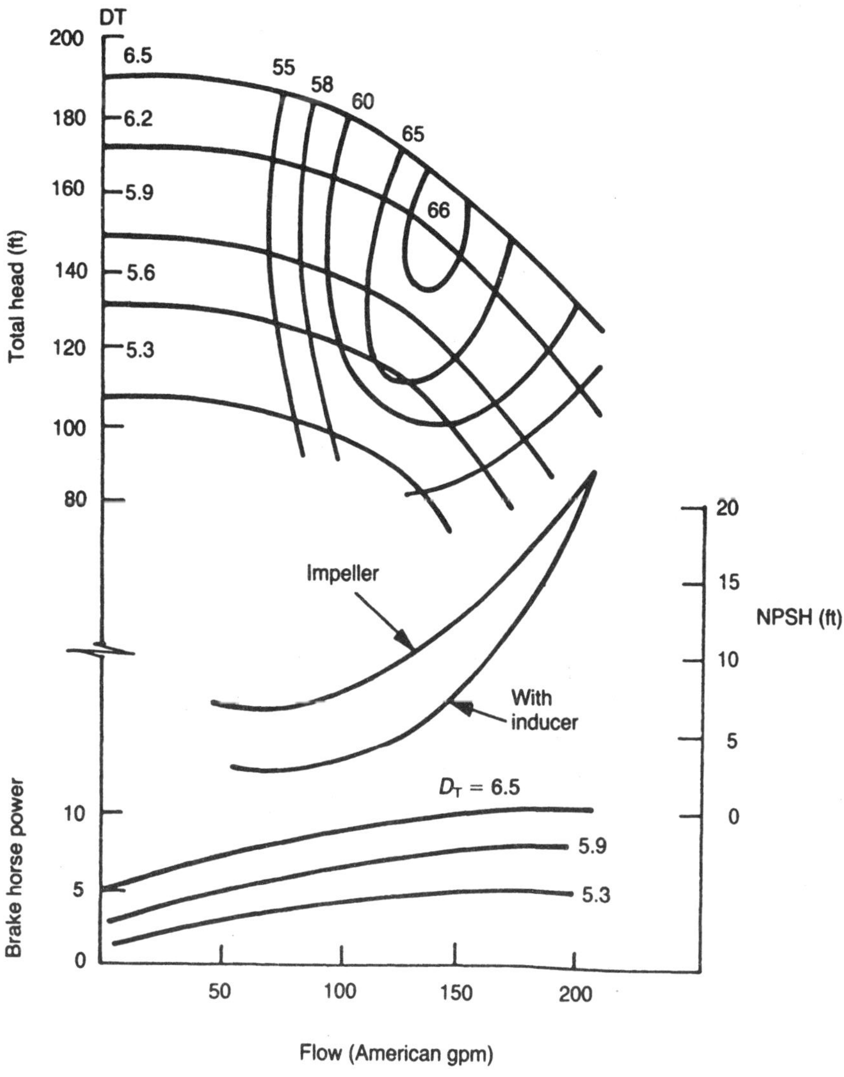


**Figure 3.17** Characteristic of a pump fitted with an inducer.

If the water is at 25 °C determine the $NPSE_A$ when the free-surface pressure is $10^5\,N\,m^{-1}$ and $1.3 \times 10^5\,N\,m^{-2}$.

3.3 If in the system described in Exercise 3.2 a limiting NPSE of $100\,J\,kg^{-1}$ is required by the pump and the losses rise to $100\,J\,kg^{-1}$, calculate the necessary pressure on the free surface.

3.4 A system to deliver *n*-butane has a tank with a free surface at 5.5 m above the pump. Calculate for a suction system loss of $50\,J\,kg^{-1}$ the $NPSE_A$ when the suction tank free surface is at 3.5 bar gauge, and when it is at vapour pressure. Assume fluid density is $0.56\,kg\,m^{-3}$, vapour pressure is 3.59 bar absolute, and atmospheric pressure is $0.95 \times 10^5\,N\,m^{-2}$.

3.5 If the tank level in Exercise 3.4 falls to 4 m, and the free surface is at vapour pressure determine the maximum suction system loss that can be allowed if the $NPSE_A$ is not to fall below $10\,J\,kg^{-1}$.

3.6 A pump is to deliver water at the rate of $80\,m^3\,h^{-1}$ against a system resistance of 50 m when rotating at 1450 rpm. The losses in the suction line are estimated to be $25\,J\,kg^{-1}$ at the specified flow rate. Assuming a suction lift suggest the maximum level difference from sump level to pump that can be sustained. Assume the water is at 20 °C and the margin between $NPSE_R$ and $NPSE_A$ is to be a minimum of $10\,J\,kg^{-1}$.

# 4 Principles of axial flow machines

## 4.1 Introduction

Axial flow pumps, fans, compressors and turbines have, as a common feature, blades which interact with the fluid involved. Typical machine flow paths are shown in Fig. 4.1, and in the two-dimensional approach that will be discussed it is assumed that the stream surfaces are symmetrical with the axis of rotation. Both the isolated aerofoil and cascade data are described, with the limitation just outlined. The three-dimensional problems are then introduced, and the concepts of radial equilibrium, the actuator disc approach, and Howell's work done factor are related to the common secondary flow problems.

The basic equations and shapes of the ideal velocity triangles and their relation with the Euler equation and the concept of reaction were discussed in section 1.2. Figure 4.2 illustrates a set or cascade of blade profiles, and relates blade angles to fluid angles by introducing incidence and deviation. The figure also demonstrates the difference between the arrangement of blades used in a compressor, which increases the fluid energy, and in a turbine, which extracts energy. Also illustrated are the other common terms used to describe rows of blades, and the conventional definitions of the coefficients in general use are defined in Table 4.1.

## 4.2 Wing theory

It is assumed that individual blades in a machine interact with the fluid as if they were wings, and the presence of other blades is ignored. This allows lift and drag data obtained from wind tunnel research on wing sections to be used.

The principles relating life and drag forces with the velocity directions for a static blade will first be discussed using Fig. 4.3. It is assumed that the lift

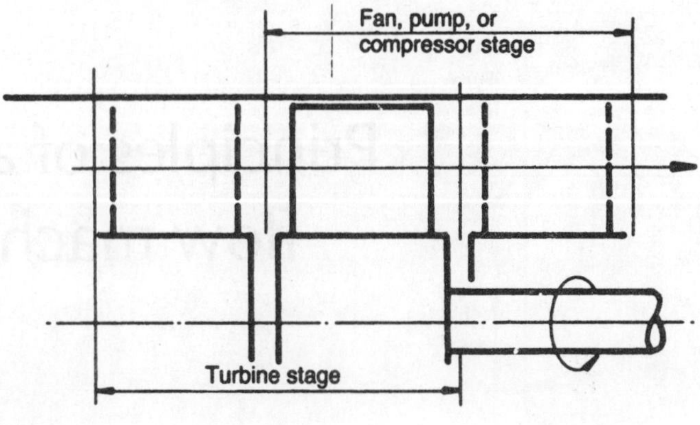


**Figure 4.1** Simple axial machine stages.

**Table 4.1** Definition of terms used with cascades of blades (to be read in conjunction with Fig. 4.2)

| Term | Static | Moving |
|---|---|---|
| Fluid inlet angle | $\alpha_1$ | $\beta_1$ |
| Fluid outlet angle | $\alpha_2$ | $\beta_2$ |
| Blade inlet angle | $\alpha'_1$ | $\beta'_1$ |
| Blade outlet angle | $\alpha'_2$ | $\beta'_2$ |
| Blade camber angle | $\theta[=(\alpha'_1-\alpha'_2)]$ | $[=(\beta'_1-\beta'_2)]$ |
| Stagger angle | $\gamma$ or $\zeta$ | |
| Deflection | $\varepsilon[=(\alpha_1-\alpha_2)]$ | $[=(\beta_1-\beta_2)]$ |
| Incidence angle | $i$ | |
| Deviation angle | $\delta$ | |
| Chord | $c$ | |
| Distance to point of maximum camber | $a$ | |
| Spacing or pitch | $s$ | |
| Blade height | $h$ | |
| Space : chord ratio | $s/c$ | |
| Solidity | $c/s$ | |
| Aspect ratio | $h/c$ | |
| Lift coefficient $C_L = L/\frac{1}{2}\rho V^2$ (projected area) | | |
| Drag coefficient $C_D = D/\frac{1}{2}\rho V^2$ (projected area) | | |
| Pitching moment $M = LX$ | | |

The projected area for a blade is usually ($c \times$ unit height). The reference direction used here is the axial or normal, but often, particularly in pumps, fans and water turbines, the tangential direction is used.

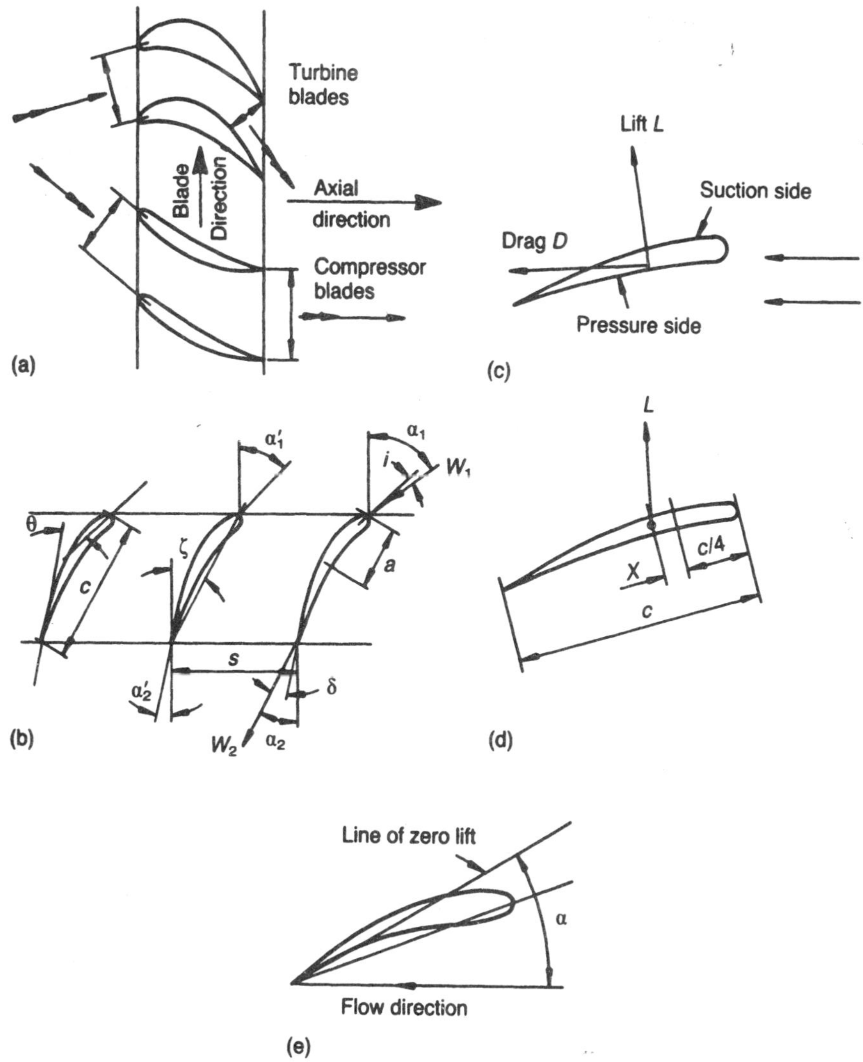


**Figure 4.2** (a) Typical cascades of blades; (b) static cascade of blades; (c) forces on a blade; (d) definition of pitching moment; (e) definition of angle of attack.

force $L$ acts at right angles to, and drag $D$ along, the mean direction. Thus the tangential force on the blade may be written as

$$F_t = L\sin\beta_m + D\cos\beta_m$$

and the axial force as

$$F_a = L\cos\beta_m \quad D\sin\beta_m$$

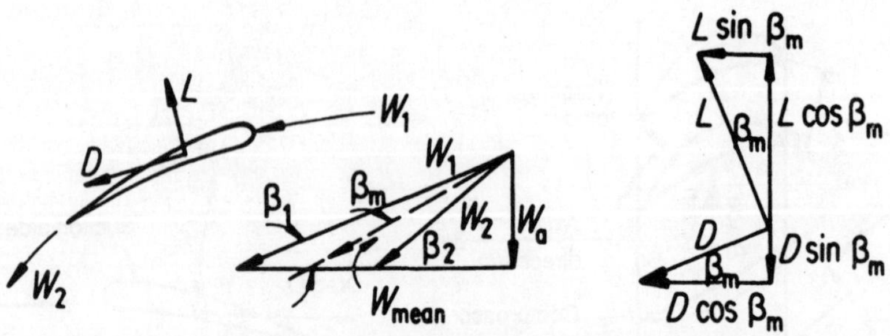


**Figure 4.3** Relative velocity triangles and forces acting on a static blade.

Since the lift coefficient $C_L = L/\frac{1}{2}\rho V^2$ (projected area) and the drag coefficient $C_D = D/\frac{1}{2}\rho V^2$ (projected area), and the projected area is the blade chord times the unit height,

$$F_T = \rho\frac{V_a^2}{2}c\left(\frac{C_L\sin\beta_m + C_D\cos\beta_m}{\sin^2\beta_m}\right) \tag{4.1}$$

or

$$F_T = \rho V_a^2 s(\cot\beta_1 - \cot\beta_2) \tag{4.2}$$

From equations (4.1) and (4.2) it can be shown that

$$C_L = 2s/c(\cot\beta_1 - \cot\beta_2)\sin\beta_m - C_D\cot\beta_m \tag{4.3}$$

Since $C_D/C_L$ is approximately 0.05 for many blade sections, equation (4.3) may be reduced to a commonly used equation by suppressing the term containing $C_D$:

$$C_L = 2s/c(\cot\beta_1 - \cot\beta_2)\sin\beta_m \tag{4.4}$$

Since in machines blades move as well as being static, a blade moving across the flow field at a velocity $U$ is shown in Fig. 4.4. Also shown are the velocity triangles at inlet and outlet, it being assumed that the fluid relative velocity directions are the same as the blade angles at inlet and outlet. Since the blade is moving and doing work, work done per second = $F_t U$. Thus, since conventionally a work or specific energy coefficient is much used (defined as $\psi = gH/U^2$), the work done equation may be rearranged to read

$$\psi = \frac{V_a}{2U}C_L\frac{c}{s}\operatorname{cosec}\beta_m(1 + C_D/C_L\cot\beta_m) \tag{4.5}$$

In many gas turbine textbooks and literature the axial direction is used as reference for angles, and so equation (4.5) may be rewritten in the form

$$\psi = \frac{V_a}{2U}C_L\frac{c}{s}\sec\beta_m(1 + C_D/C_L\tan\beta_m) \tag{4.6}$$

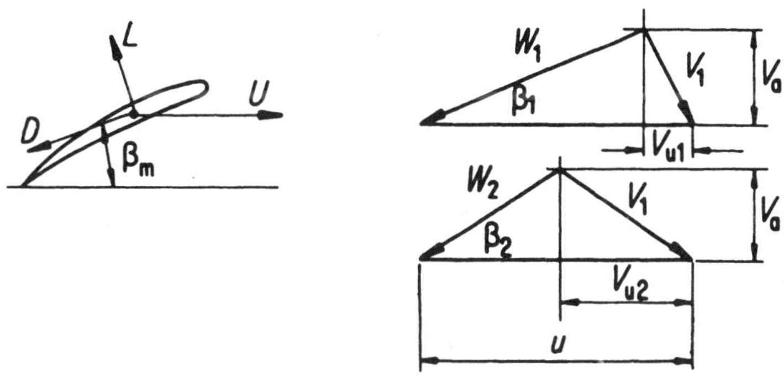


**Figure 4.4** Moving blade, the relative velocity triangles, and the forces acting.

Since $C_D/C_L$ may be neglected, equations (4.5) and (4.6) may be written

$$\psi = \frac{V_a}{2U} C_L \frac{c}{s} \operatorname{cosec} \beta_m \tag{4.7}$$

or

$$\psi = \frac{V_a}{2U} C_L \frac{c}{s} \sec \beta_m \tag{4.8}$$

Thus, by using these equations and the ideal equations outlined in section 1.2, it is possible to construct the velocity diagrams, obtain the fluid angles and determine the blade angles. By selecting a blade profile using the appropriate lift coefficient, the probable solidity, blade numbers and stagger angle may be computed. Before considering approaches to design in this way the data available for blade profiles must be discussed.

## 4.3 Isolated aerofoil data

The simplest section that may be used is the flat plate, for which Weinig (1935) produced a relation between angle of attack and lift:

$$C_L = 2\pi \sin \alpha \tag{4.9}$$

This compares quite well for low angles of attack with experimental data presented by Wallis (1961; Fig. 4.5). Also shown is some information for the cambered circular arc plate. Simple plates of this sort are easy to manufacture, and some low duty fans use them without marked loss of performance (Fig. 4.6) compared with an aerofoil of identical camber line shape but provided with a profile. In most cases strength considerations dictate some profiling and a number of sections have been used. Some of them were

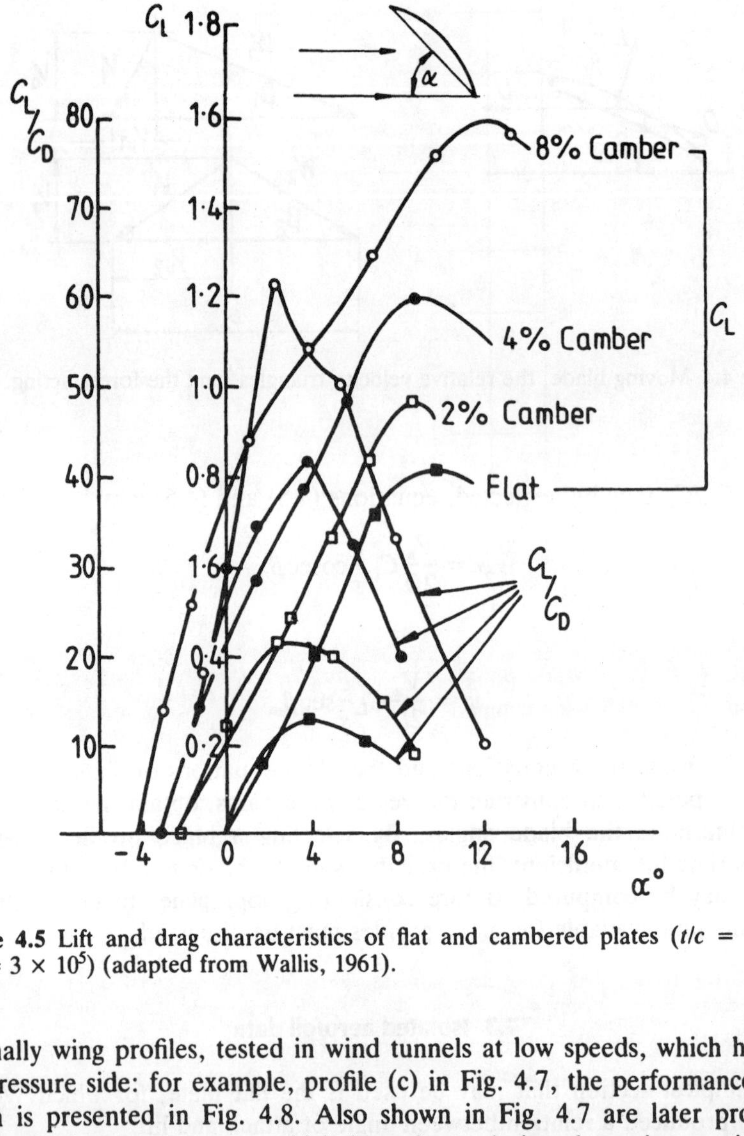


**Figure 4.5** Lift and drag characteristics of flat and cambered plates ($t/c$ = 0.02, $R_{eC} \approx 3 \times 10^5$) (adapted from Wallis, 1961).

originally wing profiles, tested in wind tunnels at low speeds, which had a flat pressure side: for example, profile (c) in Fig. 4.7, the performance for which is presented in Fig. 4.8. Also shown in Fig. 4.7 are later profiles developed for compressors, which have been designed to give particular characteristics. Many families of profiles available are detailed in standard texts like Riegels (1961), and Abbot and Doenhoff (1959) (which concentrates on the NACA profiles). Undergraduate texts deal with the hydrodynamic theory which underlies the 'design' approach to profiles. Carter (1961) discusses this approach, and Fig. 4.9 from this source illustrates the effect of choosing a 'concave' velocity distribution (a) on the variation of drag with incidence, and contrasts this with a 'convex' distribution (b). As can be seen, there is a marked difference in behaviour. Carter commented

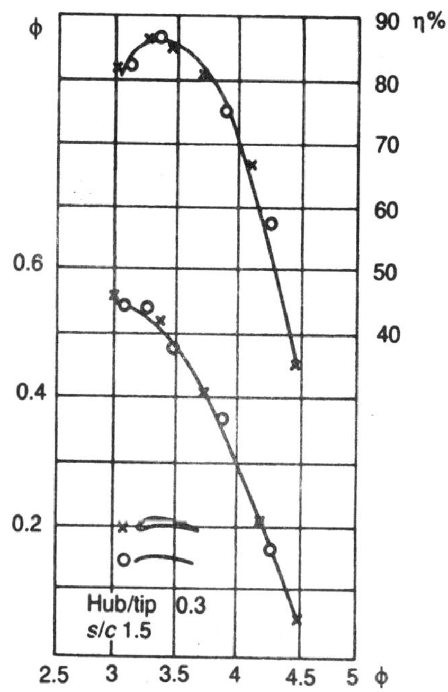


**Figure 4.6** Eckert's results for an axial fan ($R_c = 3.25 \times 10^5$, same camber and rotational speed) showing comparison between the performance of cambered plate and profiled blades.

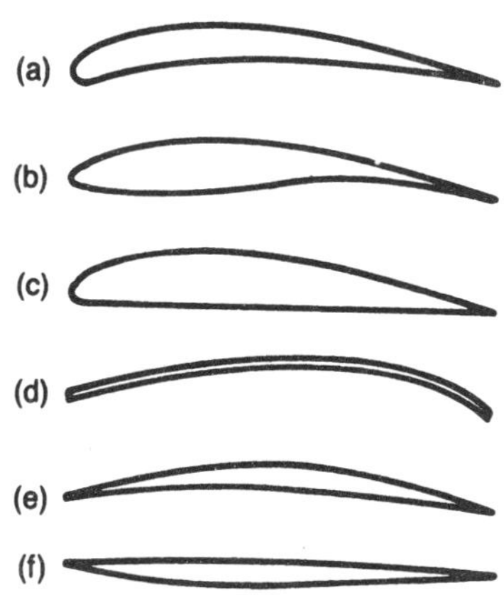


**Figure 4.7** Selection of profiles in use in axial machines: (a) axial compressor (British); (b) axial compressor (American); (c) fan profile; (d) curved plate; (e) transonic section; (f) supersonic section.

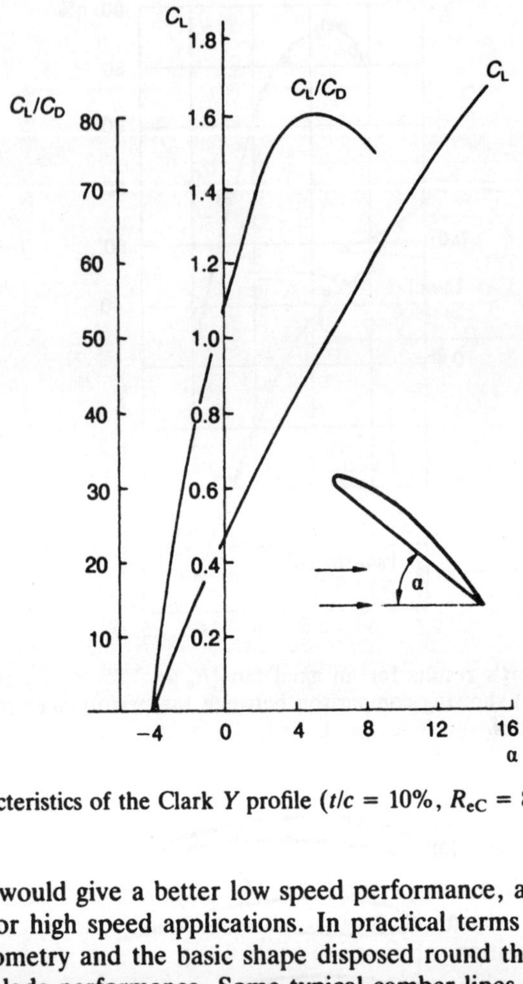


**Figure 4.8** Characteristics of the Clark *Y* profile ($t/c$ = 10%, $R_{eC} = 8 \times 10^6$).

that option (a) would give a better low speed performance, and option (b) will be better for high speed applications. In practical terms the choice of camber line geometry and the basic shape disposed round the camber line determine the blade performance. Some typical camber lines are illustrated in Fig. 4.10. It may be commented that if the point of maximum camber is towards the leading edge (LE) (well forward), the largest surface curvature and hence surface velocities will occur near the leading edge on the upper surface, producing a generally concave velocity distribution; a more rounded velocity distribution results if maximum camber is further towards the trailing edge (TE). It follows that profiles in the first category have a good low speed performance, a high stalling $C_L$ and a wide range of operation but are susceptible to cavitation in pumps, and those in the second have a lower range of operation and lower stalling $C_L$ but better cavitation behaviour.

In practice the simplest camber line is a circular arc, but if the maximum camber is placed away from mid-chord a parabolic arc is needed. Carter suggested 40% of chord for maximum camber in fans and pumps, and 50% for compressors. The maximum thickness should be as small as practicable,

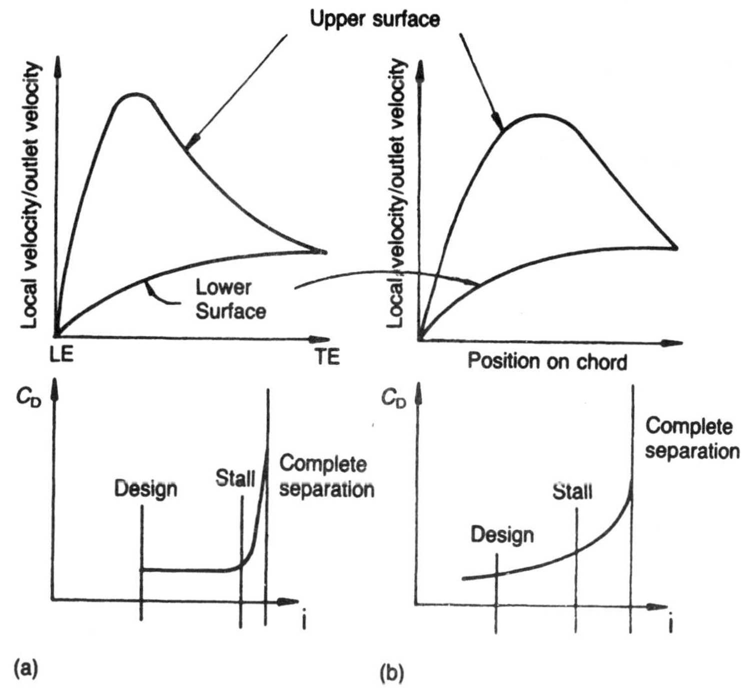


**Figure 4.9** Effect of velocity distribution on the profile drag coefficient: (a) 'concave' velocity distributions, (b) 'convex' velocity distributions (Carter (1961) courtesy of the Institution of Mechanical Engineers).

and from data referred to by Carter it appears that thicknesses up to 12.5% of chord do not affect performance, and thicknesses above this dictated by strength considerations will result in a drastic fall-off in performance. Varying the position of maximum thickness has a similar effect to the movement of maximum camber, and for many sections the maximum thickness is located at 30% of chord from leading edge, as in the C4 section. The NACA 65 series all have the maximum thickness at 40% of chord. Table 4.2 summarizes the variation in thicknesses for some commonly used profiles. It will be seen that the leading edge and trailing edge radii tend to be small, although the leading edge for low speed use tends to be fairly large to allow a tolerance of incidence changes typical of fans and pump usage, and in compressors the leading edges tend to be sharper.

## 4.4 Cascade data

When blades are in close proximity or 'cascade', the individual blade behaviour is affected (Fig. 4.11) since the passages formed by adjacent blades

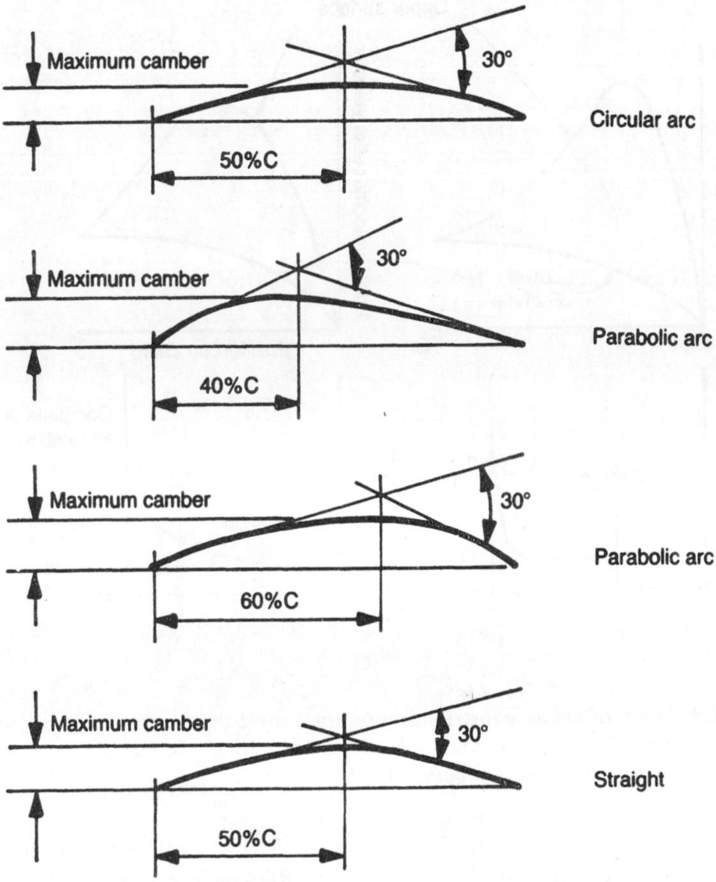


**Figure 4.10** Common camber-line shapes.

dominate the flow. A one-dimensional theoretical correction for this effect is to insert a correction factor in the Euler equation so that it reads, for a pump,

$$gH = C_H(V_{u2}u_2 - V_{u1}u_1) \tag{4.10}$$

This factor is shown in Fig. 4.12, and is a theoretical statement of the effect of blade spacing and blade angle. Weinig (1935) studied the two-dimensional problem by deriving relations for thin aerofoils which approximate to flat plates, and produced a lattice coefficient $K$ used to correct the flat plate equation, equation (4.9):

$$C_L = 2\pi K \sin\alpha \tag{4.11}$$

Figure 4.13 illustrates the effect on $K$ of blade angle and spacing.

**Table 4.2** Thickness profiles of NA CA 6510, C4 and T6 sections

| % chord | *t*/2 NACA 65 series | *t*/2 C4 | *t*/2 T6 |
|---|---|---|---|
| 0 | 0 | 0 | 0 |
| 0.5 | 0.772 | | |
| 0.75 | 0.932 | | |
| 1.25 | 1.169 | 1.65 | 1.17 |
| 2.5 | 1.574 | 2.27 | 1.54 |
| 5.0 | 2.177 | 3.08 | 1.99 |
| 7.5 | 2.674 | 3.62 | 2.37 |
| 10 | 3.04 | 4.02 | 2.74 |
| 15 | 3.666 | 4.55 | 3.4 |
| 20 | 4.143 | 4.83 | 3.95 |
| 25 | 4.503 | | |
| 30 | 4.76 | 5.0 | 4.72 |
| 35 | 4.924 | | |
| 40 | 4.996 | 4.89 | 5.0 |
| 45 | 4.963 | | |
| 50 | 4.812 | 4.57 | 4.67 |
| 55 | 4.53 | | |
| 60 | 4.146 | 4.05 | 3.70 |
| 65 | 3.682 | | |
| 70 | 3.156 | 3.37 | 2.51 |
| 75 | 2.584 | | |
| 80 | 1.987 | 2.54 | 1.42 |
| 85 | 1.385 | | |
| 90 | 0.81 | 1.60 | 0.85 |
| 95 | 0.306 | 1.06 | 0.72 |
| 100 | 0 | 0 | 0 |
| LE radius | 0.687% of chord | 12% max. *t* | 12% max. *t* |
| TE radius | 'sharp' | 6% max. *t* | 6% max. *t* |

Performance of cascades of blades was obtained using cascade tunnels sketched in Fig. 4.14 and by determining pressure and velocity changes at the mid-height of the centre blade and its associated passages. The tunnels had variable geometry to produce variations in incidence, and wall boundary layer control to ensure a two-dimensional flow field for the section under study. Figure 4.15 illustrates the typical performance of a cascade of NACA 65 series blades. The design lift coefficient $C_{L0}$ is related to camber for this foil shape in the way shown in the figure.

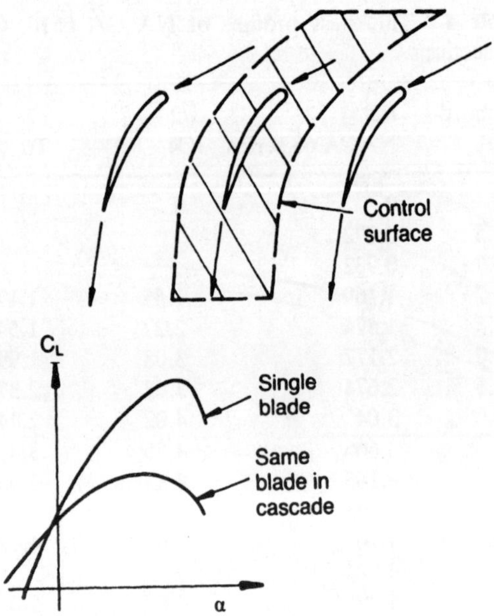


**Figure 4.11** Blades in cascade and the effect of proximity on blade performance.

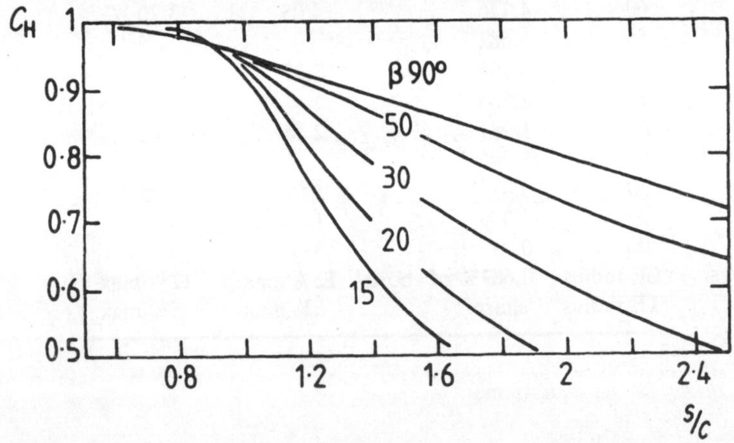


**Figure 4.12** Plot of the head correction factor to setting angle and space : chord ratio (adapted from Wisclicenus, 1965).

Since blades behave in groups it is more convenient to present information in a different form from the isolated foil equations in section 4.2. For example, the loss in total pressure for a compressor cascade may be written

$$\frac{\Delta p_0}{\rho} = \frac{p_1 - p_2}{\rho} + \tfrac{1}{2}(V_1^2 - V_2^2) \tag{4.12}$$

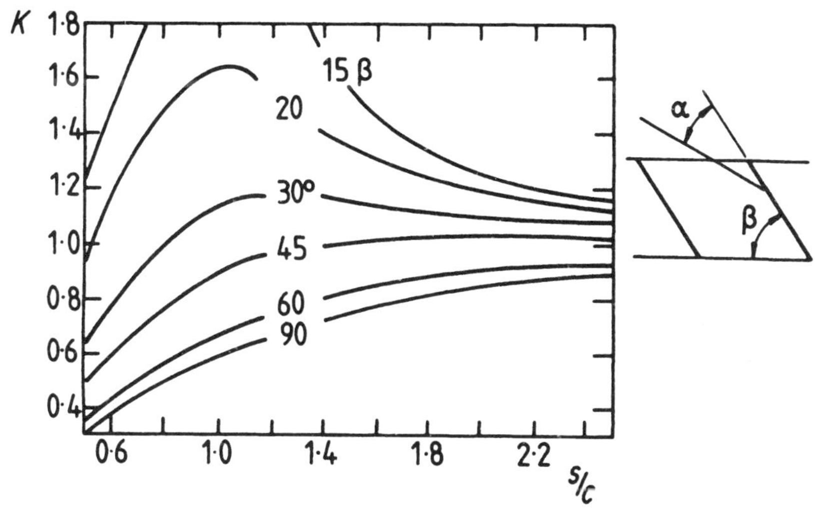


**Figure 4.13** Weinig's lattice coefficient for a simple vane cascade (after Wisclicenus, 1965).

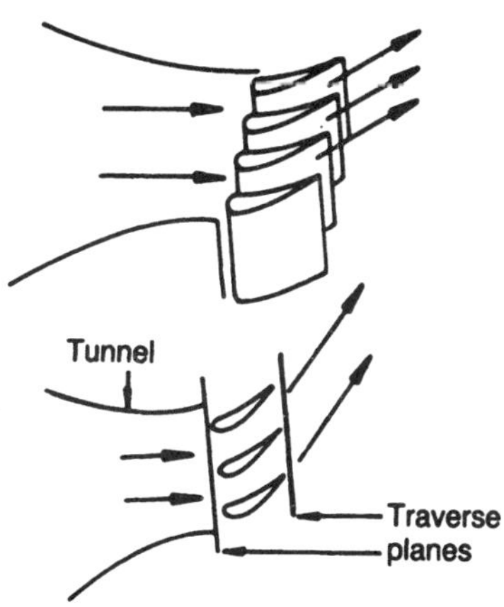


**Figure 4.14** Simple cascade tunnel.

This loss is usually expressed non-dimensionally in two forms:

$$\zeta = \frac{\Delta p_0}{\frac{1}{2}\rho V_a^2} \tag{4.13}$$

$$\bar{\omega} = \frac{\Delta p_0}{\frac{1}{2}\rho V_1^2} \tag{4.14}$$

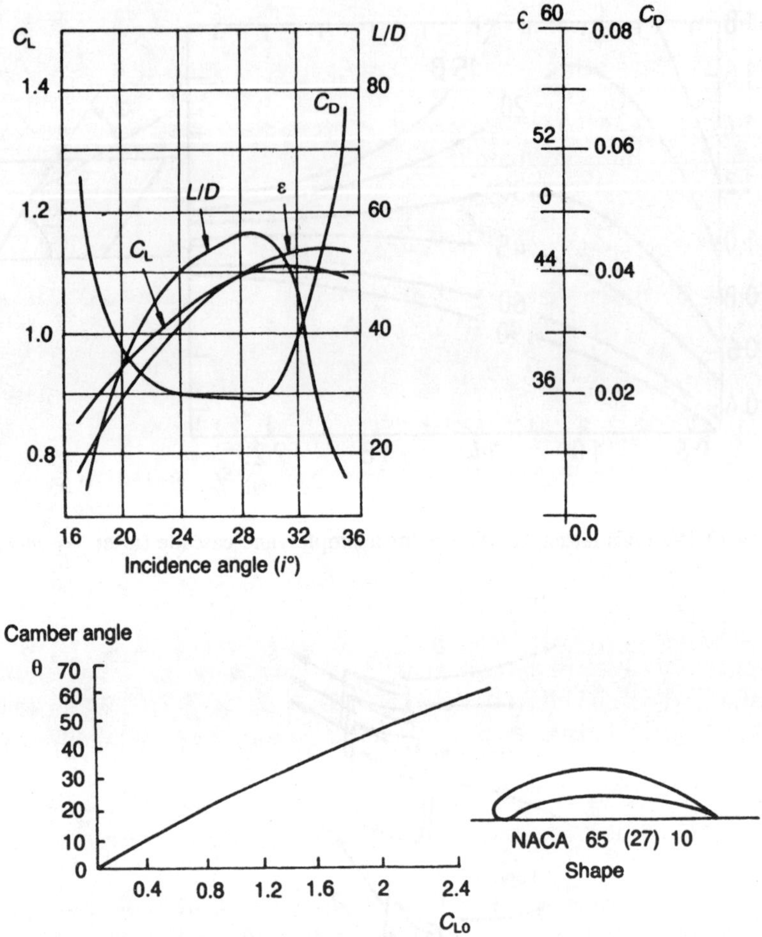


**Figure 4.15** Data for an NACA 65 (27) 10 foil in cascade ($\beta_1 = 45°$, $s/c = 1.0$). (Note that 65 (27) 10 means a 65 series foil, $C_{L0} = 2.7$, $t/c = 10\%$; $C_{L0}$ is design lift coefficient, related to camber angle $\theta$ as sketched.)

A pressure rise coefficient is often used:

$$C_p = \frac{p_2 - p_1}{\frac{1}{2}\rho V_a^2} \tag{4.15}$$

and a tangential force coefficient using $Y$ as tangential force:

$$C_f = \frac{Y}{\frac{1}{2}\rho V_a^2} \tag{4.16}$$

(quoted as $\psi_T$ by Zweifel (1945)). From these,

$$C_p = C_f \cot \beta_m - \frac{\Delta p_0}{\frac{1}{2}\rho V_a^2} \tag{4.17}$$

Howell (1945) introduced a diffuser efficiency, defined as

$$\eta_D = \frac{p_2 - p_1}{\frac{1}{2}\rho(V_1^2 - V_2^2)} \tag{4.18}$$

and it can be shown that

$$\eta_D = 1 - \frac{2C_D}{C_L \cos 2\beta_m} \tag{4.19}$$

Howell also showed (assuming $C_L/C_D$ is constant, which is approximately true for many profiles over a range of incidence) that $\beta_m$ optimum is 45° and that

$$\eta_{Dmax} = 1 - \frac{2C_D}{C_L} \tag{4.20}$$

This suggests that $\eta_D$ is maximum for a compressor row where $\beta_m$ is 45°. Howell illustrates how $\eta_D$ varies with $\beta_m$ and the lift:drag ratio, and demonstrates the small effect of variation in $C_L/C_D$ for a conventional cascade.

Figure 4.16 gives cascade performance plots presented non-dimensionally,

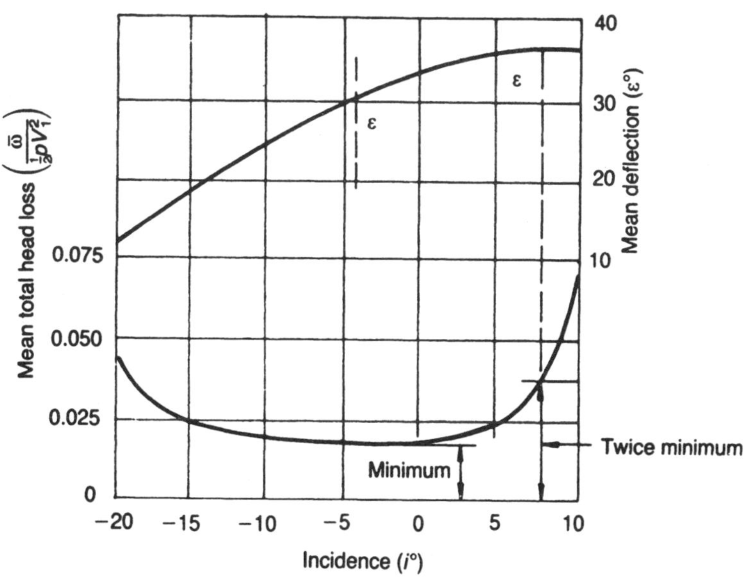


**Figure 4.16** Cascade data for a typical English cascade (after Howell (1945), courtesy of the Institution of Mechanical Engineers).

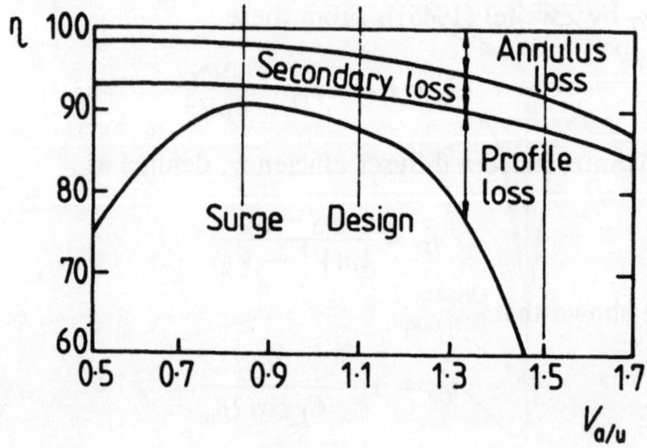


**Figure 4.17** Variation of losses with flow coefficient for an axial flow compressor.

and allows profile loss to be assessed over a range of flows and corresponding incidence angles. The other losses vary with flow rate, as Fig. 4.17 illustrates for a typical compressor stage. Cascade testing has revealed strong two- and three-dimensional flow patterns, the principal effects being illustrated in Fig. 4.18. Boundary layer and tip clearance effects have a strong effect on the overall efficiency of a blade row and, owing to the secondary and wake flows affecting blades in succeeding rows, the performance of the whole machine. An example of this interaction is shown in Fig. 4.19, where time averaged axial velocity profiles are illustrated for four stages in a compressor. The effect is to cause a variation in work capacity of a blade along its length, and on the loading of successive rows. A much used correction in compressors is Howell's work done factor $\Omega$, defined as used in

$$C_p \Delta T = \Omega u \Delta V_u \tag{4.21}$$

This factor is shown in Fig. 4.20.

The flows in Fig. 4.18 show that fluid angles are likely to differ from actual blade angle, so that there is always a deviation angle to estimate which is affected by blade geometry. Howell proposed a formula for the deviation angle $\delta$:

$$\delta = m\theta\sqrt{(s/c)} \tag{4.22}$$

The well known plots of $m$ against stagger angle for circular arc and parabolic camber lines due to Carter (1948) are shown in Fig. 4.21. Carter's discussion of blade profile behaviour has been referred to in section 4.2, and some comments relevant to cascade behaviour which extend the isolated foil now need to be made. Figure 4.9 shows a comparison of the behaviour of two types of velocity profile in cascade. Carter further discusses the effect of these profiles on loss (which is small) and shows that the effect of stagger on

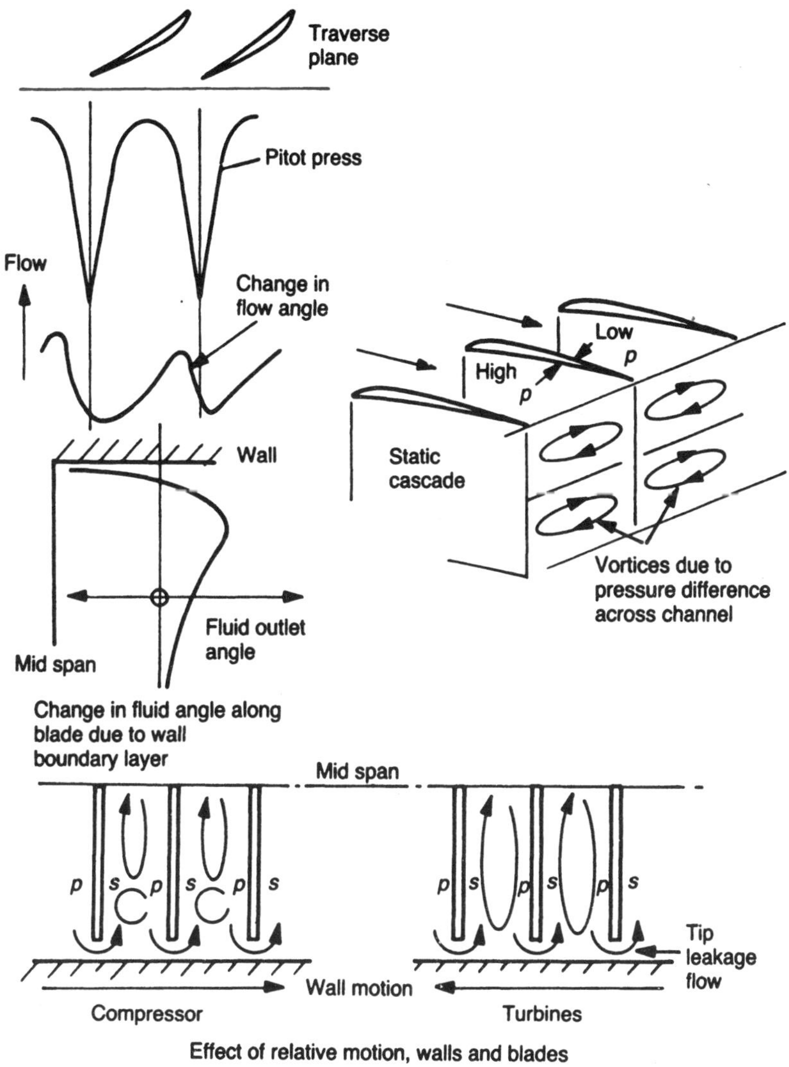


**Figure 4.18** Secondary flow effects in a compressor cascade.

blade velocity distribution is small. Figure 4.22 is a reminder that as stagger increases, relative velocities must increase for the same flow rate and camber angle in compressor and turbine cascades. The basic difference between compressor (diverging) and turbine (converging) cascade passages is also illustrated.

Figure 4.23 shows a zero reaction (impulse) blade cascade (b), illustrating the constant area passage and the blade shapes which follow. Also shown is

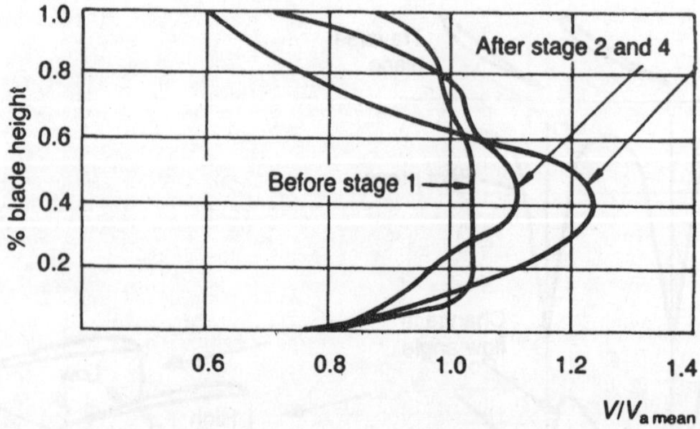


**Figure 4.19** Velocity profile changes in a four-stage compressor (adapted from Howell, 1945).

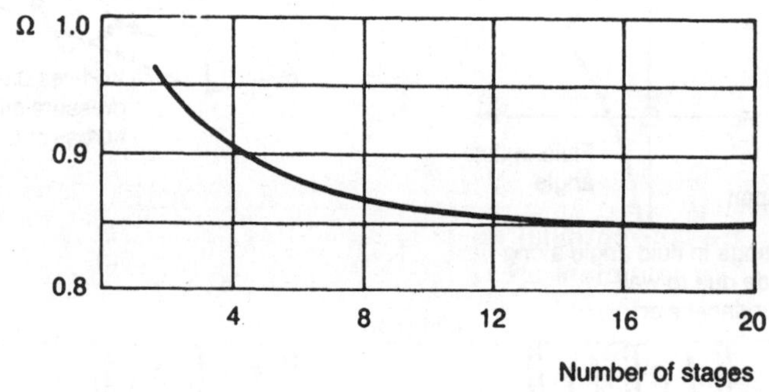


**Figure 4.20** Work done factor proposed by Howell (after Horlock, 1958).

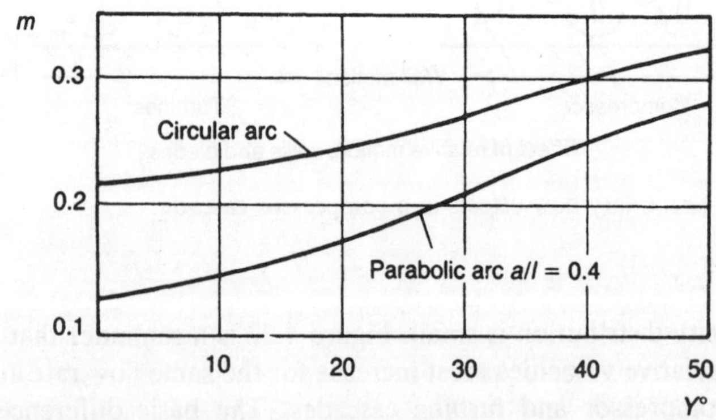


**Figure 4.21** Deviation rules proposed by Howell and plotted by Carter (1948) (courtesy of the Institution of Mechanical Engineers).

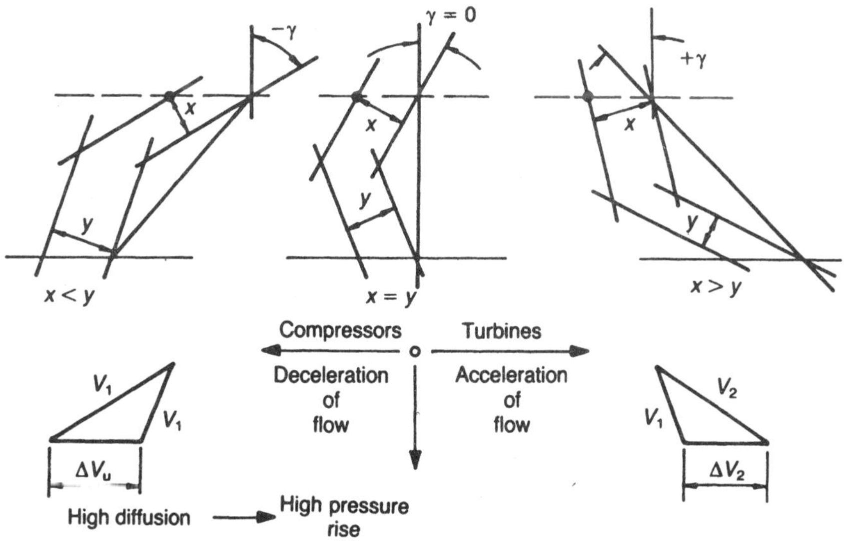


**Figure 4.22** Simplified presentation of the effect of stagger on cascade layout and the velocity triangles.

a passage of a conventional reaction design (a), with a gradual contraction to the throat or minimum area. In Fig. 4.23(d) typical profile coefficients for impulse and reaction blades are illustrated, indicating the loss penalty imposed by the profile shapes, and the restriction in incidence range involved when impulse profiles are used. Also shown in Fig. 4.23(d) is the small variation in gas angle with incidence, demonstrating how the throat geometry has a great influence on outlet angle, a distinct difference from compressor cascades. A simple empirical rule much used by steam turbine designers is, following Fig. 4.24

$$\alpha_2 = \cos^{-1} o/s \qquad (4.23)$$

where $o$ is the cascade opening or throat. This is a good approximation to actual gas angles for high Mach numbers, but Ainley and Matheison (1957) comment that at low Mach numbers ($0 < M_n \leqslant 0.5$)

$$\alpha_2 = [f(\cos^{-1} o/s)] + 4s/e \qquad \text{(in degrees)} \qquad (4.24)$$

where $f(\cos^{-1} o/s) = -11.15 + 1.154\cos^{-1} o/s$, and $e$ is the mean radius of the flank leading up to the opening $o$. They also modified equation (4.23) by replacing $o$ with throat area and $s$ with annulus area just downstream from the trailing edges.

The basic relations used in compressor and fan technology have been introduced in previous sections, and are in coherent groups, since in compressors both stator and rotor blades are treated as behaving in the same

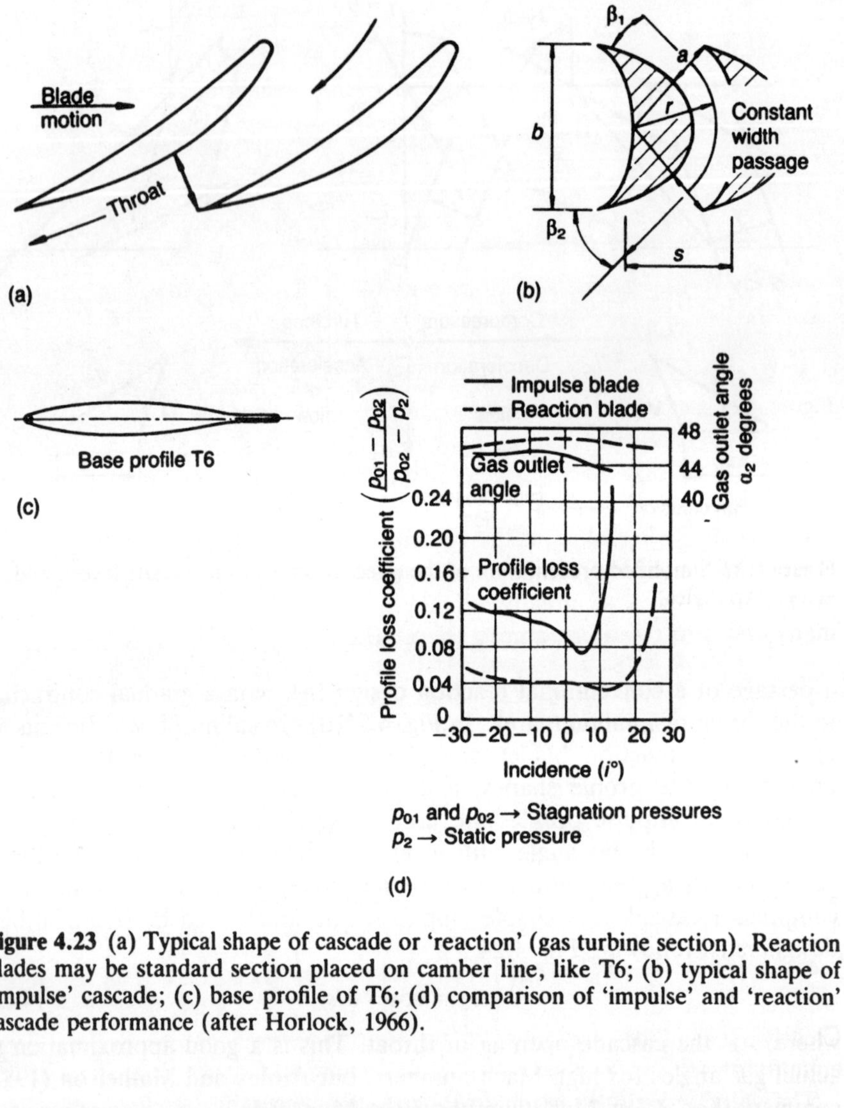


**Figure 4.23** (a) Typical shape of cascade or 'reaction' (gas turbine section). Reaction blades may be standard section placed on camber line, like T6; (b) typical shape of 'impulse' cascade; (c) base profile of T6; (d) comparison of 'impulse' and 'reaction' cascade performance (after Horlock, 1966).

way after allowance for rotor motion is made. In turbines, however, owing to the very different histories of steam and gas turbines development, a considerable range of relations exist, which need to be treated before considering turbine design (section 8.3).

A large range of equations has been used to express irreversibility in practical terms. A simple velocity ratio

$$K = \frac{V}{V_{\text{isen}}} \tag{4.25}$$

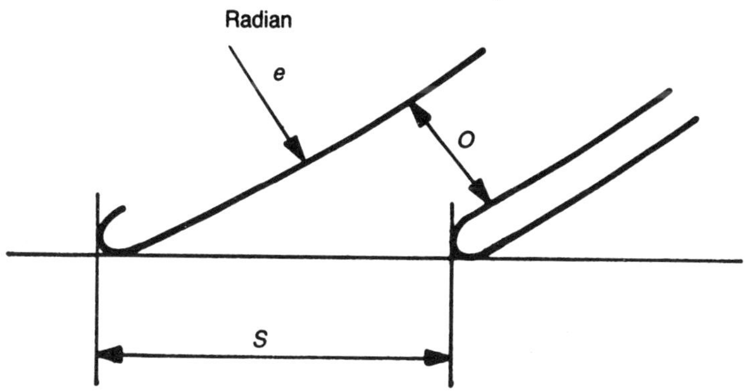


**Figure 4.24** Empirical rule used by steam turbine designer.

is much used, where $V_{isen}$ is the velocity that could be expected if expansion is isentropic to the prevailing back pressure. Kearton (1958) quotes three nozzle efficiencies, as already defined in section 1.7. Horlock (1966) quotes many different relations, among them Markov's (1958) loss coefficients:

$$\xi_N = \frac{h_2 - h_{2s}}{\frac{1}{2}V^2_{2isen}} \qquad \xi_R = \frac{h_3 - h_{3s}}{\frac{1}{2}W^3_{2\,isen}} \tag{4.26}$$

He also introduces stagnation pressure loss coefficients

$$Y_N = \frac{p_{01} - p_{02}}{p_{02} - p_2} \qquad Y_R = \frac{p_{02\,rel} - p_{03\,rel}}{p_{03\,rel} - p_3} \tag{4.27}$$

and shows that $Y_N$, for example, is given by

$$Y_N \approx \xi_N\left(1 + \frac{KM_n^2}{2}\right) \qquad \text{for } M_n < 1 \tag{4.28}$$

Clearly such indices are only valid if used when flow conditions give rough equality of $M_n$.

The velocity ratio $\sigma_{is}$ is much used in steam turbine design for fixing the optimum efficiency point:

$$\sigma_{is} = \frac{u}{\sqrt{[2(h_{01} - h_{03ss})]}} \tag{4.29}$$

Gas turbine engineers tend to use $\psi$:

$$\psi = \frac{\eta_{TT}}{2\sigma_{is}^2} \tag{4.30}$$

For $R = 0$, $\psi = 1/2\sigma_{is}^2$; for $R = 50\%$, $\psi = 1/\sigma_{is}^2$.

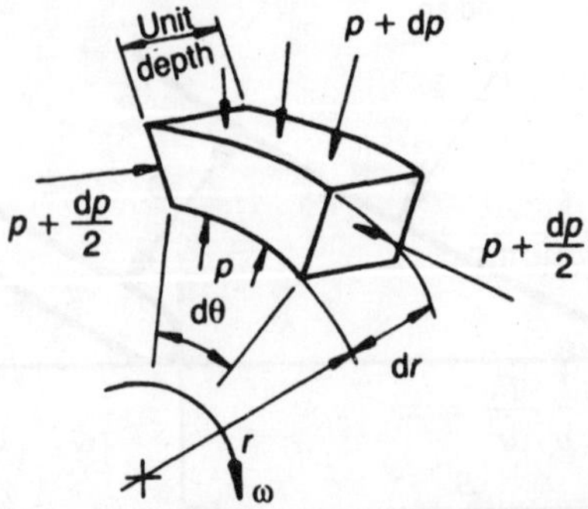


**Figure 4.25** Forces on a fluid element in a rotating field of flow.

For an isentropic turbine $\eta_{TT}$ is unity, but $\eta_{TS} < 1$, and Horlock (1966) gives 'design' charts that may be used in situations where often $\eta_{TS}$ is a true reflection of turbine effectiveness.

## 4.5 Radial equilibrium theories

Theory and experimental information was based on two-dimensional test data, and corrections for three-dimensional flow have been outlined in previous sections, but one effect that has not been discussed is the effect of the strong centrifugal forces exerted by blades. The centrifugal field distorts the flow velocity profiles considerably, and particles tend to move outwards rather than passing along cylindrical stream surfaces as classically assumed, particularly in low hub : tip ratio designs. An approach known as the radial equilibrium method, widely used for three-dimensional design calculations, assumes that flow is in radial equilibrium before and after a blade row, and radial adjustment takes place through the row.

Consider a small element of fluid (Fig. 4.25) rotating in a pressure field at constant tangential velocity $V_u$. If the angular velocity is $\omega$, the centrifugal force is $(\rho r\,\mathrm{d}r\,\mathrm{d}\theta)\omega^2 r$ on the element. Since $V_u = \omega r$, the centrifugal force is $\rho V_u^2\,\mathrm{d}r\,\mathrm{d}\theta$ on the element.

The pressure force on the element, ignoring products of small quantities, is given by

$$\text{pressure force} \approx r\,\mathrm{d}p\,\mathrm{d}\theta$$

If the two forces are the only ones acting (viscous and other effects neglected), the particle will move at constant radius if:

$$\frac{\mathrm{d}p}{\mathrm{d}r} = \rho\frac{V_u^2}{r} \tag{4.31}$$

or

$$\frac{dp}{\rho} = V_u^2 \frac{dr}{r} \tag{4.32}$$

If the concept of stagnation pressure $p_0 = p + \frac{1}{2}\rho V^2$ is introduced, it can be shown that

$$\frac{1}{\rho}\frac{dp_0}{dr} = V_a \frac{d(V_a)}{dr} + \frac{V_u}{r}\frac{d(rV_u)}{dr}$$

and if $p_0$ does not vary with radius, the usual radial equilibrium concept follows:

$$V_a \frac{d(V_a)}{dr} + \frac{V_u}{2}\frac{d(rV_u)}{dr} = 0$$

or

$$\frac{d(V_a^2)}{dr} + \frac{1}{r}\frac{d(rV_u^2)}{dr} = 0 \tag{4.33}$$

Although many approaches used in axial compressors are available as solutions to equation (4.33) the two of interest in pumps are the free and forced or 'solid' vortex solutions. If the ideal flow situation is considered, Euler's equation will apply, which with equation (4.32) gives after integration:

$$V_u r = \text{constant} \tag{4.34}$$

This is the free vortex law, and substitution in equation (4.33) results in $V_a$ = constant.

In a solid or forced vortex,

$$\frac{V_u}{r} = \text{constant } (c) \tag{4.35}$$

and equation (4.33) becomes

$$\frac{d(V_a)^2}{dr} + \frac{1}{r^2}\frac{d(c^2r^4)}{dr} = 0$$

a solution of which is

$$V_a = \sqrt{(\text{constant} - 2c^2r^2)} \tag{4.36}$$

The free vortex solution is simple and, since $V_a$ is constant across the swept annulus, allows easy comparison of velocity triangles. The sections are laid out easily. As Horlock (1958) shows, the change in blade angles with radius is considerable, and Fig. 4.26 is a sketch of such a blade showing the 'twist' in the blade length.

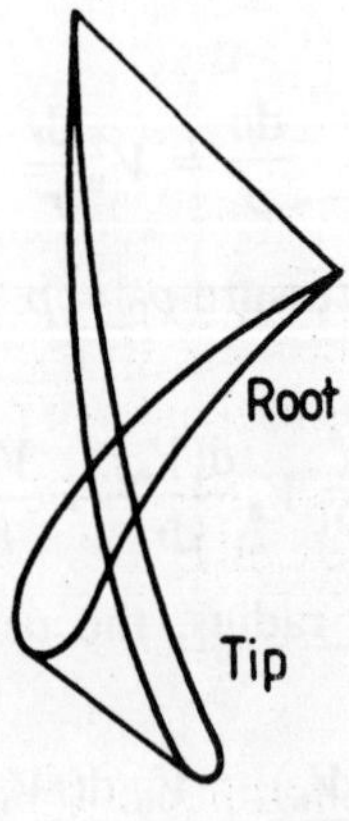


**Figure 4.26** Root and tip sections of a typical compressor rotor blade illustrating the degree of twist.

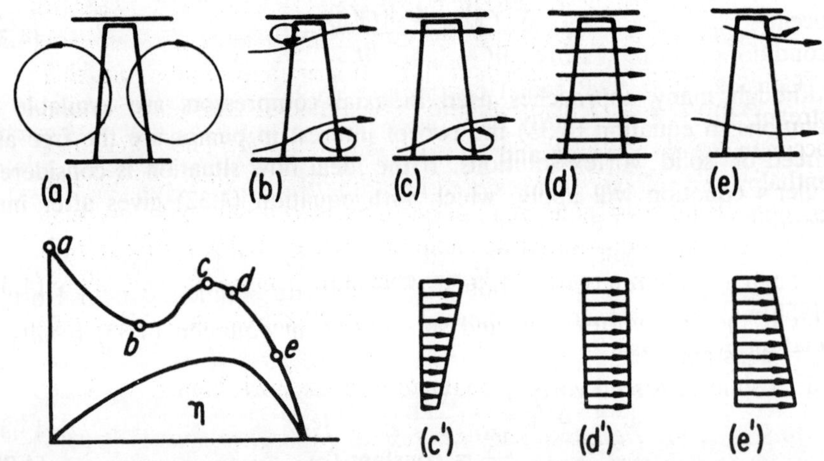


**Figure 4.27** Typical vortex and velocity distributions in an axial flow fan as flow rate changes (after Eck, 1973).

It must be said that though this approach works at design flow, considerable changes can occur in flow, particularly for long blades. Figure 4.27 following material reported by Eck (1973) for fans, illustrates the flow changes likely.

## 4.6 Actuator disc approach

The assumption of radial equilibrium before and after each blade row is only valid if the blade aspect ratio is small; for long blades it ceases to apply. The

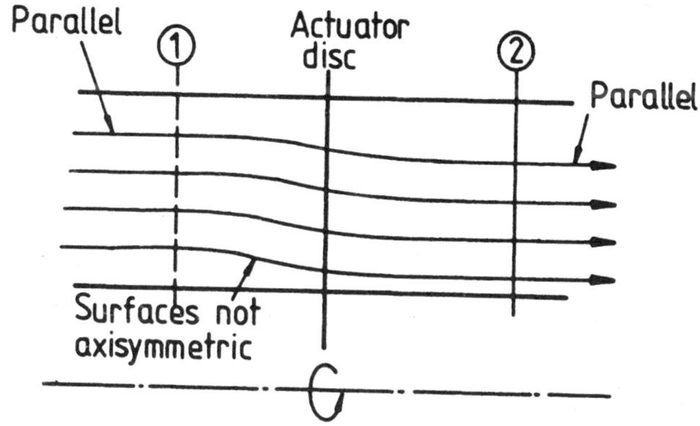


**Figure 4.28** Actuator disc concept.

idea of the actuator disc where all the turning of the flow is assumed to occur in a very small axial distance appears to approximate to actual flow conditions in compressors and turbines.

Consider Fig. 4.28: radial equilibrium applies far upstream and downstream, but not necessarily in between, and changes in tangential velocity occur in the disc. Bragg and Hawthorne (1956) derived equations relating enthalpy and velocity changes, and Ruden (1944) extended the treatment by assuming that the radial displacement of streamlines is small, and that the $V_u$ term in equation (4.33) is small. The detailed theory may be followed through in the references, but an approximate solution due to Horlock (1958) will be discussed. He showed that the axial velocity changes follow a relation of the form

$$\Delta V_a = \pm\left(\frac{V_{a2} - V_{a1}}{2}\right)e^{(-\pi Z)/h} \tag{4.37}$$

where $h$ is the blade height and Z is the distance upstream or downstream from the disc. The axial velocity at the disc is

$$V_a = \frac{V_{a1} + V_{a2}}{2} \tag{4.38}$$

Thus, following Horlock's convention, $V_a$ far upstream is given by

$$V_a = V_{a1} + \left(\frac{V_{a2} - V_{a1}}{2}\right)e^{(\pi Z)/h} \tag{4.39}$$

and downstream by

$$V_a = V_{a2} - \left(\frac{V_{a2} - V_{a1}}{2}\right)e^{(-\pi Z)/h} \tag{4.40}$$

Interference between rows is developed by Bragg and Hawthorne (1956) and Horlock (1958), and their contributions should be consulted for further information and the actuator disc technique.

## 4.7 Stall and surge effects

### 4.7.1 Introduction

The phenomenon known as 'stall' affects axial flow pumps, fans and compressor stages when flow breaks away from the suction side of the blade aerofoil, an event associated with incidence. A typical characteristic is shown in Fig. 4.29. Surge occurs as an unstable and usually large fluctuation of the flow through a machine, which results in a complete breakdown in the flow; however, it is not directly related to stall.

### 4.7.2 Stalling of fans and compressor stages

The operation of single stage fans and compressor stages at flow coefficients low enough to cause stall has been studied by many experimenters, one of the most detailed studies being that of Smith and Fletcher (1954). Dunham (1965) suggests that the stalling phenomenon may be a rotating or propagating stall, a band of reversed flow 'anchored' by spider supports, a so-called centrifuging effect (flow inwards at blade root, and outwards at blade tip) and a flow 'in and out' at all radii. The last two effects probably include propagating stall.

Rotating stall was observed by Whittle on the inducer vanes of a centrifugal compressor, and described by Emmons, Pearson and Grant (1955). When flow through the stage shown in Fig. 4.30 is reduced, the incidence

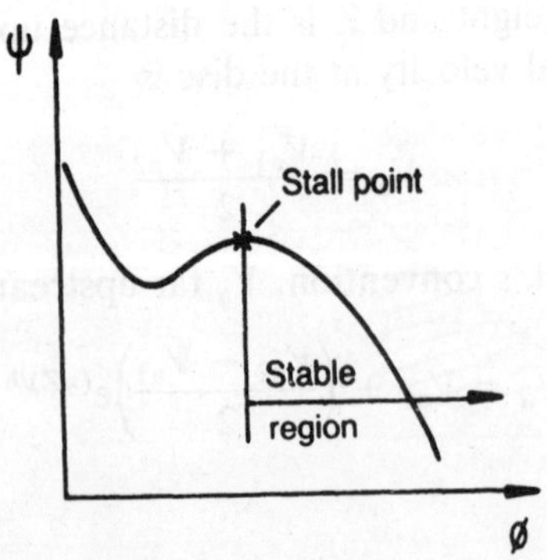


**Figure 4.29** Typical axial fan characteristic.

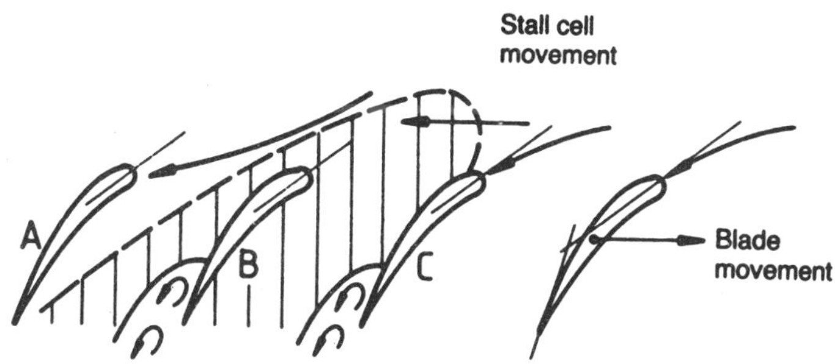


**Figure 4.30** Rotating stall cell in an axial flow rotor row.

angle increases and one or more blades may stall. Flow leaving the suction surfaces causes an effective partial blockage of a blade passage, which increases incidence on blade A; this blade will tend to stall, and blade C will come out of stall. The stall cell will thus tend to propagate round the row. To an observer **on the blades** the cell propagates in the opposite direction to blade motion at about half peripheral speed. Lakhwani and Marsh (1973) studied, among other factors, the influence of reducing blade numbers by half, and a compressor stage with 22 rotor blades and 33 stator blades. They concluded from the single rotor studies that stall cells did not move at constant speed while growing. However, once stall cells were established their angular velocity was a constant proportion of the rotor speed ($N$), but it reduced as blade numbers halved from $0.56N$ to $0.46N$. They found that adding stators downstream had little effect on stall onset. They also found that a theoretical prediction by Le Bot (1970) was fairly close for 22 blades, but not for 11 blades, indicating a need for more data. The difficulty of stall prediction, owing to the complicated cell patterns that occur, is discussed by

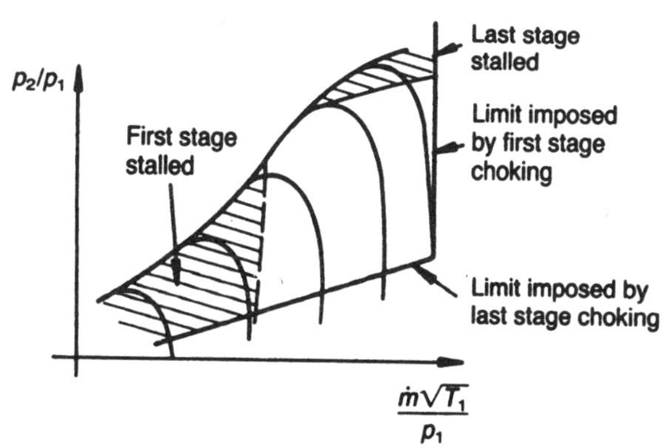


**Figure 4.31** Limits imposed by choking and stall on axial compressor performance.

Horlock (1958), where he considers the wide variations found in earlier experimental studies.

### 4.7.3 Surge and stall in compressors

The surge line on a compressor characteristic is the limit of stability, and is determined by testing. Figure 4.31 illustrates a fairly typical characteristic and also demonstrates the other limits posed on the performance envelope. Also shown is the common condition that for a wide range of lower mass flow the first stage at least is stalled, with other stages possibly so, and that at higher flows it is quite likely that the last stage will be stalled at higher pressure ratios. These phenomena are fully discussed by Horlock (1958) and Eckert and Schnell (1961) among classical texts. They argue that surge in the low speed region is associated with stalling of the first five stages, and in the high speed region with stall in the rear stages. For further information the classical paper by Huppert and Benser (1953) may be consulted.

# 5 Principles of radial and mixed flow machines

## 5.1 Introduction

This chapter is concerned with the basic principles governing the shapes of stream surfaces and of the blades for radial and mixed flow pumps, fans, compressors and turbines. In these machines flow either starts axial and changes direction as in pumps, compressors and bulb turbines, or begins by flowing radially inward and is discharged axially as in Francis and Kaplan turbines. In each case the rotor and stator systems produce the changes in direction while modifying the fluid energy level.

## 5.2 One-dimensional approach

The basic equations relating velocities, blade angles and work interchange between machine elements and the working fluid were introduced in section 1.2, and these may be simply related to machine element design by assuming that the velocities are averaged across the passage at the relevant planes in the flow path. These velocities are assumed steady, incidence and deviation are absent, and viscous and boundary effects are ignored. This approach is one-dimensional, and the 'design' problem is to relate the throughflow or meridional velocity to passage shape change.

Consider Fig. 5.1, which illustrates a typical mixed flow rotor in sectional elevation, with the bounding surfaces and two stream surfaces. Figure 5.2 illustrates the three-dimensional nature of stream surfaces, and has a velocity triangle sketched at point A. The velocity $V_n$ is simply related to the volumetric flow as described in section 1.2.

If the same type of passage is considered (Fig. 5.3), the one-dimensional method of deciding the progress of the stream surfaces, once the passage walls are established, may be illustrated. A flow net has been sketched with two potential lines and three intermediate stream lines. Each stream line

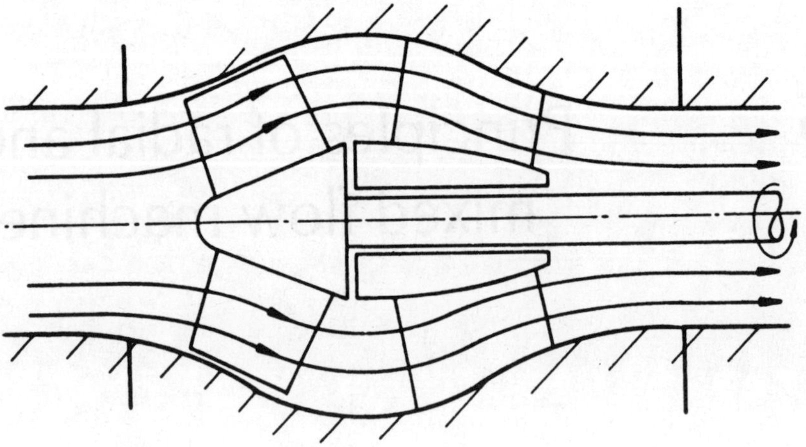

**Figure 5.1** Mixed flow or bowl pump.

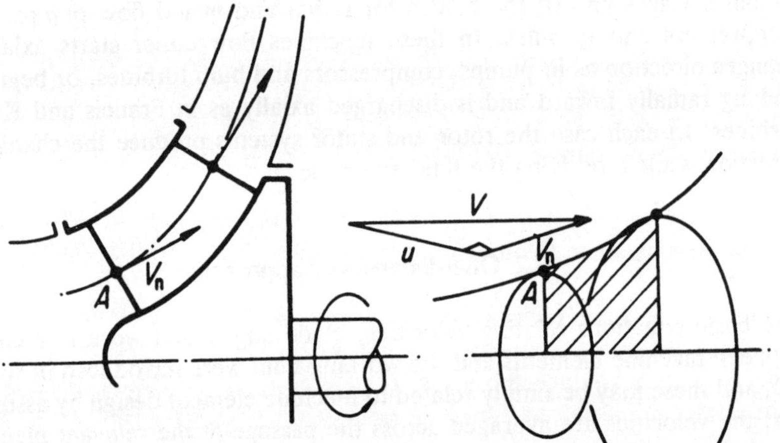


**Figure 5.2** Typical pump rotor and a stream surface.

represents a surface of revolution, as does each potential line. Noting that $V_n$ must be the same across each potential surface, it follows that

$$2\pi Rd = \text{constant} \tag{5.1}$$

for the flow field shown, along each potential surface. Applying this rule to each succeeding surface defines the progress of the stream surfaces, which are approximately true.

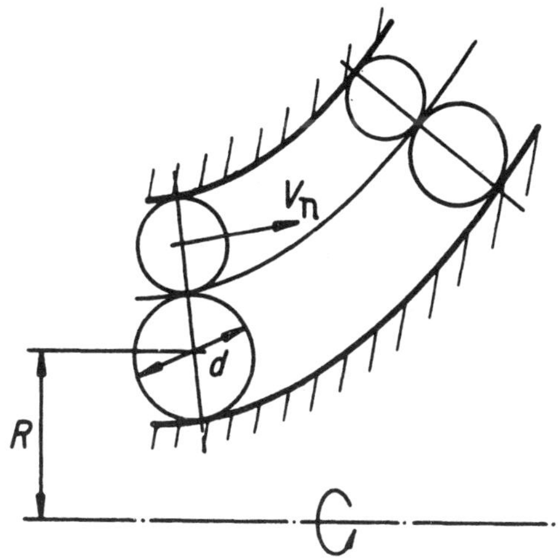


**Figure 5.3** Onc-dimensional approach to channel flow.

## 5.3 Two-dimensional approach

### 5.3.1 Passage shapes

Strictly, all flow motions will be three dimensional, and the one-dimensional approach is thus very approximate as a tool to describe them. In two-dimensional theory, flow is assumed to proceed along parallel planes, and much hydrodynamic theory may be used. This is based on the concepts of irrotational flow, where Bernoulli's theory that flow energy is constant is the connection between pressure and velocity distribution.

A relevant flow pattern is that obtained by superimposing a source and a free vortex to produce the spiral vortex of Fig. 5.4. This forms the basic model when considering flow in simple radial flow pumps, as described in section 1.4. Since both the radial and tangential velocities vary inversely with radius (the free vortex law) it follows that $\beta$ is constant, and the resulting streamlines are logarithmic spirals. These relate fairly well to flow in non-moving passages like volutes, but do not fit well with rotating blade passages, where the angle $\beta_2$ at the maximum radius is usually greater than $\beta_1$ at inlet and $V_{u2} > V_{u1}$, as Chapter 1 illustrates. Usually, blade profiles follow a modified spiral, as discussed in section 5.3.2; for more theoretical approaches see section 5.4.

Turning attention now to stream surface, we consider the contrast between Figs 5.5 and 5.6. In the first a bend is shown with streamlines based on one-dimensional considerations. In this case pressure must increase toward the outer wall, and Bernoulli's theory cannot be satisfied since

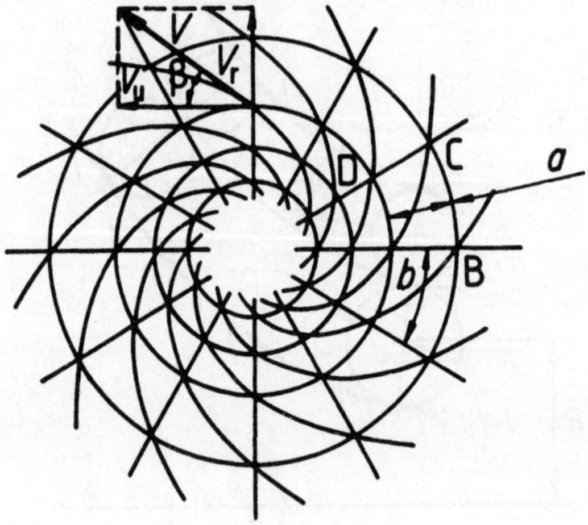


**Figure 5.4** Spiral vortex.

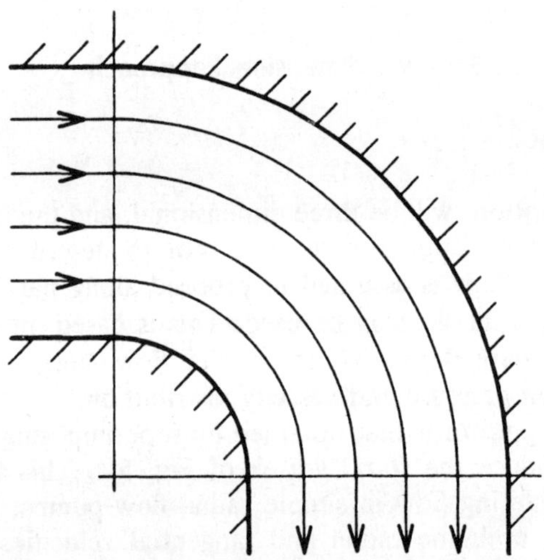

**Figure 5.5** One-dimensional stream surface solution to a bend.

velocity must decrease toward the outer wall. Figure 5.6 shows how streamlines must be adjusted to the requirements of flow net and Bernoulli principles. Acceptable stream surface shapes may be obtained using this principle, such as that sketched in Fig. 5.7 for a Kaplan machine.

5.3.2 Impeller or rotating cascade

A common technique for obtaining a radial cascade is to use conformal transformation of the linear or axial cascades discussed in section 5.4, a

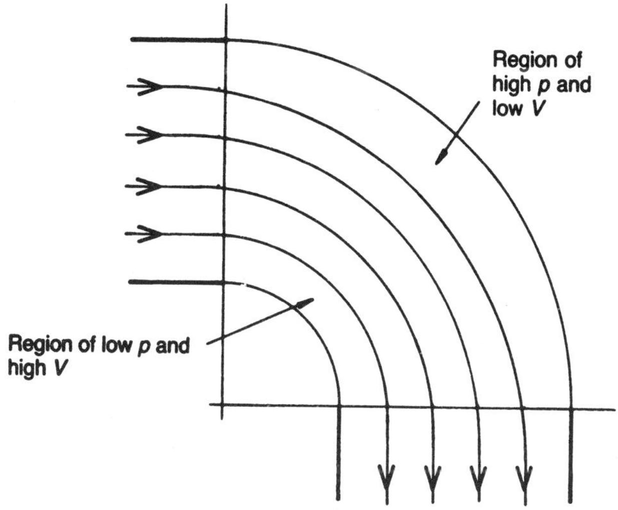


**Figure 5.6** Adjusted stream surfaces allowing for pressure and velocity distribution.

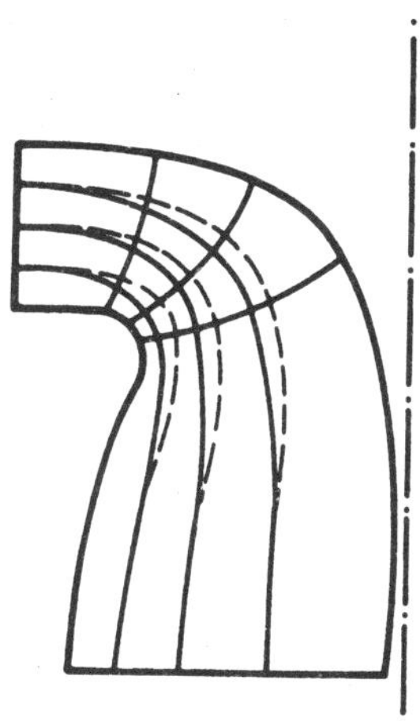

**Figure 5.7** Stream surface correction for a Kaplan turbine (dotted lines are the uncorrected surface shapes (after Pollard (1973), courtesy of the Institution of Mechanical Engineers).

simple one being illustrated in Fig. 5.8. The equation of the vane after transforming a cascade of flat vanes is

$$\psi = \cot\beta \log_e \frac{r_2}{r_1} \tag{5.2}$$

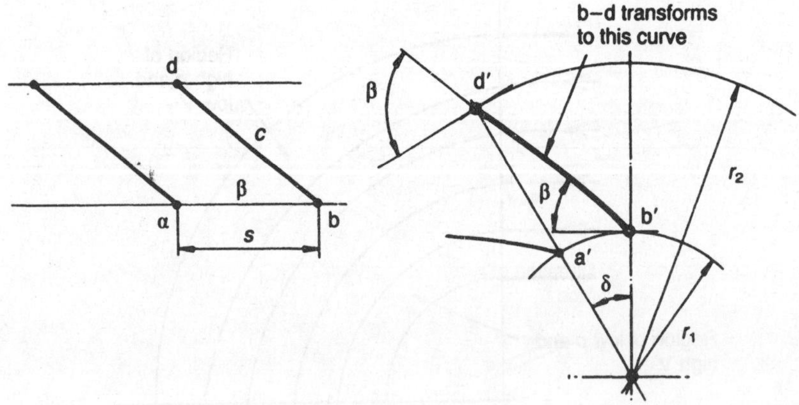


**Figure 5.8** Conformal transformation of a flat plate cascade to a circular cascade.

and

$$\frac{l}{t} = \frac{Z}{2\pi \sin\beta} \log_e \frac{r_2}{r_1} \tag{5.3}$$

The ideal performance of the transformed cascade may be predicted from the relations already described in section 5.4.

The following must be noted in comparison with the straight vane system. For centrifugal cascades, if $\beta_1 < \beta_2$, the angular momentum decreases with radius, and the curvature is less than the logarithmic spiral, the case for diffuser systems; and if $\beta_1 > \beta_2$, angular momentum increases as radius decreases and vane curvature is less than the basic spiral (inlet guide vanes). In a turbine, if $\beta_1 < \beta_2$, angular momentum decreases with radius and curvature is greater than the basic spiral (this applies to the return passages in multistage pumps and compressors).

The comments thus far relate to static cascades. In rotating cascades the irrotational concept breaks down, since flow departs considerably from the ideal shapes designed. Typical flow patterns observed are illustrated in Fig. 5.9, and the velocity distribution across a passage for one flow is shown in Fig. 5.10. It will be noted that numerous studies, particularly in compressors, suggest that this distribution varies with flow rate and is not steady, as well as being related to blade passage guidance (blade number $Z$).

A number of theoretical approaches have been made to the 'correction' of the ideal Euler-equation-based rotor outlet velocity triangle to give an actual change in $V_u$ rather than the ideal (Fig. 5.11). Busemann (1928) examined the cascade of logarithmic spiral vanes, and produced a modified Euler equation for a conventional backward curved bladed inpeller:

$$gH = h_0 u_2^2 \pm C_H u_2 V_{R2}(\cot\beta_2 - \cot\beta_1) \tag{5.4}$$

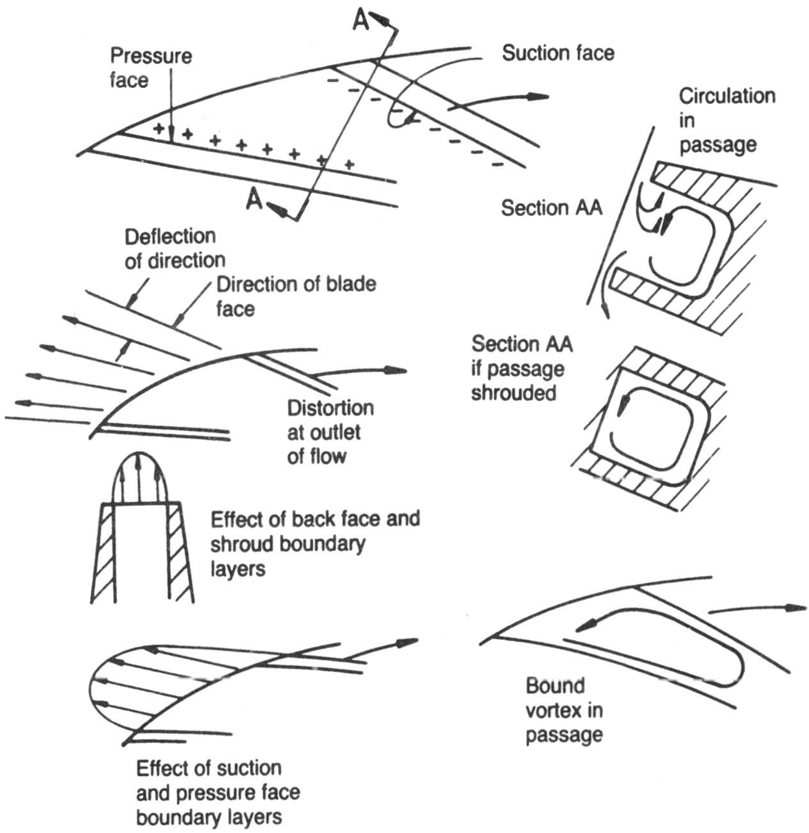


**Figure 5.9** Three-dimensional flow effects in radial passages.

Figure 5.12 is a plot of $C_H$ against $s/c$. The factor $C_H$ corresponds exactly with the factor in Fig. 4.12 and is called the throughflow factor. The term including $h_0$ is a correction for the shut valve energy transfer or displacement capacity and Busemann presented a series of plots for $h_0$ against radius ratio for a range of values of $\beta_2$ and demonstrated the effect of blade number. Weisner (1967) developed the Busemann approach during a study of the various slip theories by using the approach of Stodola (1945) who proposed that the 'slip velocity' $v$ be given by the relation

$$v = \frac{\pi \sin \beta_2 u_2}{Z} \tag{5.5}$$

and, since

$$u_2 = V_{u2} + \frac{V_{R2}}{\tan \beta_2} + v \tag{5.6}$$

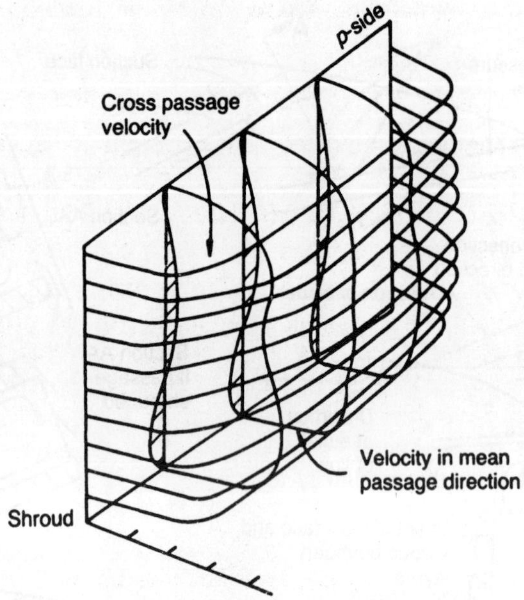


**Figure 5.10** Typical velocity distribution in flow leaving a rotor passage.

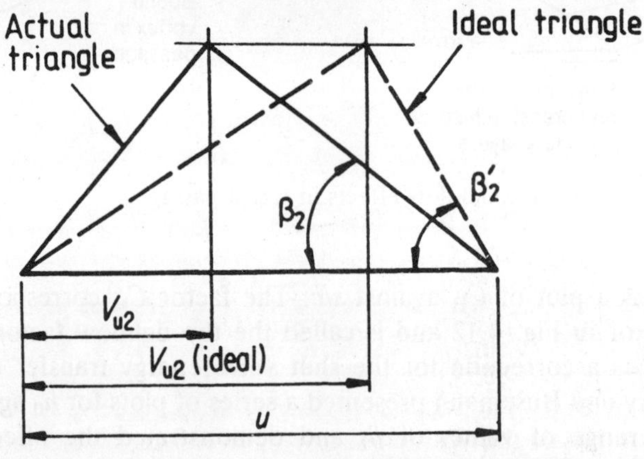


**Figure 5.11** Ideal and actual outlet velocity triangles for a radial outflow impeller.

or

$$1 = \frac{V_{u2}}{u_2} + \frac{V_{R2}}{u_2 \tan \beta_2} + \frac{\pi \sin \beta_2}{Z}$$

he proposed that

$$\sigma = \frac{V_{u2}}{u_2} + \frac{V_{R2} \tan \beta_2}{u_2} = 1 - \frac{\pi \sin \beta_2}{Z} \qquad (5.7)$$

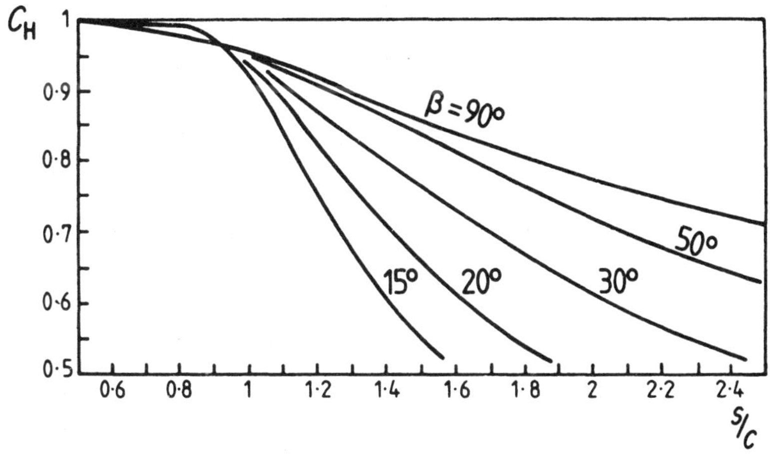


**Figure 5.12** Buscmann's throughflow factor.

He examined many correlations and compared data from over 60 pumps and compressors. From this study he proposed that the following relations applied within the limits specified:

$$\sigma = 1 - \frac{\sqrt{\pi} \sin \beta_2}{Z^{0.7}} \tag{5.8}$$

up to a limiting value of

$$\frac{R_1}{R_2} = \frac{1}{\ln^{-1}\left(\dfrac{8.16 \sin \beta_2}{Z}\right)} = \varepsilon_{\text{limit}} \tag{5.9}$$

and thereafter

$$\sigma = 1 - \frac{\sqrt{\sin \beta_2}}{Z^{0.7}}\left[1 - \left(\frac{R_1/R_2 - \varepsilon_{\text{limit}}}{1 - \varepsilon_{\text{limit}}}\right)^3\right] \tag{5.10}$$

Figure 5.13 is a plot of the Weisner factor plotted against the blade number $Z$ and the radius ratio for an outlet angle of 20°. The full lines are the Busemann factor and the dotted lines are the Weisner factor. The limit of the radius ratio given by equation (5.9) is shown also in the figure. The dotted lines are those given by equations (5.8) and (5.10).

Some empirical simple formulae which may be used as an approximation approach are as follows.

- Pfleiderer (1961):

$$\frac{gH^*_{\text{Euler}}}{gH_{\text{Euler}}} = 1 + C_p \tag{5.11}$$

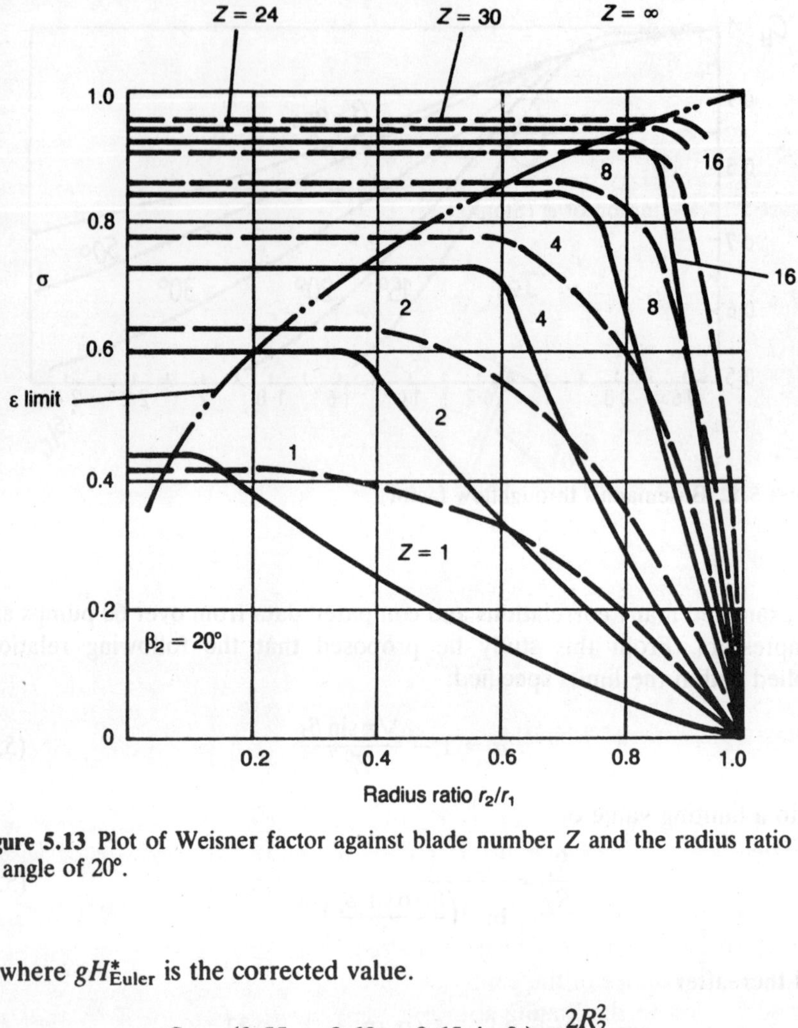


**Figure 5.13** Plot of Weisner factor against blade number $Z$ and the radius ratio for an angle of 20°.

where $gH^*_{\text{Euler}}$ is the corrected value.

$$C_{\text{p}} = (0.55 \rightarrow 0.68 + 0.65 \sin \beta_2) \frac{2R_2^2}{Z(R_2^2 R_1^2)}$$

- Stanitz (1952):

$$\mu(\text{or } K) = \frac{V_{\text{u2}}}{u_2} = 1 - 0.315\left(\frac{2\pi}{Z} \sin\frac{\phi}{2}\right) \tag{5.12}$$

where $\phi$ is defined in Fig. 5.14. For a radial machine where $\phi/2 = 90°$ equation 4.6 reduces to

$$\mu = 1 - 0.63/Z \tag{5.13}$$

- Stodola (1945) suggested an approximation to the slip velocity which has already been quoted as equation (5.5).

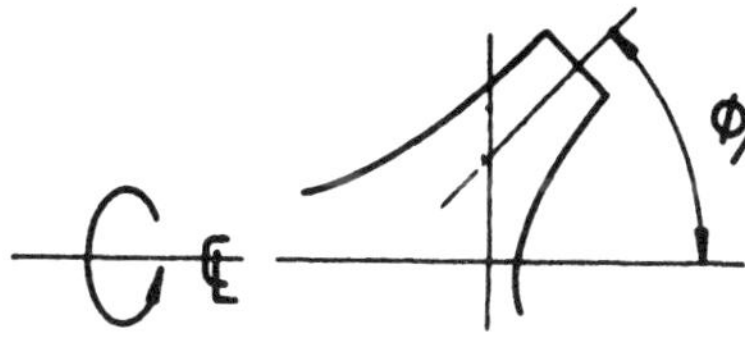


**Figure 5.14** Definition of $\phi$ (Stanitz).

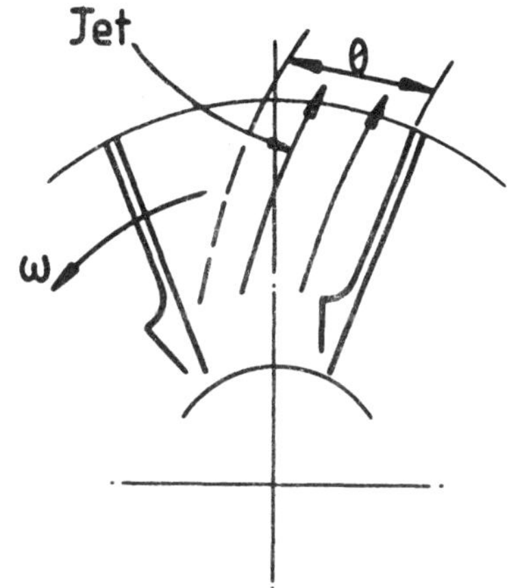


**Figure 5.15** Sketch of Whitfield's jet model.

There are discrepancies between these approaches as Stahler (1965) has demonstrated. His article proposes an alternative relation that does not correlate with the equations quoted above.

A more recent approach based on turbocharger design is that by Whitfield (1974) based upon radial bladed compressor impellers, draws attention to the lack of guidance of the emerging flow, proposes a jet model and suggests a modification to the Stanitz approach:

$$K_s = \frac{V_{u2}}{u_2} = 1 - K_1[1 - K_s^*]$$

where $K_s^* = 1 - 0.63\theta/2$ ($\theta$ defined as in Fig. 5.15. A relation of $K_1$ with blade number is given in Fig. 5.16. Whitfield's work supported the correlation, and further refinement is detailed in Whitfield, Doyle and Firth (1993).

## 5.4 Three-dimensional problem

Figure 5.9 indicates pronounced three-dimensional flow effects, some of which are a product of boundary layer behaviour on the passage walls. If the

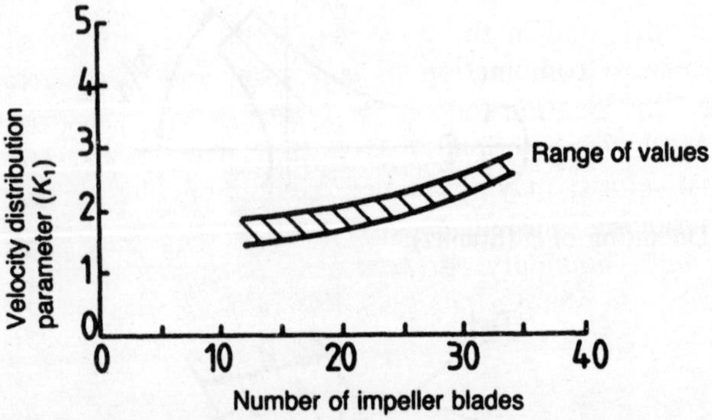


**Figure 5.16** Plot of the velocity distribution factor proposed by Whitfield (1974) (courtesy of the Institution of Mechanical Engineers).

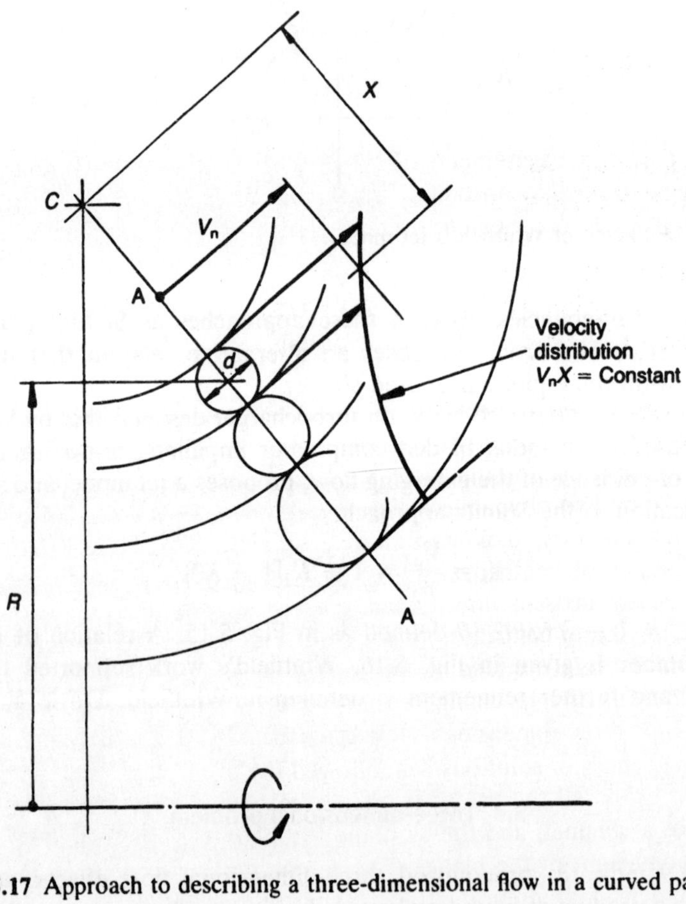


**Figure 5.17** Approach to describing a three-dimensional flow in a curved passage.

mixed flow model used in the previous section is used again, and flow is considered to be a combination of rotational and meridional effects but without vanes, the rotation follows the law of constant angular momentum, and meridional flow is described by the stream surfaces as before. However, the meridional velocity may not be constant but may be affected by the flow curvature. Assuming constant energy, and considering the passage in Fig. 5.17, where both boundary surfaces are concentric about centre C, the velocity distribution along a potential line such as AA will follow the free vortex law:

$$V_n X = \text{constant} \tag{5.14}$$

The constant depends on the line of constant potential. If continuity is applied, the total flow rate is given by

$$Q = \Sigma\, V_n 2\pi r d \tag{5.15}$$

or if a defined level at a distance $y$ is $V_{ny}$,

$$Q = \Sigma\left(\frac{V_{ny}y}{X}\right)2\pi r d \tag{5.16}$$

It will be noted that increments of flow between adjacent streamlines must be the same: $V_n 2\pi r d$ = constant. Thus the streamlines cannot be concentric. This, although an improvement on the earlier methods described, must also be modified to allow for the boundary layer with an injection of an assumed velocity profile close to the walls. Practice appears to be that the real fluid effects are ignored and the approximate approaches outlined are used.

## 5.5 Discussion of theoretical approaches to analysis and design

The idealized approach to machines has been to assume inviscid steady flow and to apply the hydrodynamic equations. Actual flow is three dimensional, unsteady and viscous, making rigorous mathematical solution extremely difficult. Studies of centrifugal and mixed flow machine theory began by assuming steady inviscid flow on asymmetric stream surfaces before and after the rotor. Later contributions sought to modify the flow patterns round critical surfaces, like the leading and trailing edges, to allow for boundary layer separation, and to change stream surface curvatures to incorporate corrections for three-dimensional flow effects.

One large school of solutions has followed the principles discussed by Wu (1952) who proposed the use of two surfaces – the S1 plane from blade to blade across a channel, and the S2 plane from hub to shroud (that is, along the length or height of the blade). The solution then proceeds on the basis of a two-dimensional stream surface shape through the machine following the meridional approach already outlined, with a pseudo three-dimensional correction to the surfaces on the basis of a loss theory. The blade sections are then designed along each surface in turn from hub to tip section. An early English example of this type of solution is that due to Marsh (1966), which was based upon inviscid flow; this was subsequently improved upon by Perkins (1970). A later technique introduced the effect of viscosity by incorporating a loss model in the form of a polytropic efficiency based on experimental relations. This technique was used by Pollard (1973), for example, to both analyse a number of machines and to perform design studies. He considered a typical pump and a Francis turbine, establishing first the stream surfaces, then performing blade to blade solutions on each surface in turn, utilizing a conformal transformation due to Wilkinson (1968) that produced a planar cascade. This cascade may then be used to study velocity distributions with inlet and outlet angles and pressure distributions that, when considered satisfactory, are then transformed back to the stream surface. This process, repeated for each stream surface in turn, allows the blade profiles to be built up from hub to tip. Pollard demonstrated how the predicted pressure distributions matched experimental data for a pump, and showed quite good agreement apart from the leading edge and trailing edge areas. These techniques are being continuously improved, and many contributions correlating experimental data to provide corrections to the computer-based analytical solutions have been published. Some are referred to in context in later chapters.

An example of the combination of an analytical solution approach with experimentally obtained data is that due to Stirling and Wilson (1983). The design of a mixed flow pump, illustrated in Fig. 5.18, is described in some detail. They discuss their earlier work using a mainframe computer at the

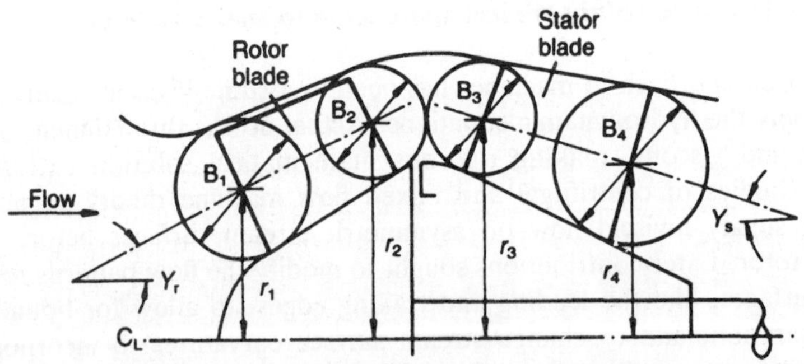


**Figure 5.18** Mixed flow pump (adapted from Stirling, 1982).

NEL, and demonstrate how a desktop computer system may be used to produce the same design. Their method first determines the casing profiles and the most suitable positions of the blade leading and trailing edges $B_1$, $B_2$, $B_3$ and $B_4$; the spacing of blade rows are about a half-blade chord in practice, and the shapes are optimized to give good flow characteristics. The method then proceeds to predict the velocity distribution without blades, and moves on to select blade energy distribution and to look at the blade solution for the rotor, including losses based on experimental and theoretical correlations taken from many sources. The rotor blades are then designed, and the stator blades are designed in similar routines, and the complete machine drawings may then be produced for manufacture. They demonstrate the correlation between prediction and test for pump designs having dimensionless specific speeds of 1.8, 2.5 and 3.2. The validity of the approach, compared with the experimental and empirical approaches, was demonstrated when pumps were designed by several engineers and quite close agreement found between the techniques deployed. These are described in the papers by Bunjes and Op de Woerd (1982), Richardson (1982), Thorne (1982) Stirling (1982) and Chiappe (1982).

A survey undertaken by Hughes, Salisbury and Turton (1988) reviewed commercial computer-based methods of analysing flow through machines pump and compressor passages, and also approaches to design. There are now a number of manufacturer-produced software systems that interface with computer-based manufacturing systems, and also give user friendly method of pump selection. These are reviewed at intervals in such journals as *World Pumps*.

# 6 Centrifugal machines

## 6.1 Introduction

Centrifugal machines of the pressure increasing type are covered here: these include pumps, fans, and compressors or blowers. In all these designs the fluid enters axially and is discharged by the rotor into a static collector system and thence into a discharge pipe. Pumps and fans have a common feature in that the fluid is considered incompressible, and in the compressors and blowers there is usually a considerable density change.

Many machines are single stage designs as sketched in Figs 6.1–6.4, and may either have an axial inlet or a ducted inlet as shown in the double entry pump shown in Fig. 6.2. The single stage pump is the workhorse of the process and petrochemical industries, producing specific energy rises from about $50\,\mathrm{J\,kg^{-1}}$ to over $1200\,\mathrm{J\,kg^{-1}}$ (head rises from 5 to 125 m of liquid) and delivering flow rates from 6.3 to $400\,\mathrm{m^3\,h^{-1}}$. Single stage compressors in turbochargers, for example, give pressure ratios up to at least 4:1 but more conventionally will give about 2.5:1; fans used for ventilation duties give pressure rises equivalent to a maximum of around 500 mm water gauge. Multistage machines like that shown in Fig. 6.5 are used for boiler feed and similar duties, and may be called on to deliver an energy rise of $300\,000\,\mathrm{J\,kg^{-1}}$ (or head rises up to about 3000 m of liquid).

Since the flow paths are the same for the different machines, the main considerations involved in the suction area, the impeller and the stator delivery system, are discussed in that order, with some attention being paid to the design choices open to the engineer.

The chapter concludes with a discussion of radial and axial thrust loads, and a brief discussion of the additional problems posed by the mixed flow machines.

## 6.2 Inlet or intake systems

The simplest inlet system found in pumps, fans and compressors is the straight pipe coaxial with the impeller centreline, as seen in Fig. 6.6, but

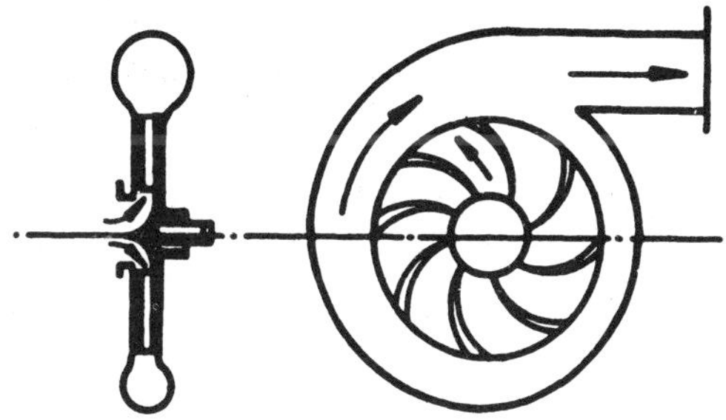

**Figure 6.1** Simple centrifugal pump.

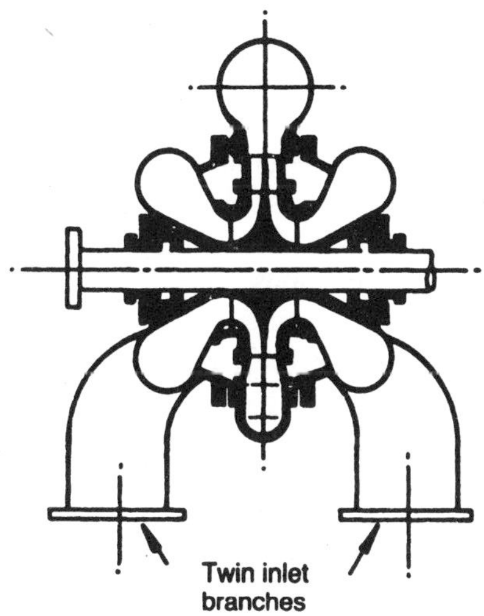


**Figure 6.2** Double suction centrifugal pump.

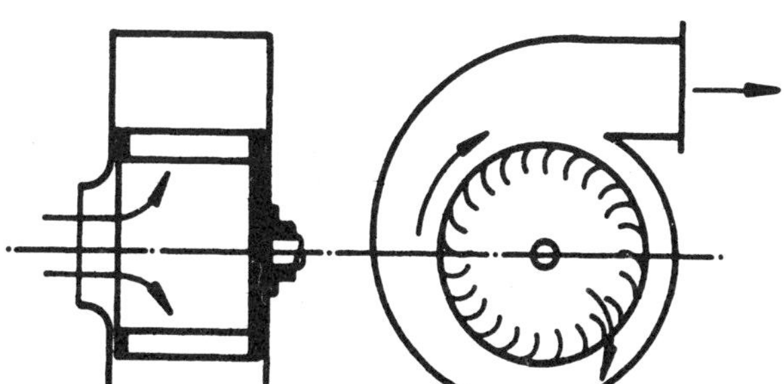


**Figure 6.3** Centrifugal fan.

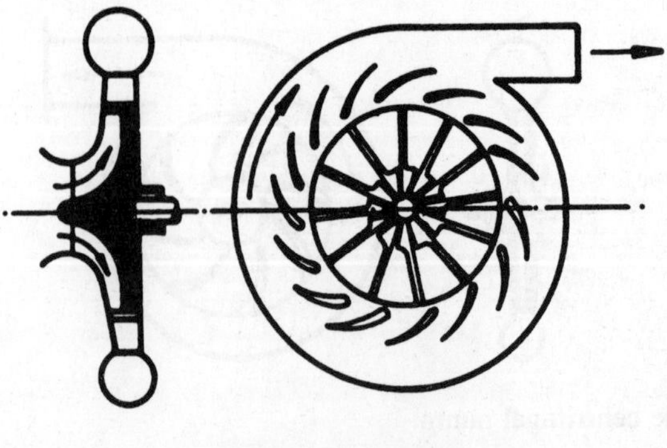


**Figure 6.4** Centrifugal compressor.

space and suction system layout frequently requires a ducted inlet as shown in Fig. 6.7. These can cause undesirable inlet velocity profiles and strong three-dimensional flow patterns in the eye of the impeller. This is particularly true in double suction machines where the shaft passes through the flow on both sides of the impeller, and a good treatment of this problem is found in the paper by Matthias (1966). However good are the design flow patterns, there will be problems at part flow with vortex formation in the straight suction pipe which penetrates far into the suction from the suction flange in the manner shown in Fig. 6.6.

## 6.3 Impeller

### 6.3.1 Eye or inducer section

As the fluid approaches the impeller it slowly takes up its rotation so that the absolute velocity begins to increase with a resulting drop in the static pressure (Fig. 6.8). This process continues till the fluid reaches the leading edges of the rotor vanes, and then has to pass round the nose and into the impeller passages. The pressure may tend to drop, depending on the incidence angle, and may in pumps reach the local liquid vapour pressure with resulting cavitation problems. It is conventional to make allowance for incidence angles up to 6° in design in order to allow for these effects.

In compressors, which tend to have radial blade sections at the outlet, the inducer or inlet section is given a large amount of twist, as indicated in Fig. 6.9, to accommodate the relative velocity direction imposed by the inlet motion of the gas. The incidence angle rule just noted for pumps is used but great attention often has to be given to the shape of the nose. This is

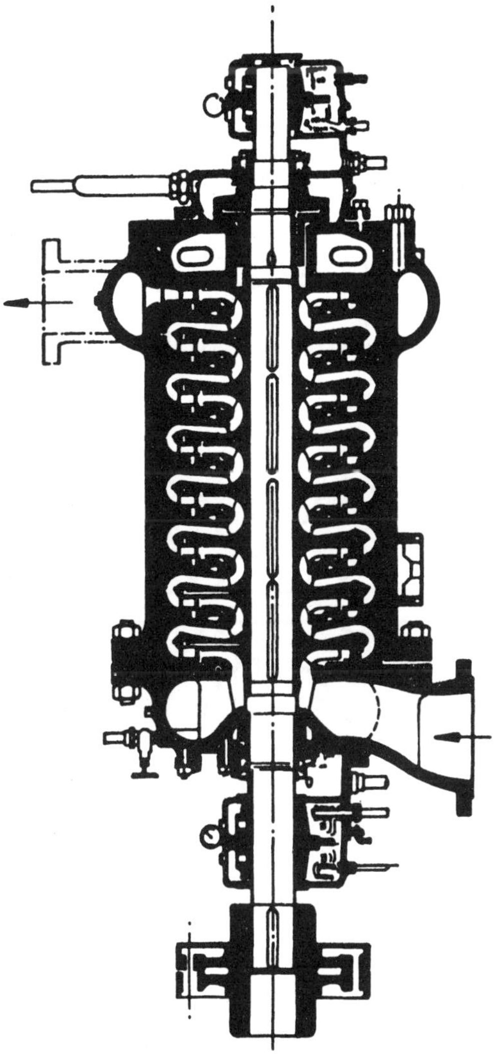

**Figure 6.5** Multistage centrifugal pump.

illustrated in Fig. 6.9, which shows the effect of three nose profiles and their effect on the local Mach number, shape 1 being preferred to allow smooth local velocity changes.

The axial velocity is also an important parameter, and in pumps is kept as low as possible to avoid cavitation problems. The size of this velocity is related to the eye or suction diameter, and empirical rules will be found in Stepannof (1957b) and Turton (1994) among other texts. The relations

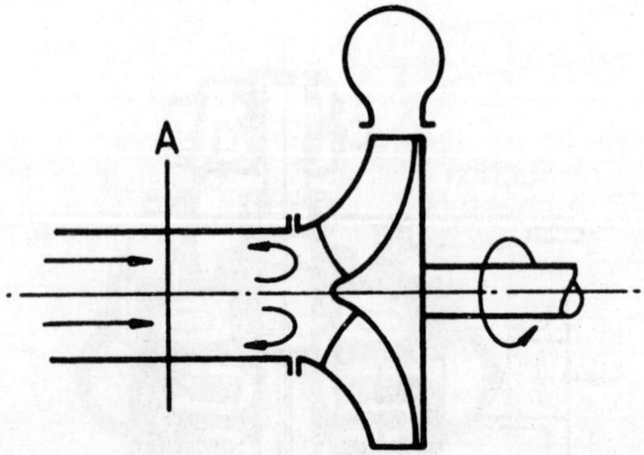


**Figure 6.6** Simple suction pipe.

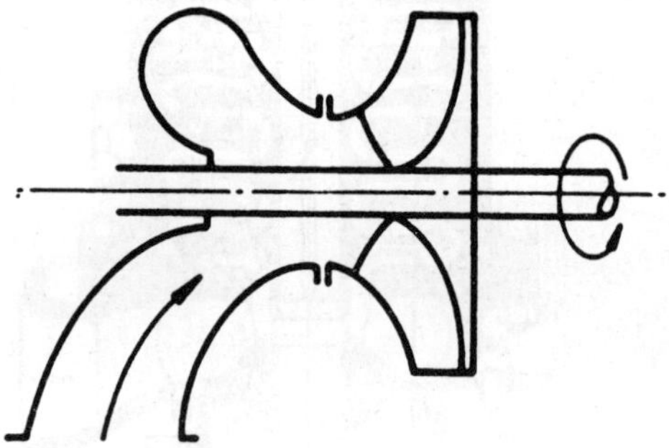

**Figure 6.7** Ducted entry suction.

between axial velocity and suction diameter were discussed in section 3.4, and the risks of oversizing were outlined. In compressors the need to control the velocity at the maximum diameter of the suction zone to avoid Mach number problems is paramount. Figure 6.10 illustrates the probable design relations that can be involved and used as design guidelines.

To illustrate the basic approach, consider the following compressor example. A compressor is to deliver air at the rate of $9\,kg\,s^{-1}$ drawn from a plenum chamber where the conditions are $10^5\,N\,m^{-2}$ and 295 K. The plenum conditions are stagnation. Thus, if it is assumed as the ideal case that there is zero inlet whirl for the impeller and an acceptable axial velocity component of $75\,m\,s^{-1}$, the static conditions in the inducer, and the density and other parameters, may be found. Assuming the stagnation conditions remain the same along the inlet pipe (approximately true if it is short),

$$295 = T_1 + 75^2/2C_p$$

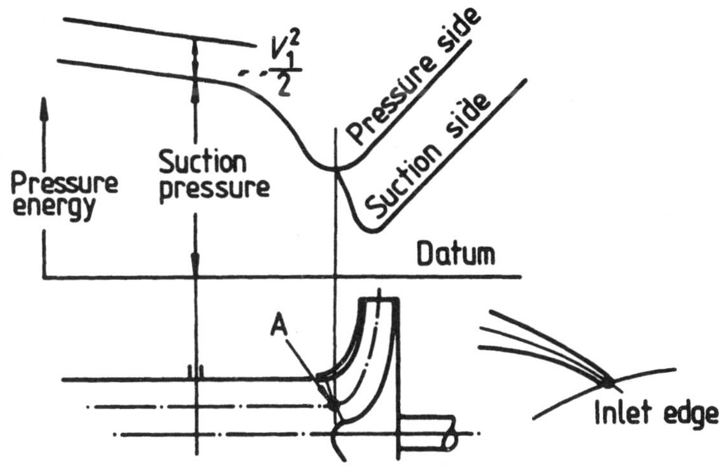


**Figure 6.8** Pressure changes on a stream surface in the suction of a centrifugal machine.

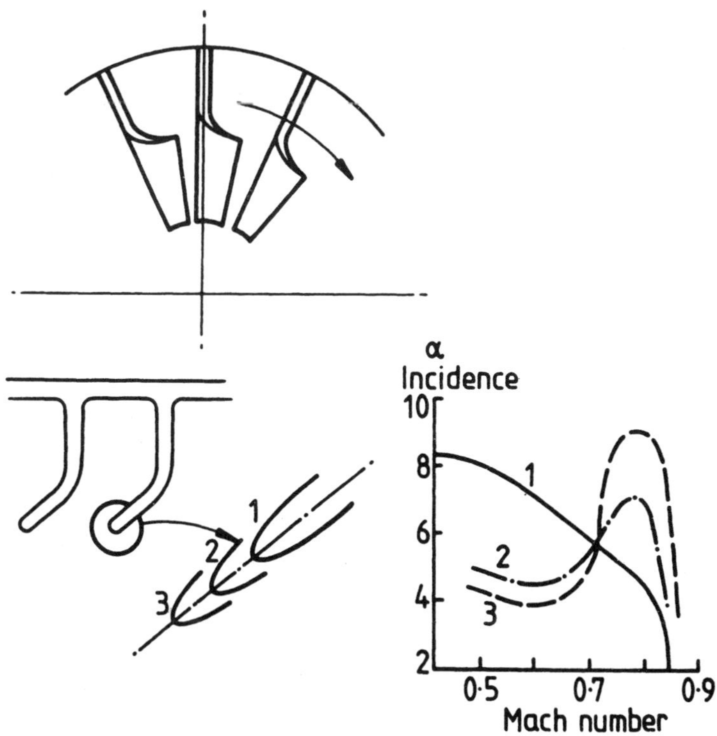


**Figure 6.9** Inducer of a centrifugal compressor.

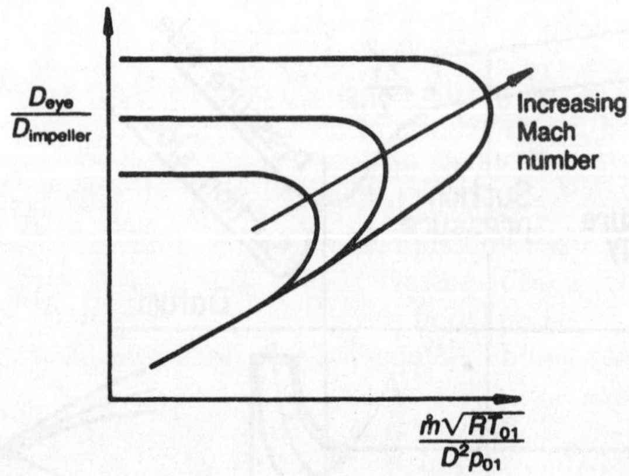


**Figure 6.10** Design plot for a compressor/inducer.

thus

$$T_1 = 292.2\,\text{K}$$

Following the gas laws,

$$p_{01}/p_1 = (T_{01}/T_1)^{1/3.5}$$

Therefore

$$p_1 = 0.997 \times 10^5\,\text{N m}^{-2}$$
$$\rho = 1.189\,\text{kg m}^{-3}$$

Assuming a hub to tip diameter ratio of 0.35,

$$\frac{9}{1.189 \times 75} = \frac{\pi}{4}(D_{tip}^2 - D_{hub}^2)$$

$$D_{tip} = 0.385\,\text{m}$$

$$D_{hub} = 0.135\,\text{m}$$

The acoustic velocity at the inducer section is given by

$$a = \sqrt{(\gamma RT)} = 342.65\,\text{m s}^{-1}$$

The velocity triangle is drawn for the maximum diameter of the inducer, using the assumption of zero whirl and a driving speed of 15000 rev min$^{-1}$, i.e. a tip peripheral velocity of 302.38 m s$^{-1}$. This gives $W_1 = 311.54\,\text{m s}^{-1}$, a Mach number based on $W_1$ of 0.91, and $\beta_1 = 13.93°$ (with incidence probably 16°). Similarly, for the hub diameter, $W_1 = 129.87\,\text{m s}^{-1}$ and $\beta_1 = 35.27°$.

### 6.3.2 Impeller design

The inducer (or eye) of the impeller has been discussed, and attention may now be devoted to the choice of blade or passage design.

As in all turbomachines there is a need to balance control with friction loss due to guiding surface. It is possible using recent computer techniques based on loss data to obtain blade numbers, which can then be fed into the suites of programs for passage shapes. Empirical formulae for pumps and compressors based on 'good practice' will be quoted here.

For pumps (quoted by Stepannof (1957b), for example) blade number is

$$Z = \beta_2/3 \tag{6.1}$$

Another, quoted by Pfleiderer (1961) and others, is

$$Z = 6.5\frac{D_2 + D_1}{D_2 - D_1}\sin\beta_m \tag{6.2}$$

In these equations, $D_1$ and $D_2$ are inlet and outlet diameters of the impeller, respectively, $\beta_2$ is the impeller outlet angle, and $\beta_m$ is the arithmetic mean of the inlet and outlet angles of the impeller, quoted with respect to the tangential direction.

Equally simple rules exist for compressors, such as

$$Z = \beta_2/3 \tag{6.3}$$

Another, quoted by Eckert and Schnell (1961), is

$$Z = \frac{2\pi\sin\beta_m}{(0.35 \text{ to } 0.45)\log_e D_2/D_1} \tag{6.4}$$

Pump design conventions indicate backward curved blades with outlet angles $\beta_2$ in the range 15° to 35° or 40°, the lower values being found typically in high specific speed designs. Compressors have tended for stress reasons to have outlet angles of 90°, but some modern designs have been provided with backward curved vanes; McCutcheon (1978) quoted an angle of 33° for a 4.4:1 pressure ratio machine with a rotational speed of 70 000 rev min$^{-1}$.

Varley (1961) studied the effects of blade number on pump performance, with the effects of blade angle and surface roughness as additional factors. The machine used was a double entry design with an impeller diameter of 244 mm, designed to give a duty of 16 m at 6.44 l s$^{-1}$ when rotating at 1400 rev min$^{-1}$. Figure 6.11 indicates that high numbers of blades resulted in the head characteristic falling to zero flow, a condition known as unstable. The best shaped curves apply to five and six blade designs, compared with the seven suggested by equation (6.1) on the basis of an angle of 27°. Figure 6.12 demonstrates that higher outlet angles gave unstable curve shapes, that

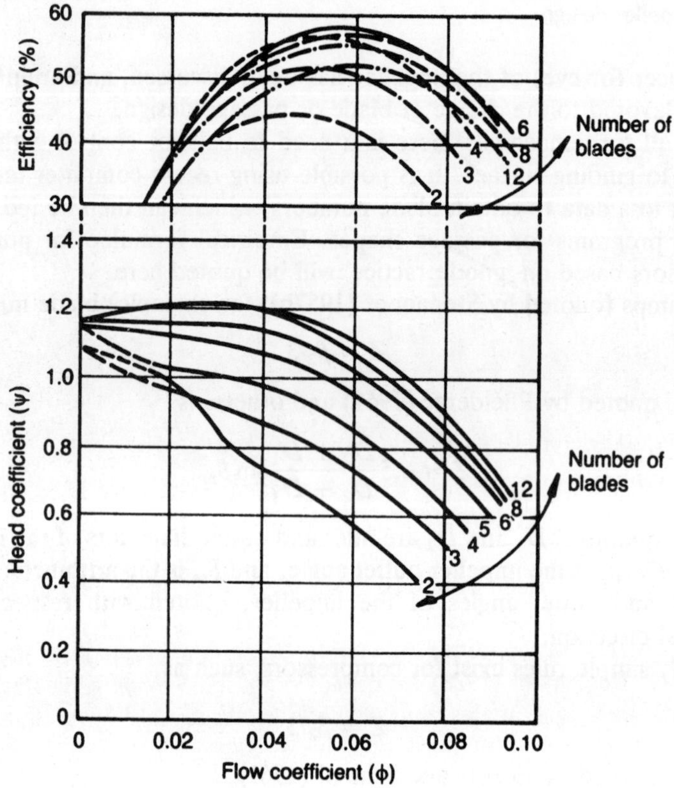


**Figure 6.11** Pump characteristics with varying blade number (after Varley (1961) courtesy of the Institution of Mechanical Engineers).

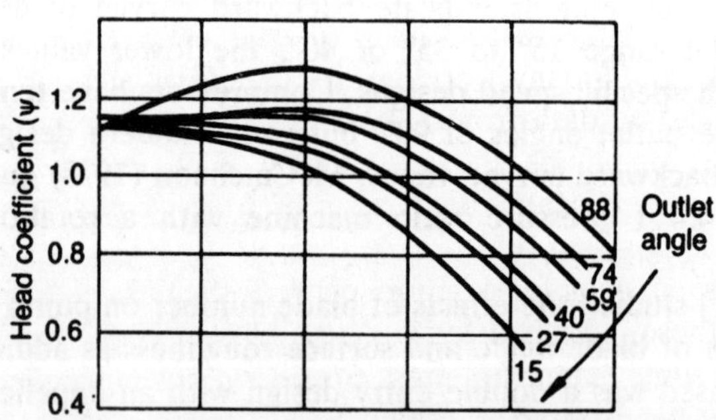


**Figure 6.12** Effect of outlet angle on pump characteristics (after Varley (1961) courtesy of the Institution of Mechanical Engineers).

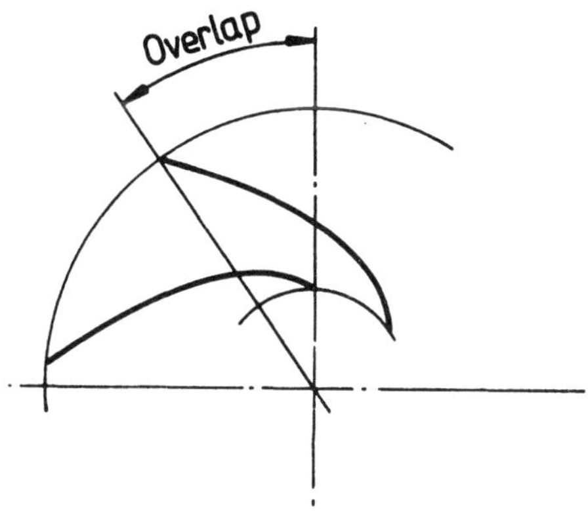


**Figure 6.13** Angle of blade overlap.

27° gave a characteristic which rose steadily to shut valve, regarded as a very desirable feature, and that this angle gave the best efficiency. Varley studied the slip correlations outlined in section 5.3.2, and found that the Stodola (1945) and Busemann (1928) formulae compared well with his data for low blade numbers and angles, but deviated considerably as both angle and blade number rose, particularly that due to Stodola.

Another design consideration related to blade number is the angle of overlap, illustrated in Fig. 6.13 for a pump design. Texts that deal with pump design suggest an angle of 45° as a good compromise for pumps of conventional designs. Centrifugal fans tend to have more but shorter blades; Eck, in his classic text (1973), demonstrated this for the usual circular arc blades, and showed the discontinuity in characteristic curves due to local stall effects. He gives many working formulae for blade and pitch based on good practice. Some data from work by Myles (1969; Fig. 6.14) indicates that for fans, as for pumps, the lower angles give rising characteristics. This paper also gives useful data about losses and their codification in a manner similar to the compressor diffusion factor.

Optimum proportions for compressors follow the same principles already outlined, but these have had to be modified to account for compressibility effects, and to some extent for stress considerations in high speed machines. As has already been stated, the blades tend to be radial at outlet, so that equation (6.3) indicates 30 blades. Although this will give good control at outlet and reduce slip, there will be too much metal in the inducer area, so that the usual arrangement is to make every other blade a half blade as shown in Fig. 6.15. McCutcheon (1978) describing the backward curved impeller, showed half blades; he reported that empirical rules for blade number were checked by finding blade loadings as a part of the design

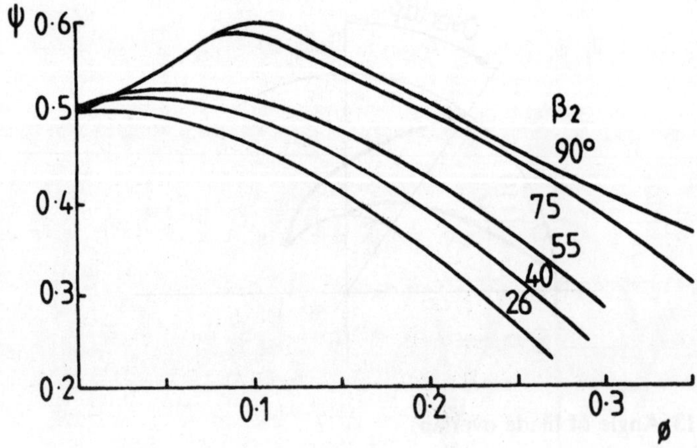


**Figure 6.14** Effect of outlet angle on fan characteristics (after Myles (1969) courtesy of the Institution of Mechanical Engineers).

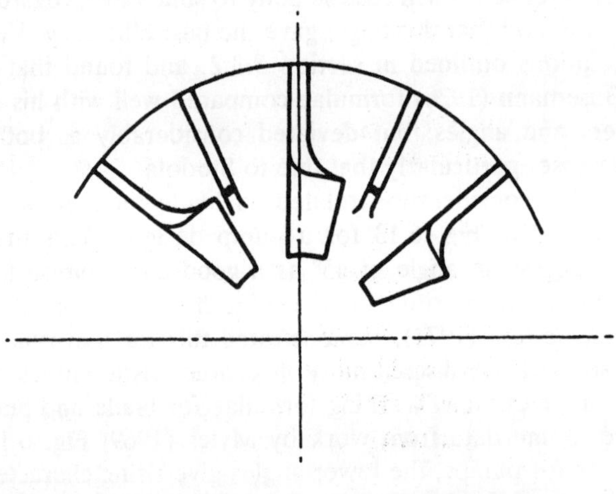

**Figure 6.15** Provision of half vanes in a centrifugal compressor.

process using blade to blade methods, and he found quite close agreement with the empirical approach.

Another important consideration is the shape of the blade passages. The side elevations found in good pump practice are shown in Table 6.1, and the placing of the inlet edge in elevation is seen to vary with specific speed. In blowers the leading edges tend to be parallel to the axis of rotation; the flow is high but pressure rise relatively small. In the majority of compressor designs the tendency is to provide inlet edges close to the plane at right angles to the axis of rotation, allowing for as long a passage length as

**Table 6.1** Pump side elevations

| Non-dimensional $k_s$ | Impeller profiles | Velocity triangles | Characteristics |
|---|---|---|---|
| 0.188–0.567 | 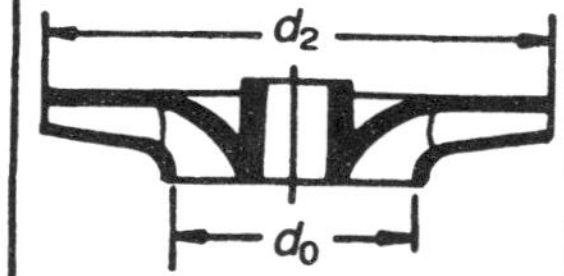  $d_2/d_0$ = 3.5–2.0 | 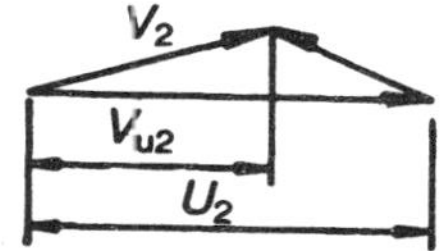  | 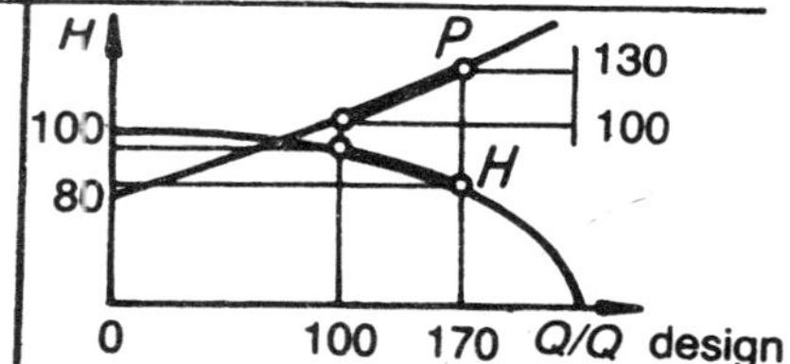  |
| 0.567–0.944 | 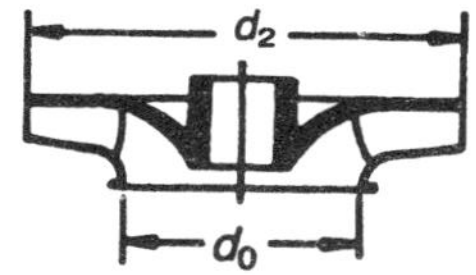  $d_2/d_0$ = 2.0–1.5 | 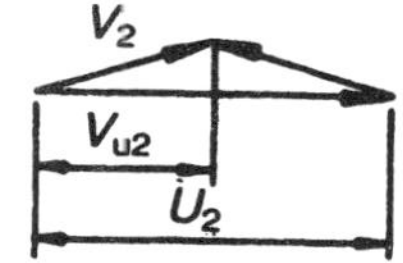  | 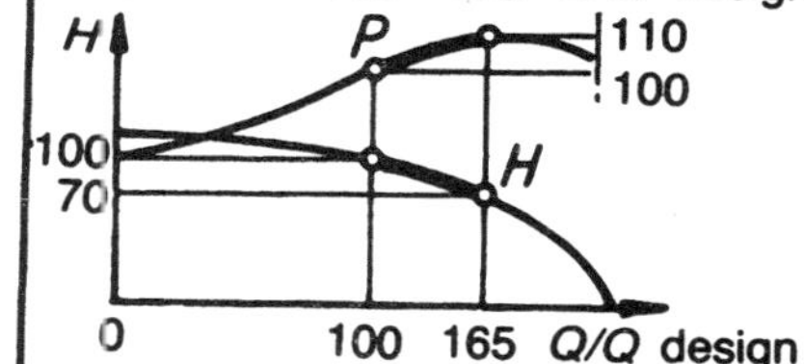  |
| 0.944–1.511 | 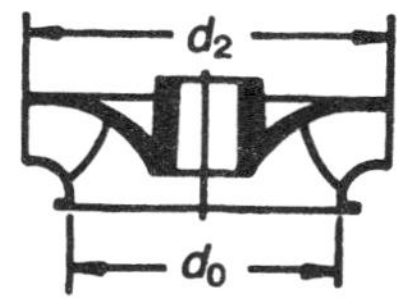  $d_2/d_0$ = 1.5–1.3 | 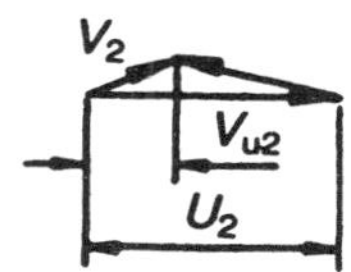  | 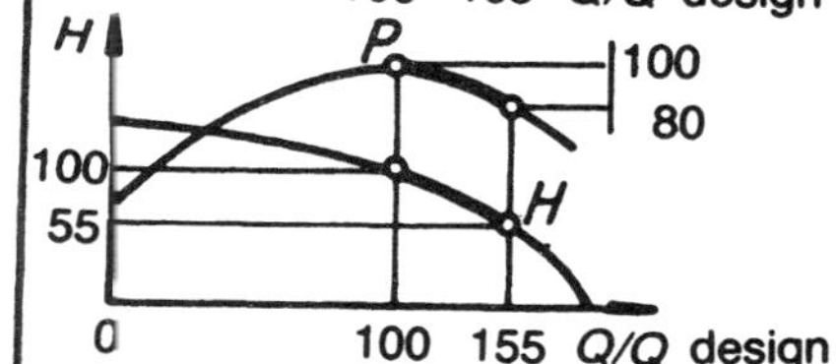  |
| 1.511–2.833 | 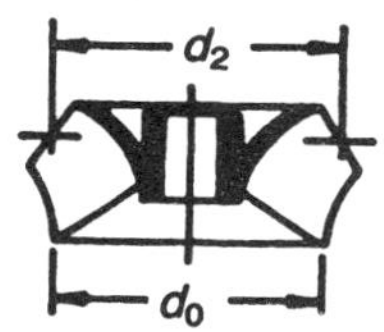  $d_2/d_0$ = 1.3–1.1 | 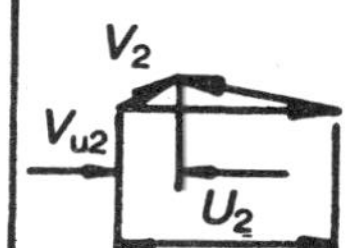  | 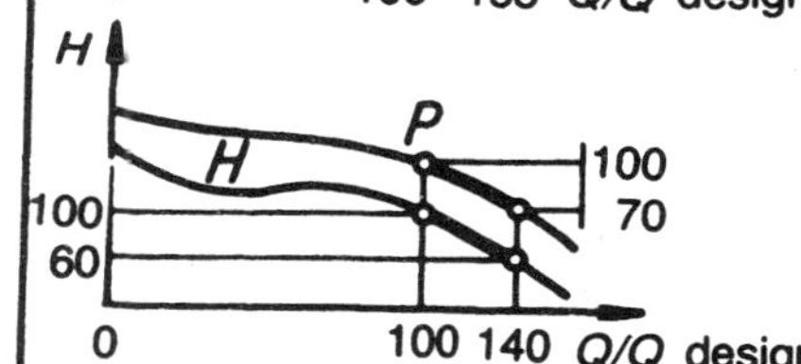  |

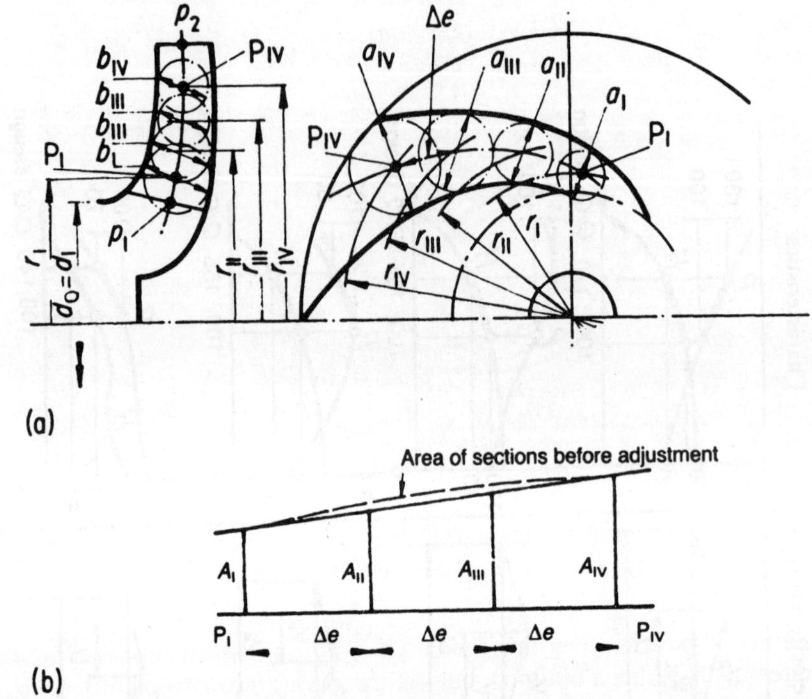


**Figure 6.16** (a) Method of establishing passage area changes (at point $P_1$, area $A_1 = b_1 a_1 2\pi r_1$). (b) Cross-sectional area of impeller passage plotted against passage length.

possible to give good flow control. Blade passages are designed to give a controlled diffusion from inlet to outlet, so that the area changes required to give this must be determined by the profiles of the shroud and the back-plate, and by the blade shapes too. The texts cited give ground rules for blade profile: in pumps and compressors the blades will follow an approximate Archimedian spiral, either using point to point programmes or combinations of radii. When the blade profiles are set and the elevation profiles determined the passage area changes must be checked as illustrated in Fig. 6.16. The principles outlined in Chapter 5 may be used, but the area changes must give smooth changes from inlet to outlet; if this is not achieved by the shapes designed, modifications need to be made and the process followed through again.

In compressors the pump approach just outlined may be followed but compressible considerations demand care. Figure 6.17, for example, shows a compressor design obtained by using meridional and blade-to-blade solutions that, by successive passes of calculation, achieve a smoother velocity variation. Detailed discussion is not possible here, but the contributions cited in Chapter 5 may be consulted as well as those given in this chapter.

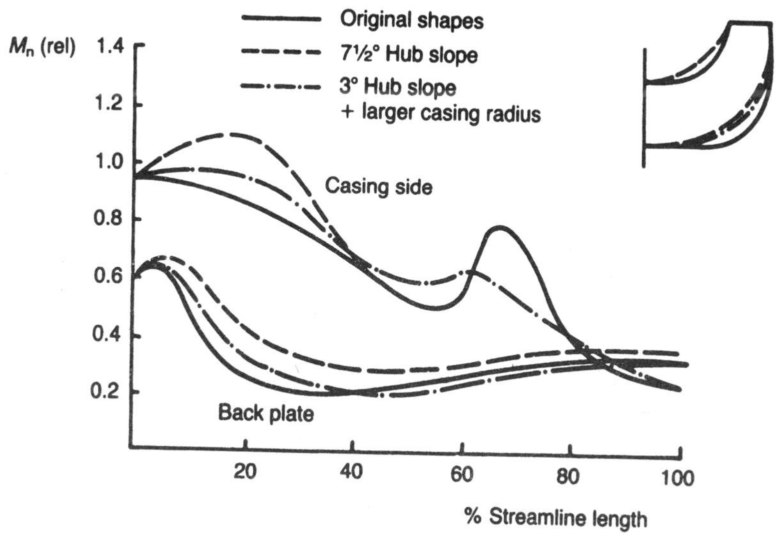


**Figure 6.17** Effect of wall shapes on local relative velocities in a compressor.

The main features governing the design of good performance pumps and compressors have been outlined, but before leaving a discussion of impellers it is of interest to consider again the paper by Varley (1961). He studied the effects of surface roughness on performance. Figure 6.18 indicates that roughness cannot be ignored as unimportant, so that cost reductions which lead to deliberately poor surface production give rise to performance penalties. This is true for pumps, fans and compressors, although the latter, being 'high technology' machines, normally have high class finishes on all flow surfaces. In fans and pumps, of course, the selling price is also a consideration that may modify material and surface finish choices. Other factors affecting performance are clearances between rotating and static surfaces, and their influence on leakage flow. This was briefly remarked upon when discussing the approach to performance prediction by Nixon, in section 2.4.1. For efficient operation clearances should be a minimum, but tolerances in manufacture, and such effects as end float in bearings and thermal gradients give rise to differential rates of expansion in many machines, which result in large cold clearances in compressors and normally wide clearances in pumps and fans. That care is needed in settling dimensions is indicated by a well known manufacturer of turbochargers, who reported a drop of 2% in performance as the side clearance between impeller and casing (expressed as a proportion of impeller diameter) increased from 0.036 to 0.143. Pampreen (1973) quoted a similar drop in a compressor passing 0.904 kg of air per second when this clearance ratio changed from near zero to 0.06.

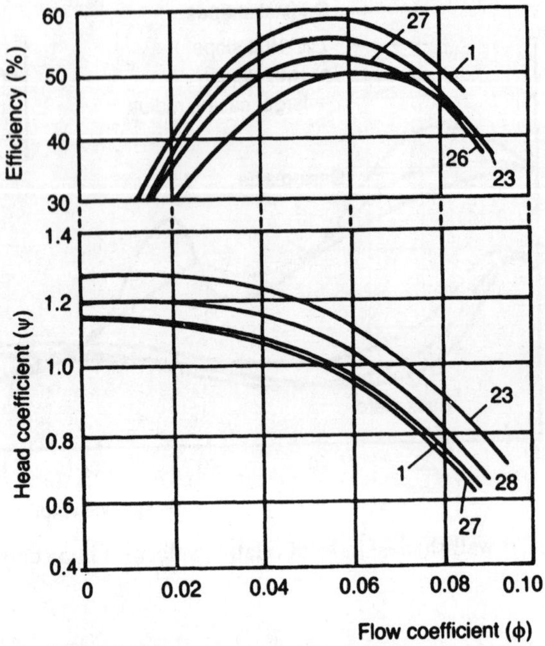


**Figure 6.18** Pump characteristics with various impeller components roughened (after Varley (1961) courtesy of the Institution of Mechanical Engineers).

Commercial considerations result, particularly for pumps, in the practice of providing a range of impellers in one casing, and also in reducing the diameter of an impeller ('turning down') up to values approaching 15% of the maximum size. Head and flow may be predicted using the so-called scaling laws:

$$\frac{P_1}{P_2} = \frac{N_1^3 D_1^3}{N_2^3 D_2^3} \tag{6.5}$$

$$\frac{Q_1}{Q_2} = \frac{N_1 D_1}{N_2 D_2} \tag{6.6}$$

$$\frac{gH_1}{gH_2} = \frac{N_1^2 D_1^2}{N_2^2 D_2^2} \tag{6.7}$$

These laws allow the likely performance to be predicted from the full size characteristics, but do not with accuracy allow prediction if the impeller is cut down to less than 75% of the full size. Another well known 'adjustment' technique is the process of filing the vane tips, which can make small but significant adjustments in performance, as Fig. 6.19 illustrates.

For further illustration the example in section 6.3.1 will be continued.

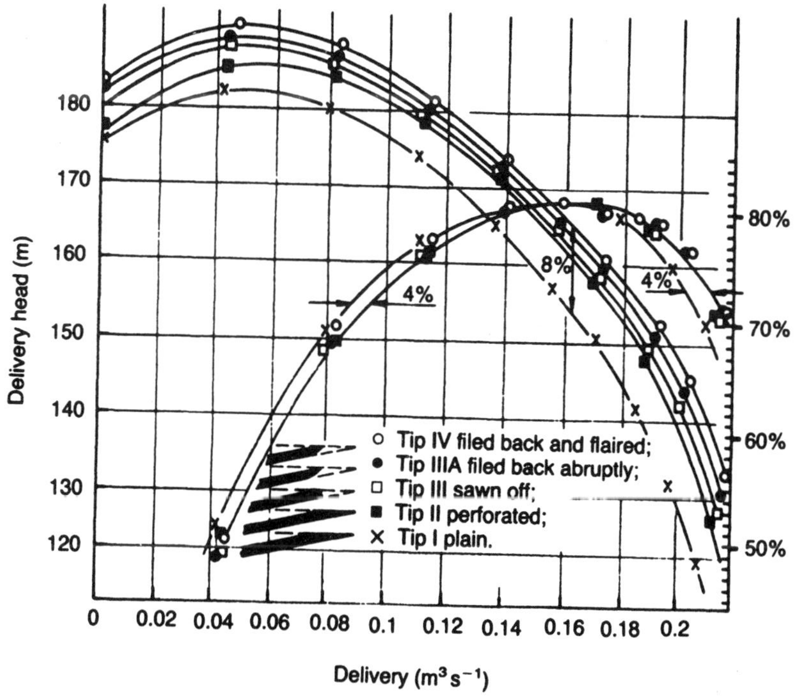


**Figure 6.19** Effect of centrifugal pump blade tip shape on performance (after Worster and Copley (1961) courtesy of the Institution of Mechanical Engineers).

Since the pressure ratio required is 3.2:1, and the efficiency suggested is 80%,

$$3.2 = (1 + 0.8\Delta T/T_{01})^{3.5}$$

Therefore,

$$\Delta T = 145.37$$
$$T_{02} = 440.37\,\text{K}$$

Assuming zero inlet whirl

$$V_{u2}u_2 = C_p\Delta T$$

Assuming as a first trial that $D_2/D_1$ is 2,

$$\beta_m = \frac{90 + 13.93}{2} = 52°$$

Then, from equation (6.4),

$$Z = \frac{2\pi \sin 52°}{0.45 \log_e 2} = 16$$

Using the Stanitz equation, equation (5.12),

$$\mu = \frac{V_{u2}}{u_2} = 1 - 0.315\left(\frac{2\pi}{16}\sin 90\right)$$

which gives

$$\mu = 0.876$$

and therefore

$$0.876u_2^2 = 1.005 \times 10^3 \times 145.37$$

Thus

$$u_2 = 408.4\,\mathrm{m\,s^{-1}}$$
$$V_{u2} = 357.76\,\mathrm{m\,s^{-1}}$$

and, for 15 000 rev min$^{-1}$,

$$D_2 = 0.52\,\mathrm{m}$$

The passage height at this diameter is now needed. The density is required as well as the outlet radial velocity, and to establish this some estimate of efficiency or loss is required. For example the outlet stagnation conditions at the discharge flange are known from the overall pressure ratio already calculated. As a first trial calculation it may be assumed that half of the loss of 20% implied by the efficiency of 80% occurs in the casing. Therefore a loss of perhaps 10% may be assigned to the impeller passages, giving an efficiency for the impeller of 90% as a guess. Thus, assuming that the stagnation conditions remain the same from impeller outlet to the discharge flange, a calculation may be made as follows:

$$p_{02}/p_{01} = (1 + 0.9 \times 145.37/295)^{3.5} = 3.75$$

Thus

$$p_{02} = 3.75 \times 10^5\,\mathrm{N\,m^{-2}}$$
$$T_{02} = 440.37 = T_2 + V_2^2/2C_P$$

Since in the limit $V_2$ must not really reach Mach 1, the maximum velocity is fixed by assuming that $V_2$ equals the acoustic velocity; $V_2 = 384\,\mathrm{m\,s^{-1}}$, and hence $T_2 = 367\,\mathrm{K}$. Using the gas laws

$$p_2 = 1.98 \times 10^5\,\mathrm{N\,m^{-2}}$$
$$\rho_2 = 1.88\,\mathrm{kg\,m^{-3}}$$
$$Q_2 = 4.79\,\mathrm{m^3\,s^{-1}}$$

If now a diffusion is assumed such that $V_{R2} = 0.95V_{R1}$,

$$V_{R2} = 71.25\,\mathrm{m\,s^{-1}}$$

Using $Q_2$ shows that an outlet area of $0.067\,m^2$ is required. Allowing for the blockage effect of 16 vanes each 2 mm thick and, if the passage height is $h$,

$$0.067 = \pi \times 0.52 \times h - 16 \times 0.002 \times h$$

which gives

$$h = 42\,\text{mm}$$

## 6.4 Outlet systems

Fluid leaves the impeller at an absolute velocity considerably higher than that in the discharge pipe. The outlet system, therefore, as well as collecting fluid from the impeller without affecting its performance, must also reduce the flow velocity efficiently and so improve the outlet pressure from the level achieved by the impeller working on its own. Figure 6.20 illustrates the basic layouts used in fans, pumps and compressors. All the alternatives have an annular space outside the impeller before the volute or diffuser ring. The vaneless diffuser continues the annular space and will be discussed first; the volute will then be described, and finally the vaned diffuser.

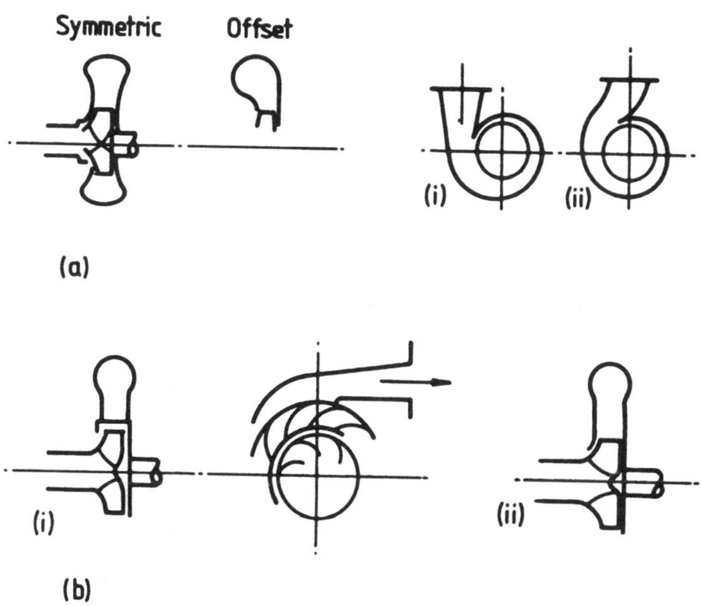


**Figure 6.20** Alternative diffuser systems used in pumps and compressors: (a) simple volute casing with (i) tangential discharge (ii) radial discharge; (b) diffuser system with (i) vaned diffuser (ii) vaneless diffuser.

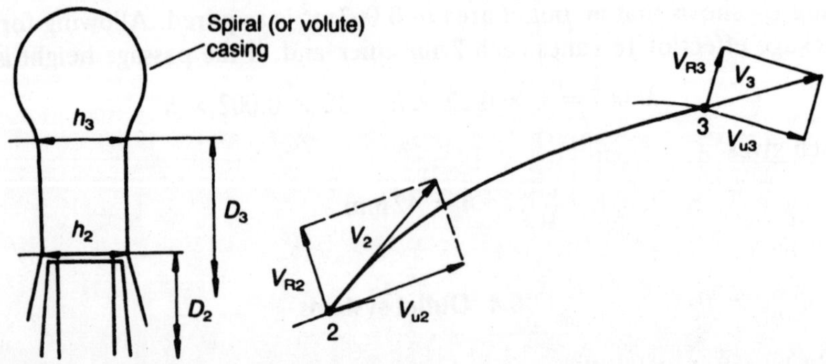


**Figure 6.21** Vaneless diffuser.

### 6.4.1 Vaneless diffuser

A two-dimensional model of the flow in a vaneless diffuser is represented in Fig. 6.21. The streamlines follow a spiral vortex path that, shown in the figure, traces a streamline from the impeller outlet at point 2 to the casing at point 3, the space between the points being the diffuser. Application of the principle of constant angular momentum and the continuity equation between points 2 and 3 gives (ignoring losses) an expression for the pressure rise as follows:

$$\Delta p_{\text{diffuser}} = p_3 - p_2 = \left(\frac{V_2^2}{2} - \frac{V_3^2}{3}\right) \tag{6.8}$$

This is the ideal pressure rise, and if flow losses between points 2 and 3 are accounted for by using the diffuser efficiency, then

$$\eta_{\text{diffuser}} = \Delta p_{\text{actual}}/\Delta p_{\text{ideal}} \tag{6.9}$$

Ferguson (1963, 1969) discussed flow in vaneless diffusers and the approaches to the estimation of losses. He demonstrated (Fig. 6.22) uncertainties in flow regime, and strong three-dimensional flow as flow rate changed. He discussed the relevance of the friction loss estimation approaches used in normal diffusers, and concluded, as did Johnston and Dean (1966) that all the available theoretical approaches have to be modified in the light of the flow instabilities found in experimental studies. Of great interest are studies published by Sherstyuk and his colleagues (1966, 1969) on the effect of profiling the side walls on the diffuser efficiency. Figure 6.23, taken from this work, illustrates the effect of quite limited changes in principal dimensions. Clearly the control of boundary layer behaviour pays dividends. More recently, Whitfield and Sutton (1989) have shown how geometry affects surge margins and flow range in compressors.

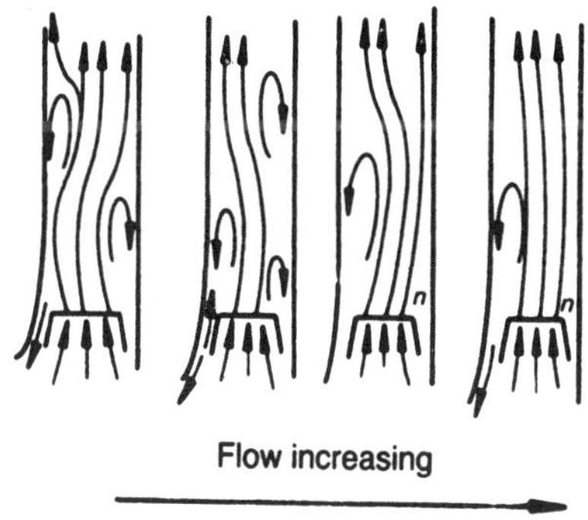


**Figure 6.22** Flow patterns in a vaneless diffuser as pump flow changes (Ferguson, 1969).

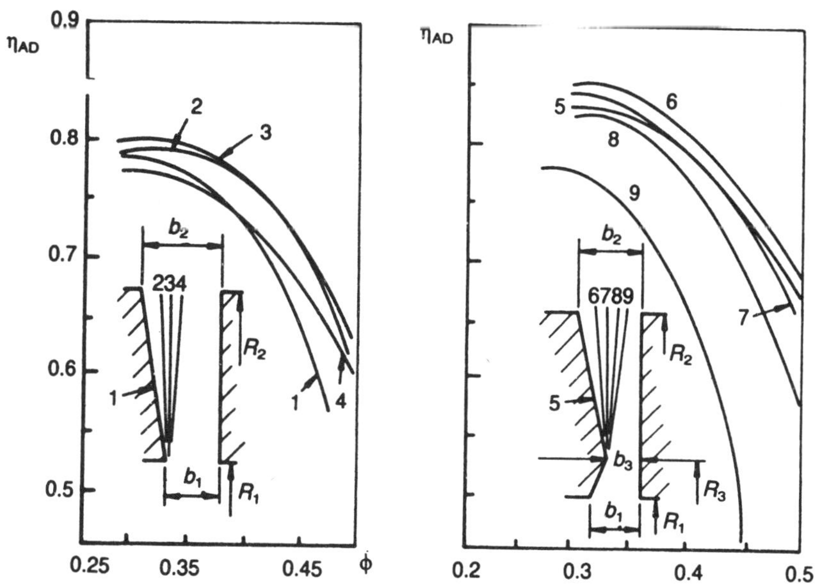


**Figure 6.23** Effect of geometry on vaneless diffuser performance (after Sherstyuk and Kosmin, 1966; Sherstyuk and Sekolov, 1969).

### 6.4.2 Volute or spiral casing

The volute surrounds the impeller, and is a spiral casing whose cross-sectional area increases from a minimum at the cutwater to the throat, which is the beginning of the diffuser to the outlet pipe, as illustrated in Figs 6.24 and 6.25. The cutwater represents the nearest part of the casing to the impeller, and is aligned into the general direction of the flow leaving the impeller. The cross-sectional area increases to take the volume leaving the

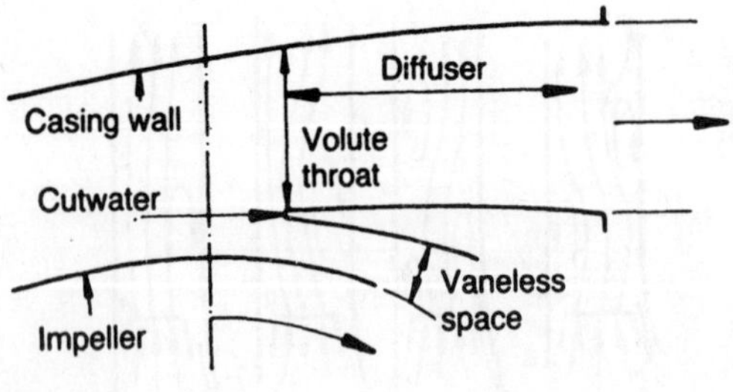


**Figure 6.24** Volute diffuser.

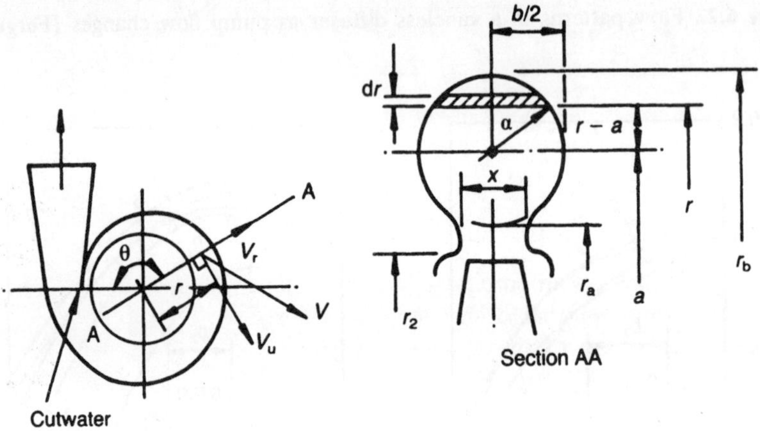


**Figure 6.25** Volute and a typical cross-section.

impeller; for example, in Fig. 6.25, if section AA is at $\theta = 90°$ from the cutwater, 25% of the pump delivery flow will be passing through the section.

If ideal flow is considered, the law of constant angular momentum will apply, and the flow will leave the impeller on a logarithmic spiral path. In many casing designs the cross-sections are circular sections, and the outer wall follows an Archimedian spiral. It is possible, using integration across each section of small elements indicated in Fig. 6.25, to schedule the area changes and get the correct areas for all designated sections like AA. The equations and solutions are covered in textbooks like those of Stepannof (1957b) and Pfleiderer (1961) though usually the effect of friction is neglected as it has been found to affect the areas by very small amounts. A much simpler approach is to size each cross-sectional area by assuming that the throat velocity is constant round the impeller circumference. Stepannof (1957b) also demonstrates this technique, based upon a velocity given by the relation

$$V = K_3\sqrt{(2gH)} \tag{6.10}$$

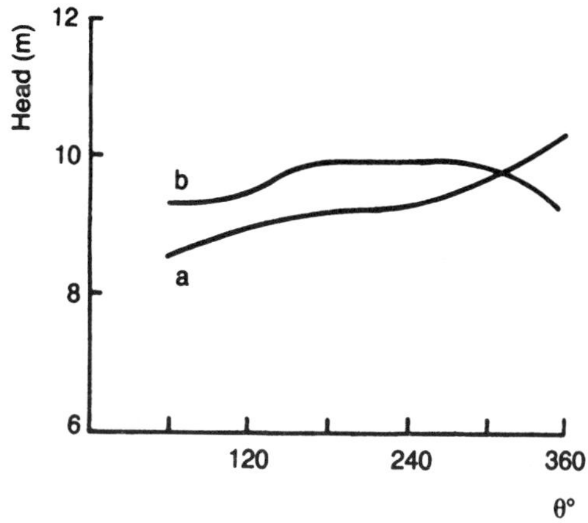


**Figure 6.26** Effect of the choice of velocity distribution on pressure head variation around the impeller: (a) constant angular momentum; (b) constant velocity (Rutschi, 1961).

In this equation $K_3$ is an empirical factor plotted against specific speed, and $gH$ is the pump nominal design specific energy rise. Rutschi (1961) presented a comparison of the angular momentum and constant velocity approaches which is shown in Fig. 6.26, and demonstrated little difference in pressure head at successive sections round the casing.

Worster (1963) more recently studied the casing and its effect on pump performance and proposed that the most important parameter was the volute throat area. He argued, as did Anderson and others referred to in his paper, that the best efficiency or design match between the impeller and the casing occurs when the impeller and volute characteristics cross, as indicated in Fig. 6.27. He based his analysis on the simple casing cross-section shown in Fig. 6.27 and maintained that, although the outlet angle $\beta_2$ and impeller diameter $D_2$ do determine performance, the correct matching of impeller and casing is a decisive factor in the design process. Figure 6.28 illustrates his point, and Fig. 6.29 is his design plot that allows the selection of the correct volute area once the impeller is designed. This approach has been used with some success, and Thorne (1979) has recently underlined its utility.

Although the technique just discussed has been applied to pumps, it is possible to argue that it could be applied to fan and blower design since the density changes are low. However, as Eck (1973) shows, the principles of constant angular momentum or constant velocity are both used with empirical factors such as equation (6.10).

Considering pump design still, Worster (1963) used the loose or large side clearance design approach, as against the more conventional tight or close

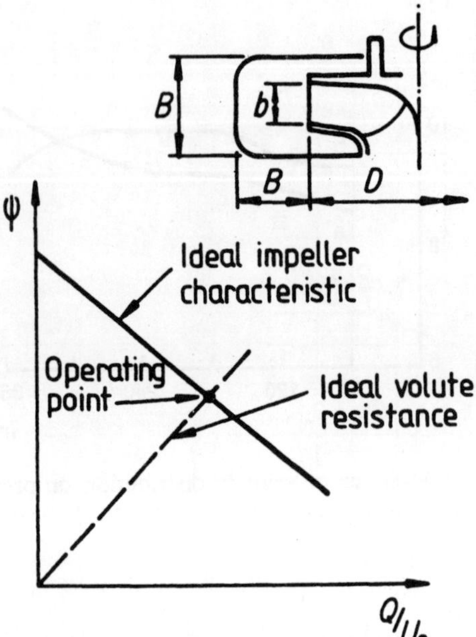


**Figure 6.27** Illustration of the approach to impeller and casing matching proposed by Worster (1963).

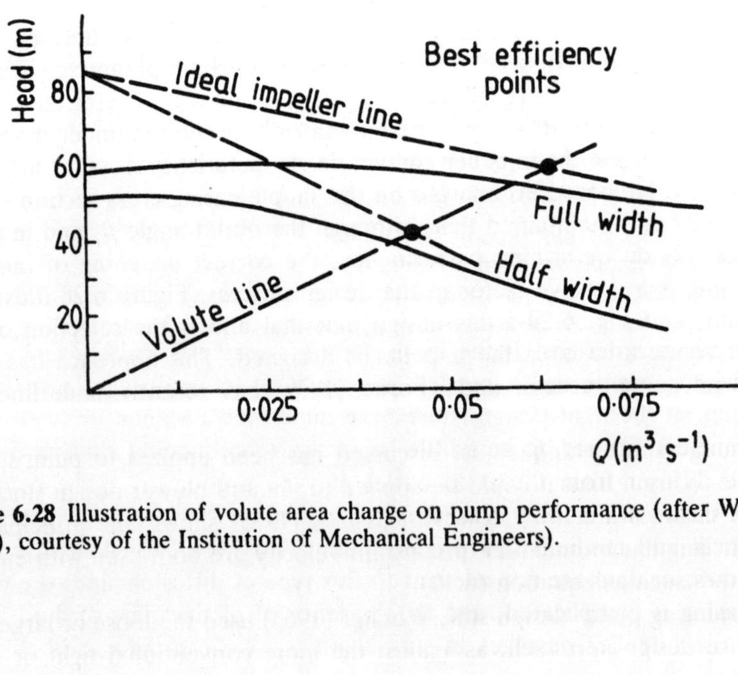


**Figure 6.28** Illustration of volute area change on pump performance (after Worster (1963), courtesy of the Institution of Mechanical Engineers).

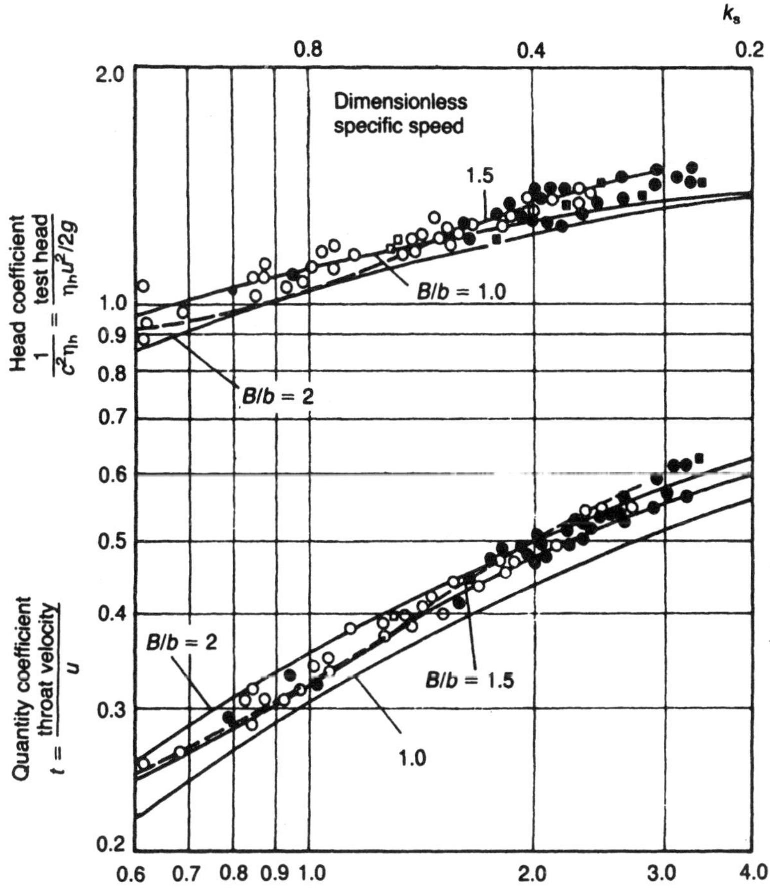


**Figure 6.29** 'Design' plot proposed by Worster (1963; modified from that quoted by Worster, the symbols are defined in Fig. 6.27).

side clearance designs. If the loose design is used the width $x$ in Fig. 6.25 will be the casing width needed to give adequate side clearance for the impeller. If the tight design is used the dimension $x$ should be fixed, as for the vaneless diffuser, to be a little larger than the impeller channel height.

The diffuser from throat to outlet may be a cone or, if the centreline outlet shown in Fig. 6.20 is used, as demanded by the current ISO standards for small and medium size process pumps, may be curved. Conventional pipe diffuser data are not relevant to this type of diffusion since the flow in the casing is complicated, and the presence of distinct jets of fluid issuing from the impeller passages in succession will be causing disturbed three-

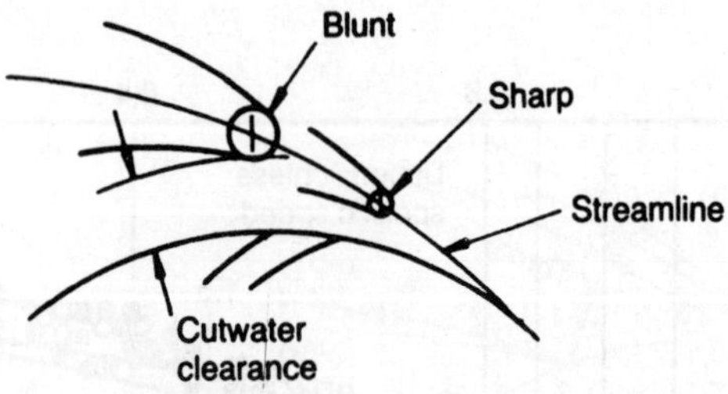


**Figure 6.30** Cutwater nose profile.

dimensional flows into the throat of the casing. The shape of the cutwater nose is also important, as indicated by information published by Turton (1966). It is also well known that the clearance between the impeller and the cutwater nose is important, as too tight a clearance gives rise to noise and vibration. A working rule is to arrange that the minimum clearance should be of the order of 5–10% of impeller diameter. Depending on casting, the cutwater nose may be blunt or sharp (Fig. 6.30), but must be aligned along the streamline as shown in the figure. A blunt nose allows for tolerance of flow misdirection and thus an accommodation for flow changes from design point.

### 6.4.3 Vaned diffuser systems

The diffusion system is now a vaneless space whose radial depth follows the same proportions as already given for the volute cutwater clearance. The fluid follows a spiral path up to the diffuser throat, is diffused over the length bc (Fig. 6.31) and discharged into the spiral casing. Ideally the diffuser vane walls should follow an Archimedian spiral but, since diffusers are often machined, the surfaces abc and ef are convenient radius approximations. Each diffuser passage has an active length $L$, which a working rule suggests can be between $3t$ and $4t$ to balance effective diffuser control and surface friction loss. Clearly the sum of throat areas will follow the same basic rule as the volute throat already discussed, and the number of vanes is always different from the number of rotor blades $Z$, a working rule often used being $Z + 1$. Many working designs that follow the rule for $L$ given above appear to have a ratio for $D_3/D_2$ of 1.3–1.5, a constant width $b$ which is a little wider than the impeller tip width, and a maximum diffusion angle of 11°. The vanes are as thin as practicable, following the type of profile shown in Fig. 6.31. Designs using aerofoil sections have been used, but are not conventional. Experiments with 'pipe' or drilled diffusers for flow that is supersonic at impeller outlet have been used but are not general, as the simple generated shapes of Fig. 6.31 are easy to produce and control.

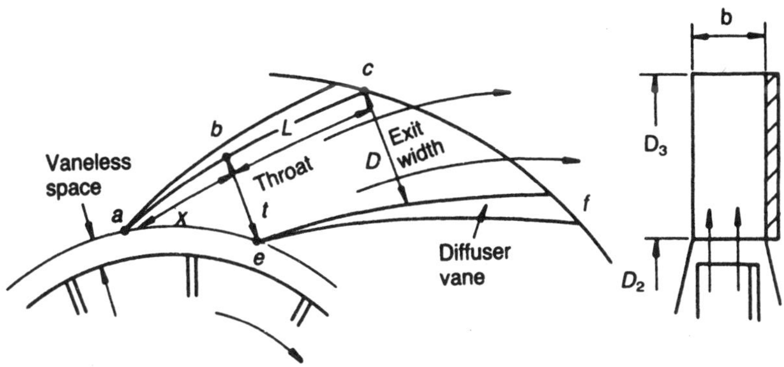


**Figure 6.31** Vaned diffuser.

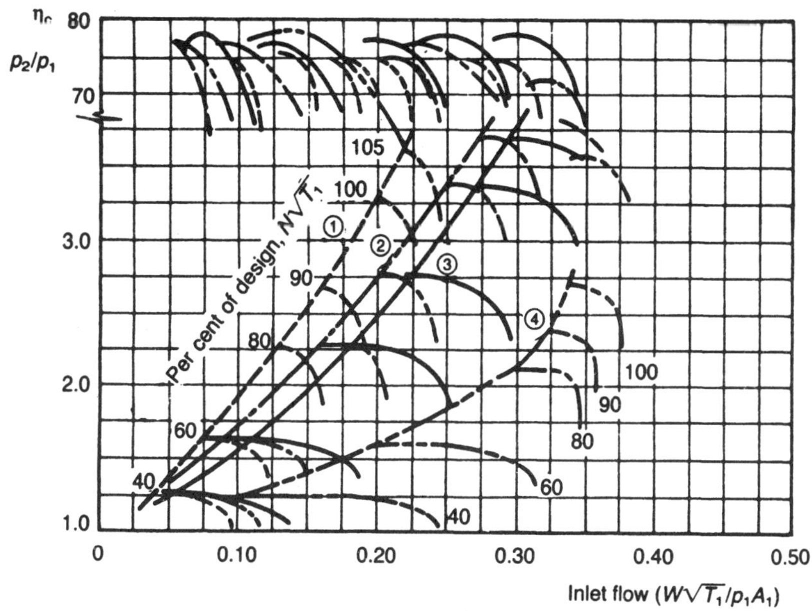


| | Vane inlet angle (degrees) | t/L (Fig. 6.31) |
|---|---|---|
| 1 ——— | 10.7 | 0.15 |
| 2 —·— | 12.8 | 0.19 |
| 3 —— | 15.0 | 0.265 |
| 4 —··— | vaneless | |

**Figure 6.32** Effect of the choice of diffuser system on compressor performance (after Petrie (1964) courtesy of the Institution of Mechanical Engineers).

Petrie (1964) reports a study for a helicopter gas turbine where four diffusion systems were investigated. Figure 6.32 illustrates the effects on performance of three vaned diffusers with different inlet angles, and a vaneless diffuser. It can be seen that the larger angle and area both moved the surge point to higher mass flows and extended the flow range before choking. However, the vaneless diffuser lowered both the maximum pressure ratio and efficiency, significantly extended the total flow range, but did not significantly improve the range of flow from surge to choking at any driving speed. The paper by McCutcheon (1978) also emphasizes the role the diffuser plays in determining the range of a compressor, particularly when the impeller tip Mach number approaches unity.

In multistage compressors and pumps the passages that guide the fluid from one impeller to the next are partly diffusers and partly simple transfer ducts. Typical shapes employed are shown in Fig. 6.33 for a multistage pump.

The example of sections 6.3.1 and 6.3.2 is now continued through the diffuser. The height of the diffuser passages (dimension $b$ in Fig. 6.31) will

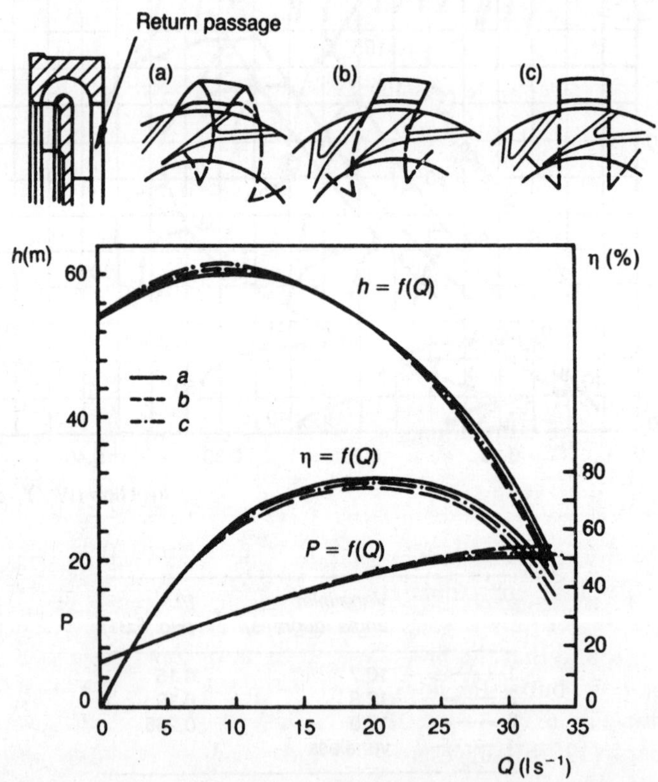


**Figure 6.33** Comparison of the effect of crossover duct geometry on pump performance.

be 0.045 (larger than the tip width $h$ already found), and the number of passages will be 17 (16 + 1). Diameter $D_2$ in Fig. 6.31 will be 0.52 + 0.05 = 0.57 m, and the radius of the mean width of the throat will be 0.35 as an estimate.

Assuming free vortex flow in the vaneless space,

$$357.76 \times 0.52 = V_u 0.57$$

Thus the tangential component of flow velocity at 0.57 m diameter is 326.38 m s$^{-1}$. The radial component of velocity now must be found. One approach is to assume constant density and then correct, or to guess the radial velocity. Assume that $V_R$ = 55 m s$^{-1}$. Thus the absolute velocity = $\sqrt{(55^2 + 326.38^2)}$ = 330.98. Assume that stagnation temperature remains constant:

$$440.37 = T + 330.98^2/2C_p$$

$$T_{static} = 385.87\,K$$

$$p_2 = p_{02}(385.87/440.37)^{3.5} = 2.36 \times 10^5\,N\,m^{-2}$$

The density thus becomes 2.13 kg m$^{-3}$, and

$$V_R = 9/2.13(\pi \times 0.045 \times 0.57) = 52.37$$

Thus another iteration is really needed, but the solution will be left at this point.

The throat width $t$ in Fig. 6.31 must now be determined. The radius of its centre is fixed, as a guess, at 0.35. The process is followed again, with the assumptions that the stagnation temperature remains the same and that $V_R$ at the throat radius is 38 m s$^{-1}$; this gives the width $t$ = 18 mm. The assumption of a diffuser efficiency then allows the solution. Alternatively a suitable change in area of the passage may be designed, and the design checked to give a pressure rise. The solution will stop at this point.

## 6.5 Thrust loads due to hydrodynamic effects

### 6.5.1 Radial thrust forces

In vaned diffuser designs the pressure distribution round the impeller periphery remains approximately uniform for a range of flows, and to a reasonable degree this is so for vaneless diffusers. For the volute type of design it is found that the pressure distribution is approximately constant at the design flow, but as the flow departs from design the distribution changes as shown in Fig. 6.34.

Stepannof (1957b) suggests for pumps a simple equation for the radial thrust force:

$$F_r = K\rho gHD_2B_2 \qquad (6.11)$$

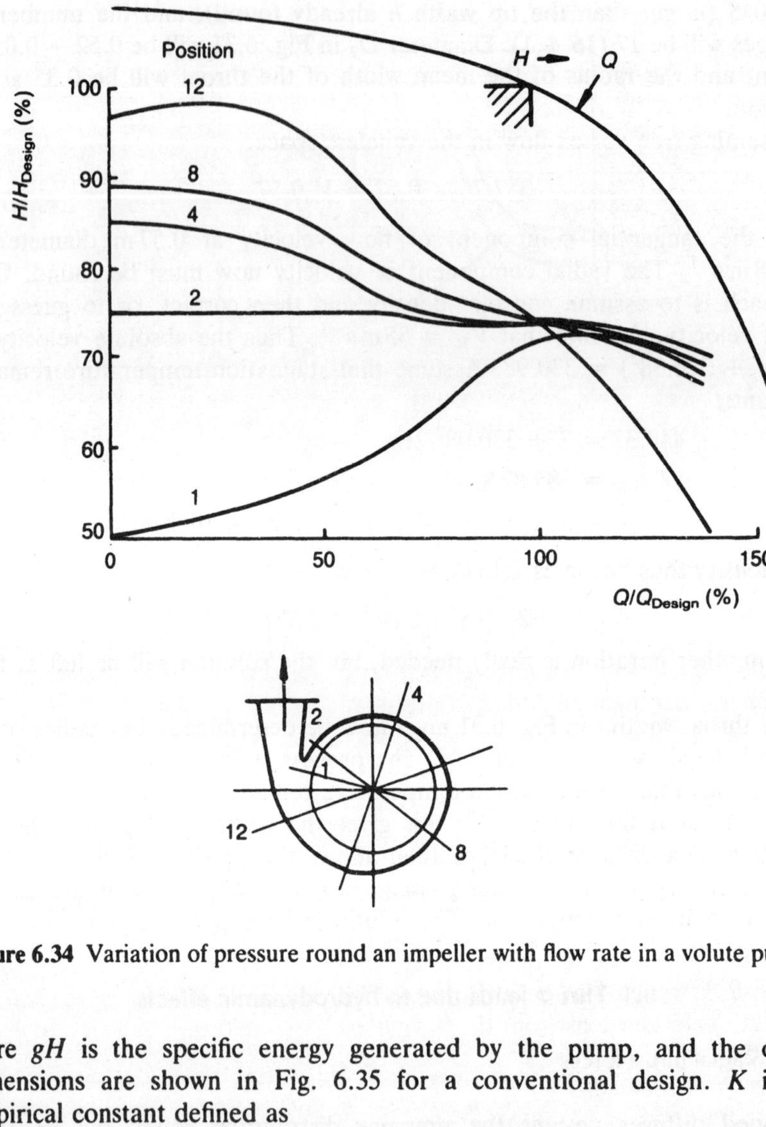


**Figure 6.34** Variation of pressure round an impeller with flow rate in a volute pump.

Here $gH$ is the specific energy generated by the pump, and the other dimensions are shown in Fig. 6.35 for a conventional design. $K$ is an empirical constant defined as

$$K = 0.36[1 - (Q/Q_{\text{design}})^2] \tag{6.12}$$

Other texts give more sophisticated equations that allow for recirculating flow effects in the spaces between impeller and casing. All indicate that the resulting radial loads to be absorbed by the bearings supporting the shaft tend to be large, and in all cases may not be neglected. The formulae are used to size shafts and bearings, but it is found in many pumping installations that even though the shaft has been carefully sized, with a margin for error, shafts do break. These events may be due to poor materials or to

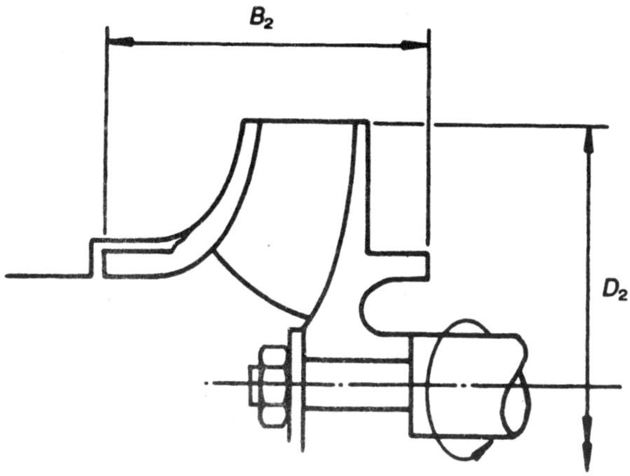


**Figure 6.35** Definition of dimensions used in radial force calculation.

misuse or poor fitting. In other cases the failures are due to effects not allowed for in the formulae, which are based on steady-state flow conditions. This circumstance was studied and many researchers agree that, since flow conditions are not steady, there is a substantial fluctuating force component that must be added to the steady value. In papers by Turton and Goss (1982, 1983) it is shown that the fluctuating forces may be of the same size as the steady load, and this factor as well as the fluctuating nature of the load effects gives rise to vibration and bearing and shaft failure in some cases. The reason for the fluctuating components is felt to be the varying velocity distribution leaving the impeller and interacting with the casing.

One device for reducing the radial loads is the double volute; two flow paths are provided in the volute so that each receives half the flow from the impeller, and the radial thrust loads cancel out. The penalty is increased friction loss, which may result in probably a 2% reduction in efficiency. A few other solutions exist, but the only tried method is the double volute.

### 6.5.2 Axial thrust loads

Axial hydraulic thrust is the summation of unbalanced forces acting on the impeller in the axial direction. Reliable thrust bearings are available so that this does not present problems except in large machines, but it is necessary to calculate the forces. These forces arise due to the distribution of pressure in the spaces between the impeller and the casing. In Fig. 6.36 the assumption is illustrated for the discharge pressure acting over the backplate and front shroud in the single entry pump, and down to the wear rings for a double entry impeller; the suction pressure is assumed to be distributed across the area up to the wear ring diameter in both designs. It can be

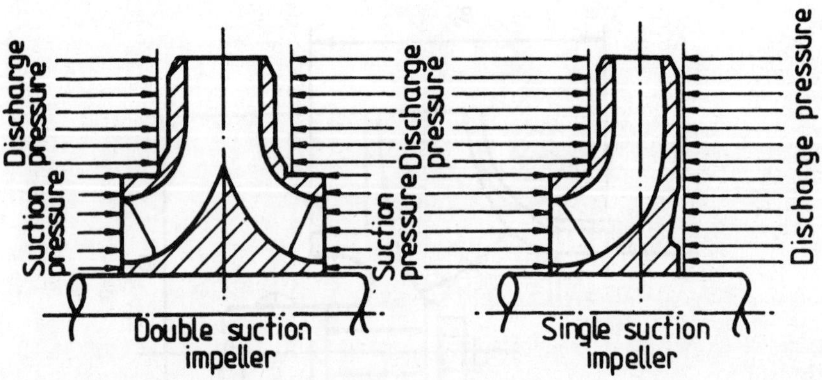


**Figure 6.36** Pressure distributions leading to axial force.

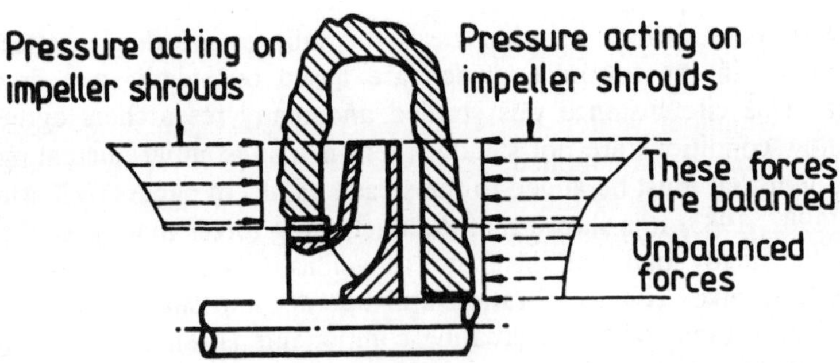


**Figure 6.37** Unbalanced force due to pressure distribution on a centrifugal impeller.

argued that the pressure forces are balanced in the double suction design, and this is approximately the case in practice; the single suction impeller clearly has an unbalanced force. In practice there is always an unbalanced force acting on a double entry impeller due to such factors as unequal flow distributions in the two entry passages, and interference due to external features such as bends in the suction line, so that thrust bearings are always needed. Figure 6.37 illustrates the probable pressure distribution on the impeller surfaces. The actual pressure variation will depend upon surface roughness of the pump surfaces on the side clearances, and on leakage flows through any wear rings fitted, so that front and back net forces may vary from the design conditions assumed. The classic solutions assume that the fluid in the clearance space rotates at about half the impeller speed as a solid mass, and the texts by Stepannof (1957b) and Pfleiderer (1961) may be consulted for the equations. The net result of the pressure forces and the

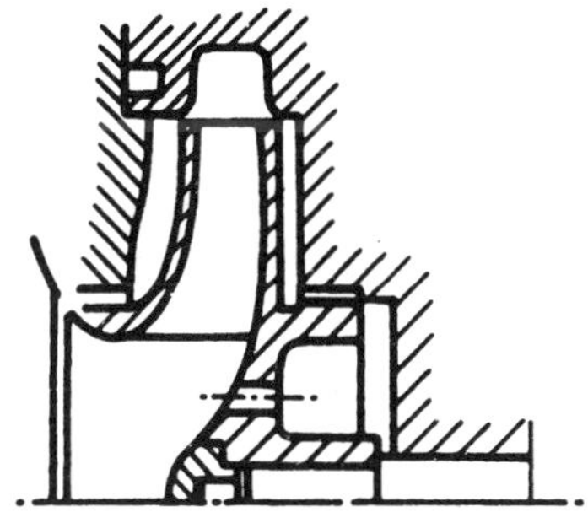

**Figure 6.38** Rear wear ring, balance chamber and balance holes method of axial force balance.

fluid change of momentum is a force towards the suction flange; the force magnitude depends on size, outlet pressure, rotational speed, and whether the impeller is provided with a shroud, as in many pumps, or is open, as in many compressors. The most common way of reducing the axial thrust load is shown in Fig. 6.38. A wear ring is formed on the impeller backplate, thus creating a chamber vented to the suction through 'balance holes'. Suction pressure is in this way applied to the backplate, up to the wear ring diameter, and hence reduces the delivery pressure related force. This increases the risk of cavitation inception but is tolerated because it reduces the load. Other methods connect the balance chamber to the suction by using a balance pipe, or use the so-called pump-out vanes which centrifuge fluid back to the casing. These also work quite well but there is a critical clearance : vane height ratio, as too tight an axial clearance results in vibration. In multistage machines these measures cannot be used, and in some designs impellers are used back-to-back to balance thrusts, but the usual technique is to use balance pistons or balance discs as illustrated in Fig. 6.39 for compressors. The sizes of the components, such as wear ring diameter, balance piston diameter and balance hole sizes, depend on the calculated loads and the probable leakage flows considered acceptable. A common design approach is to consider that 10% of the thrust load is taken by the bearings and the remainder absorbed by the balance system being used. The equations published in the standard texts refer only to steady loads, and the fluctuating load problems already introduced in section 6.5.1 again apply. As Turton and Goss (1982, 1983) among others have indicated, the fluctuating component in some cases tends to be twice the steady value, and the shafts and other components should be sized accordingly. A comprehensive series of papers on *Hydraulic forces in Centrifugal Single and Multistage Pumps* was given during an Institution of Mechanical Engineers Seminar in 1986, and this group of papers should be consulted for more detail on thrust causes and cures.

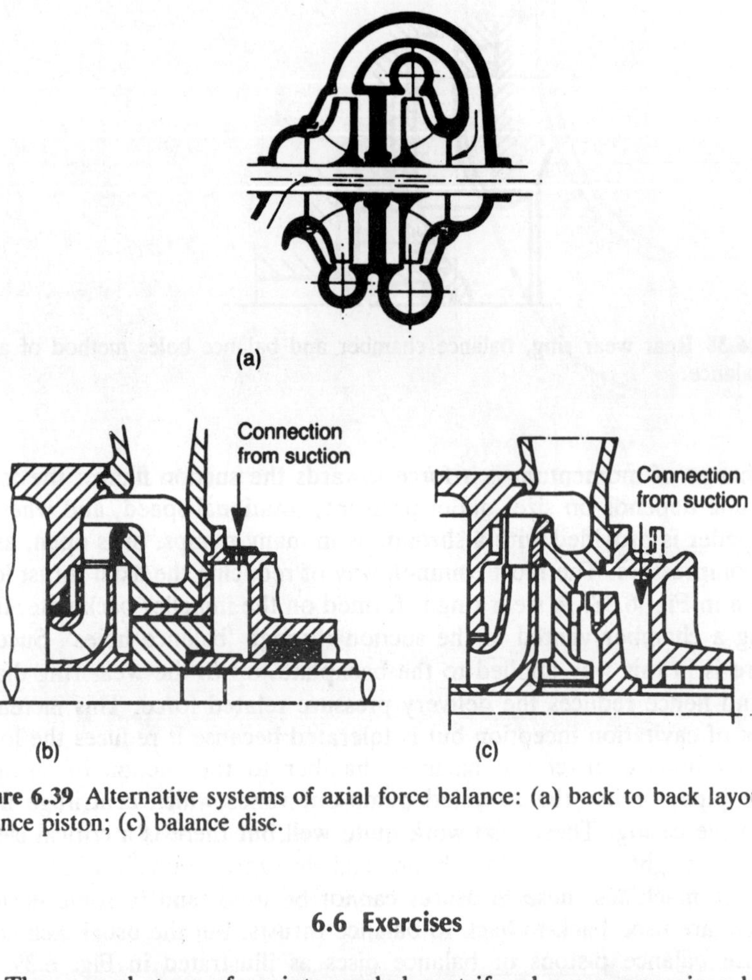


**Figure 6.39** Alternative systems of axial force balance: (a) back to back layout; (b) balance piston; (c) balance disc.

## 6.6 Exercises

6.1 The second stage of an intercooled centrifugal compressor in a helium line is illustrated in Fig. 6.40. The design rotational speed is 24 000 revolutions per minute and the static pressure and temperature are $4 \times 10^5\,\mathrm{N\,m^{-2}}$ and 300 K at inlet to the stage. If the inlet angle at point A is 30° to the tangential direction determine the mass flow rate and maximum mach number for zero incidence at point A if the absolute velocity makes an angle of 3° to the axial and its tangential component is in the same direction as the rotation of the blades. If the impeller total to total efficiency is 90%, determine the discharge stagnation pressure.

6.2 The double suction water pump shown in Fig. 6.41 is driven at 1750 revolutions per minute and delivers $0.8\,\mathrm{m^3\,s^{-1}}$. There are seven blades per flow path of backward curved design with an outlet angle of 25°. Flow can be assumed to be divided equally between the flow paths.

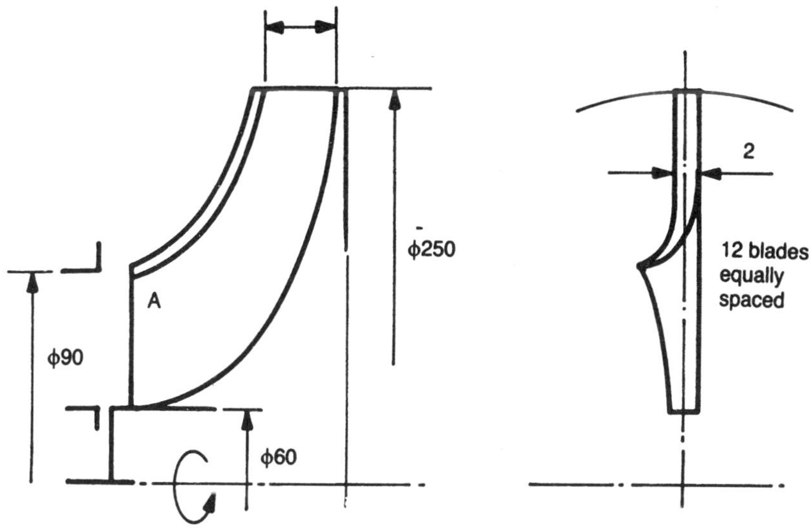


**Figure 6.40** Exercise 6.1.

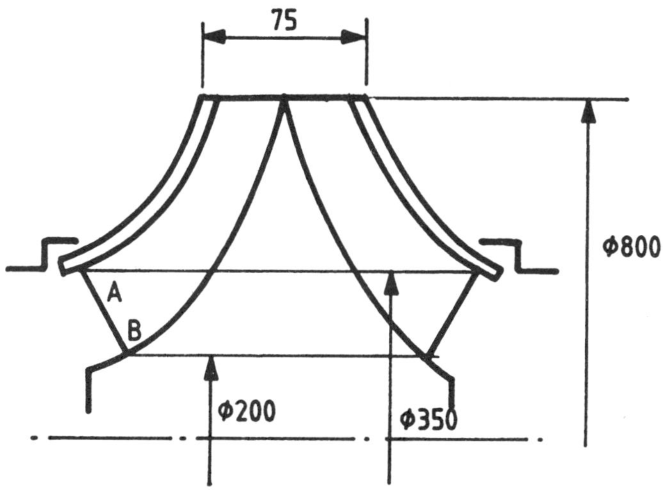


**Figure 6.41** Exercise 6.2.

Assuming zero inlet whirl, estimate the Euler specific energy rise and the blade angles at points A and B on the suction edge of the blades.

If the flow conditions change to give an inlet swirl of $5\,\mathrm{m\,s^{-1}}$ in the direction of rotation at point A, estimate the reduced specific energy rise and the fluid angles at points A and B.

6.3 A single entry centrifugal compressor provided with a vaneless diffuser delivers air at a mass flow rate of $10\,\text{kg}\,\text{s}^{-1}$ when rotating at 30 000 revolutions per minute. Air is drawn in from a large plenum chamber where the stagnation temperature and pressure are, respectively, 290 K and $1.25 \times 10^5\,\text{N}\,\text{m}^{-2}$.

The impeller diameter is 270 mm, the tip width 25 mm, the diffuser passage height is 27 mm and its outer diameter is 550 mm, there are eight impeller blades 3 mm thick radial at outlet, and the compressor efficiency can be assumed to be 79%.

If the Mach number based on impeller outlet absolute velocity is limited to 0.95, construct the outlet velocity triangle.

Calculate the static pressure and temperature at outlet from the diffuser using a diffusion factor *Cp* of 0.5, and discuss the limitations of this factor ($Cp = \Delta p/0.5\rho V^2$. Where *V* is the gas absolute velocity at outlet from the impeller).

6.4 A single stage single entry centrifugal pump is to deliver water at the rate of $100\,\text{m}^3\,\text{h}^{-1}$ when driven at 2900 revolutions per minute. The pump is illustrated in Fig. 6.42. There are five backward curved impeller blades each 3 mm thick, and the nominal outlet blade angle is 20° to the tangential direction.

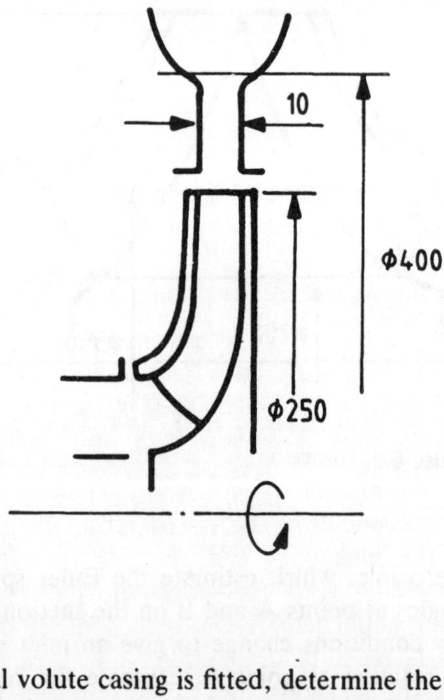


**Figure 6.42** Exercise 6.4.

If a conventional volute casing is fitted, determine the specific energy rise generated at design flow and the probable input power needed.

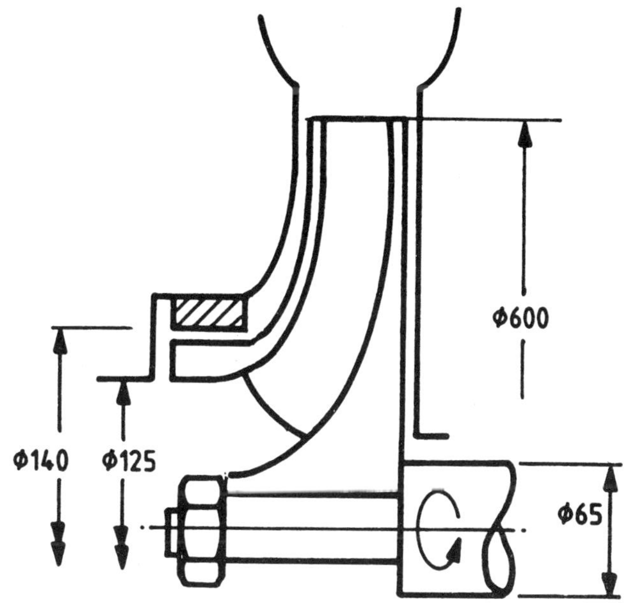


**Figure 6.43** Exercise 6.5.

If a vaneless diffuser is fitted with the same volute design placed around it, calculate the probable extra specific energy rise generated.

6.5 Figure 6.43 shows a centrifugal pump impeller. It is to deliver water at the rate of $150\,m^3\,h^{-1}$. If the value of $k_s$ is to be 0.75, suggest the specific energy rise probable. It may be assumed that the impeller has five blades, that the outlet angle is 25°, and that the pump has a conventional volute casing.

For the design shown estimate the axial thrust, if the suction pressure is $0.85 \times 10^5\,N\,m^{-2}$.

6.6 An air compressor is to be of centrifugal design, is to be driven at $1575\,rad\,s^{-1}$ and will draw directly from a plenum chamber where the stagnation conditions are $1.25 \times 10^5\,N\,m^{-2}$ and 300 K.

The impeller has an outer suction diameter of 300 mm, a hub of 100 mm, a tip diameter of 600 mm and a tip width of 20 mm.

Estimate the maximum mass flow rate if the inlet relative Mach number is limited to 0.85. Also determine the probable outer stagnation pressure if the outlet radial velocity at limited flow is $100\,m\,s^{-1}$ and the impeller has 12 radial blades. Assume the efficiency is 82%.

Suggest how usual slip factors can be improved upon and also how the flow effects can be modified.

# 7 Axial machines for incompressible flow

## 7.1 Introduction

Axial fans and pumps may be treated as incompressible flow devices, producing large flow rates and small pressure rises. Axial pumps in service deliver up to $80\,000\,\mathrm{m^3\,h^{-1}}$ against resistances equivalent to a level change of 20 m, and have been applied to irrigation, draining, dock dewatering and circulation duties. Fans have been applied to ventilation duties and cooling tower circulation, and commonly produce pressure rises of the order of $500\,\mathrm{N\,m^{-2}}$; special designs give higher pressure rises. Sizes clearly vary with duty, from kitchen ventilation fans to machines up to 4 m in diameter.

Axial turbines may have their guide vane and runner blades coaxial, as in the tubular or bulb turbine, or have guide vanes disposed in the radial plane and the runner of axial design. Hydroelectric plant using this type of machine will be working with level drops from a few metres to over 50 m, producing powers from 100 kW to over 100 MW, the maximum runner tip diameter being of the order of 8 m.

The principles underlying pumps, fans and turbines of the axial type have been introduced in Chapter 4. In the sections following, two examples are used to illustrate the way the principles are applied to fans and pumps, with the practical implications indicated.

A brief description of the axial water turbine follows, covering the essential points concerning the shapes of the hydraulic components. This section concludes with simple examples to illustrate typical velocities and velocity diagram shapes.

The chapter concludes with a discussion of the forces acting on static and rotating blades, and their complex nature is illustrated.

## 7.2 Axial flow pumps and fans

The machines described have stream surfaces that are axisymmetric, and in their simplest form consists of a rotor only; in some cases guide vanes are added downstream, and in rare applications are used upstream to provide control. In many instances the prime mover is an electric motor, so that the rotational speeds tend to be synchronous and constant; variable-speed drives are rarely used except for the largest machines.

One approach to design is outlined by using a fairly typical fan application with appropriate comments. The references cited may be consulted for other approaches and for more advanced techniques.

A fan draws air in from the atmosphere (pressure $10^5\,\mathrm{N\,m^{-2}}$ and 290 K) and is to supply at a rate of $6\,\mathrm{m^3\,s^{-1}}$ against a pressure of $1.005\times10^5\,\mathrm{N\,m^{-2}}$. The applications engineer suggests that the probable duct diameter will be 750 mm.

The hub diameter needs to be defined first. Assuming a hub tip ratio of 0.5 (a typical value), $D_h = 375$ mm, and the axial velocity through the rotor ignoring blade blockage effects is $18.108\,\mathrm{m\,s^{-1}}$. The peripheral velocities are now to be found, bearing in mind that for noise level limitation tip velocities should not exceed $70–75\,\mathrm{m\,s^{-1}}$. The driver speed may be assumed to be $1450\,\mathrm{rev\,min^{-1}}$, so the peripheral velocity at the tip is $56.94\,\mathrm{m\,s^{-1}}$ and the hub peripheral velocity is $28.97\,\mathrm{m\,s^{-1}}$.

The density of air at the conditions specified is $1.2\,\mathrm{kg\,m^{-3}}$, so the specific energy rise is $416.67\,\mathrm{J\,kg^{-1}}$. The Euler specific energy rise, assuming a fairly typical efficiency of 80% will thus be $520.8\,\mathrm{J\,kg^{-1}}$. (If the machine were a pump the overall efficiency would be found from Fig. 2.6 after the specific speed has been calculated. The text by Balje (1981) presents the same sort of picture for fans as well, and this may be referred to for additional information.)

Using the Euler equation (1.4),

$$520.8 = u\Delta V_u$$

Considering the hub section first, $\Delta V_u = 16.14\,\mathrm{m\,s^{-1}}$. The resulting velocity triangles are shown in Fig. 7.1, and the air angles relative to the axial direction are $\beta_1 = 58°$ and $\beta_2 = 31.44°$, and $\beta_m = 44.72°$. If the incidence and deviation are ignored these will be the blade angles too, so that $\theta = 26.56$. This will be assumed for the purpose of this example.

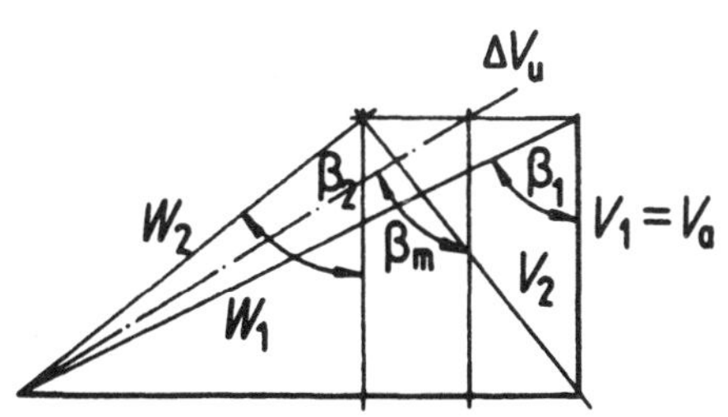


**Figure 7.1** Velocity triangles for axial flow fan example.

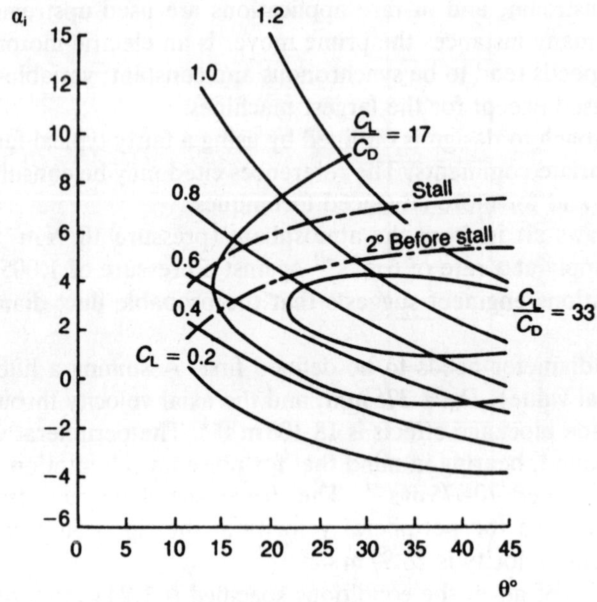


**Figure 7.2** Plot of $\alpha_i$ and blade data camber angle (Hay, Metcalfe and Reizes (1978) courtesy of the Institution of Mechanical Engineers).

A choice of profile must now be made, and in this treatment the circular arc cambered plate will be used, the data for which is given in Fig. 4.5. Cambered plates are often used in low performance fans as they are cheap but have sufficient strength. Where stress levels are higher, as in high performance fans and in pumps, the blade profiles discussed in section 4.3 must be used. Whatever profile is used, $C_L$ and $C_L/C_D$ are read from the data sheet for a suitable angle of attack that gives a good margin before stall, or are related to the maximum lift to drag ratio. These values can then be fed into equation (4.7) to find solidity, and then choices would be made of blade number, chord and stagger angle.

An alternative approach that gives an optimizing view is the paper by Hay, Metcalfe and Reizes (1978), from which Fig. 7.2 is derived. They approached the problem of blade design by producing carpet plots for several profiles of $\alpha_i$ against $\theta$, where

$$\alpha_i = \beta_1 - \gamma \tag{7.1}$$

As Fig. 7.2 illustrates, the carpet plot has two sets of curves for $C_L$ and for $C_L/C_D$. From the figure, an optimum value for $\alpha_i$ is 4.5°, with $C_L = 0.95$ and $C_L/C_D$ approximately 30.

The design approach then goes on to correct for the fact that the blades

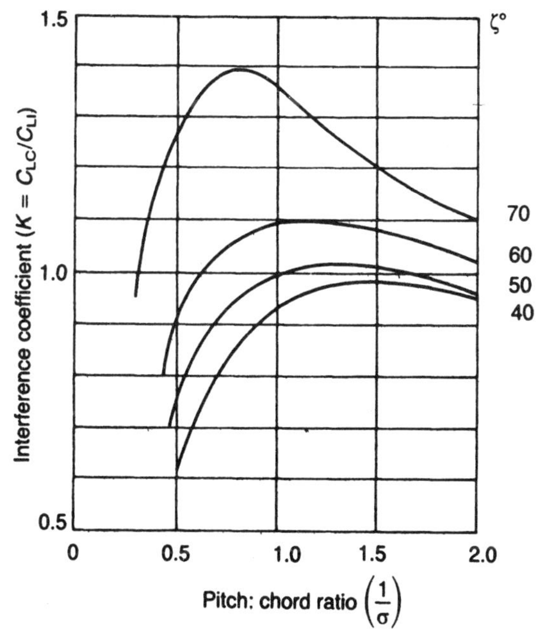


**Figure 7.3** Plot of interference coefficient against pitch:chord ratio (Hay, Metcalfe and Reizes (1978) courtesy of the Institution of Mechanical Engineers).

will be in cascade, the appropriate correction factor being shown in Fig. 7.3. The angle of incidence $\alpha_i$ is replaced by $\alpha_c$, and $\beta_1$ by $\beta_m$, with

$$\alpha_c = \beta_m - \gamma \tag{7.2}$$

They relate $\beta_2$ to $\beta_m$ and $\alpha_c$ by using the equation

$$\beta_2 = (\beta_m - \alpha_c) - \theta/2 + \delta \tag{7.3}$$

Here $\delta$ is defined by Howell ($\delta = m\theta(s/c)^n$), and for a circular arc camber line

$$m = 0.23 + \beta_2/500$$

Thus

$$\beta_2 = (\beta_m - \alpha_c) - \theta\left[0.5 - m\left(\frac{s}{c}\right)^n\right] \tag{7.4}$$

Then, assuming $n = 0.2$, using this treatment and equation (7.4) gives $c/s$ = 1.48 and $s/c = 0.676$, and $\gamma = 44.72 - 4 = 40.72°$.

From Fig. 7.3 the proximity factor $k = 0.8$, the corrected $C_L = 0.76$, $s/c$ = 0.451; from equation (7.4), $\beta_2 = 33.6°$ and $\delta = 5.724°$.

The method then goes on iterate again, but for this example the numerical treatment will stop. If iteration continued a correction to $\alpha_c$ of the order of the difference in $\beta_2$ would be made, a corrected $\theta$ obtained, $C_L$ and the other qualities found, resulting in a new value for $\beta_2$, and so on until the differences became very small.

Reverting to the example, the probable blade angle will be $\beta_2' = 27.88°$ and $\beta_1' = 58°$. With $c/s = 1.85$, using 12 blades gives $c = 182$; from this a simple calculation gives the radius of the circular arc as being 400 mm.

Similar calculations may be performed for the tip and other sections, and the blade profiles related to one another as outlined in section 7.4. The numerical values will not be given here, however, but the concept of radial equilibrium may be used to relate the basic velocity diagrams, and the iteration procedure used as needed. Alternative approaches are detailed by Eck (1973) and Wallis (1961). With the advent of suites of computer programs it has been possible to develop approaches based upon the principles outlined and to inject corrections for blockage and for secondary flow effects, as well as to build-in procedures for stacking the blade profiles on selected lines for manufacture. Such approaches are found in contributions by Myles and Watson (1964), Myles, Bain and Buxton (1965) and by Hesselgreaves and McEwen (1976), among others.

The preceding discussion has dealt with rotor blades, but the same procedures will follow for stator blades, for the absolute fluid vector angles are found in the velocity diagrams and the profiles will emerge in the same way as described above.

The mechanical problems in fans of fixing and location are discussed briefly by Eck (1973), and the pump applications are well dealt with in the standard text by Stepannof (1957b) and by other authorities. A recent conference on fans updates design principles and gives some interesting examples of applications (Woods-Ballard, 1982 and Lack, 1982)

As a further illustrative example consider an axial flow fan with a tip diameter of 1 m and a hub diameter of 0.4 m, which rotates at 1450 revolutions per minute. Five cambered plates have been used as blades, at the root section the chord is 100 mm, and the camber is 2%. For a flow rate of $10\,m^3\,s^{-1}$ estimate the probable pressure rise, the stagger angle at the hub section. Assume that the free vortex principle has been used in design of the blades, that the air density is $1.2\,kg\,m^{-3}$, and the plates that form the blades are bent to a circular arc form.

From the data given

$$\text{flow area} = \pi/4\,(0.1^2 - 0.4^2) = 0.66\,m^2$$

Therefore

$$V_A = 10/0.66 = 15.15\,m\,s^{-1}$$
$$u_H = 1450 \times \pi \times 0.4/60 = 30.37\,m\,s^{-1}$$

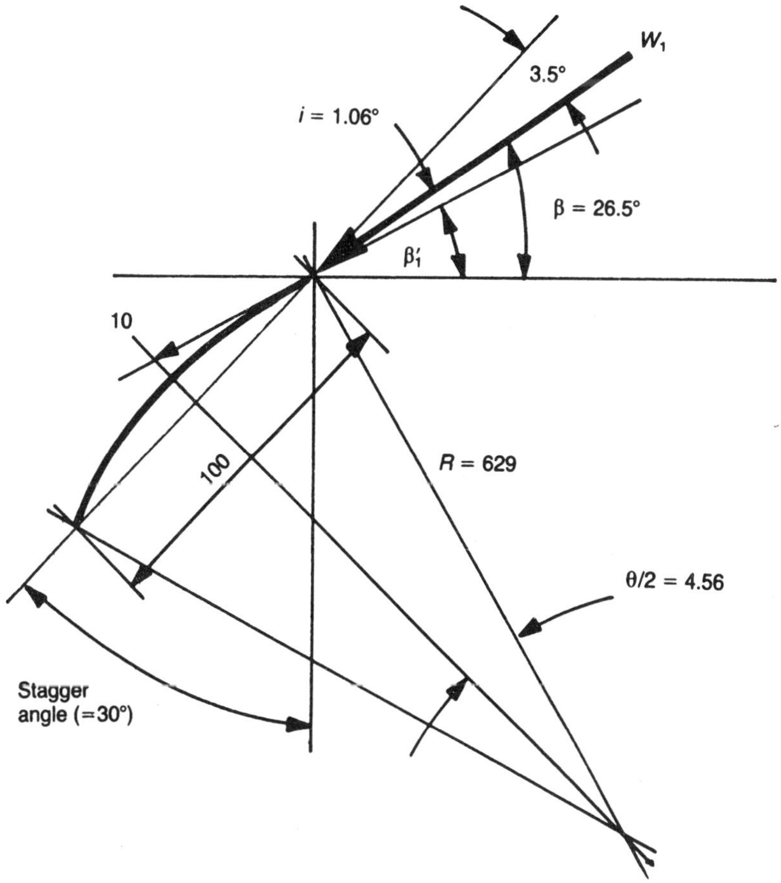


**Figure 7.4** Axial flow fan example, geometry.

The spacing at the hub section is given by

$$\pi \times 0.4/5 = 0.251$$

so that

$$c/s = 0.398$$

For a 2% cambered plate $C_L = 0.5$ at $\alpha = 3.5$ for best performance (from Fig. 4.5), and using geometry for the hub section the camber radius is 629 mm and the camber angle is 9.12° (Fig. 7.4). If zero inlet whirl is used the blade inlet angle referred to the tangential direction is $\beta_1 = 26.5°$. From Fig. 7.4 $\alpha$ is 3.5° to the chord line, so the relative velocity makes an angle of incidence of $-1.06°$.

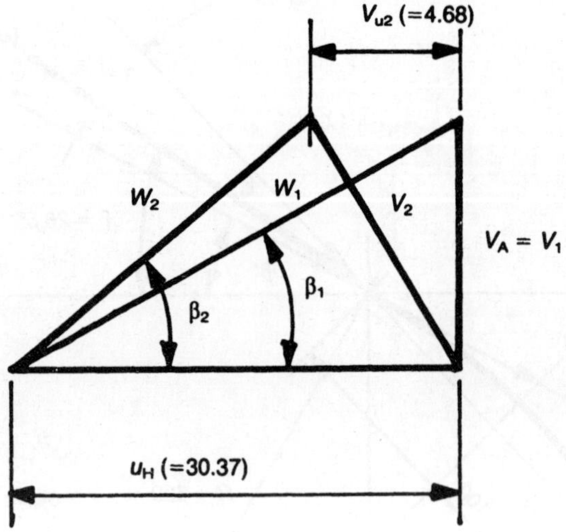


**Figure 7.5** Axial flow fan example, velocity triangles.

The stagger angle will be 90 − 30 = 60°, as shown in the figure. If zero deviation is assumed, $\beta_2 = 34.56°$ and

$$\beta_m = 26.5 + 34.56/2 = 30.53°$$

From the velocity triangles (Fig. 7.5)

$$V_{u2} = 4.68\,\mathrm{m\,s^{-1}}$$

and $gH_E = 4.68 \times 30.37 = 142.17\,\mathrm{J\,kg^{-1}}$.

## 7.3 Axial water turbines

Water turbines of this type may be either of the Kaplan propeller type, or the fully axial or bulb type (Fig. 7.6). As can be seen the turbine rotor is axial in both designs, and the guide vanes (or wicket gates) may be disposed in the radial plane or be coaxial with the rotor.

For both flow paths the problem is the same – the establishment of the flow surface shapes and the stream surfaces on which the blade profiles are to be disposed, and the design of the blade shapes themselves. The basic hydraulic problems are the same as those met in the axial pump, with the essential difference that energy is being extracted from the fluid. The meridional approach to stream surfaces has been covered in Chapter 5, and the methods outlined may be used with empirically determined proportions, based upon experience and presented against specific speed as shown in texts such as those by Nechleba (1957) and Balje (1981).

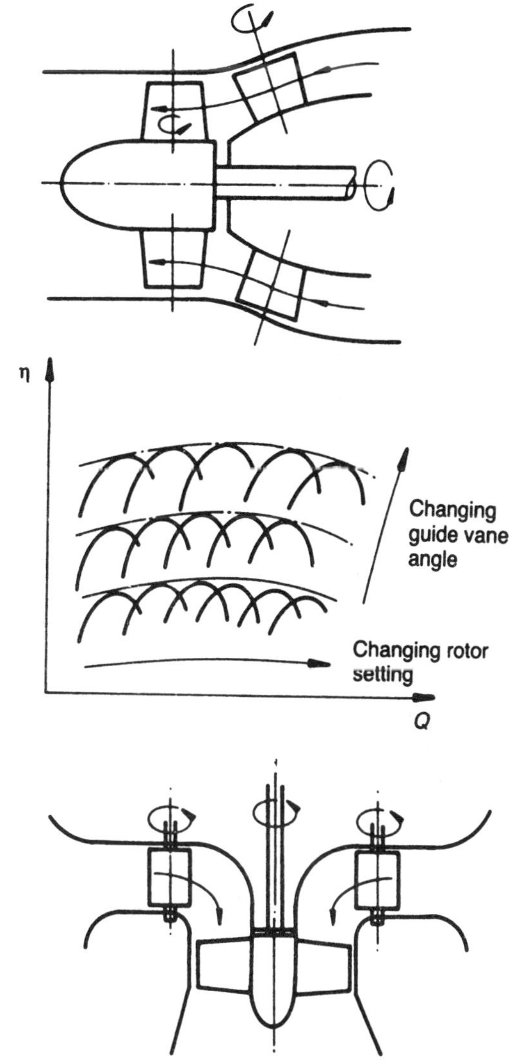


**Figure 7.6** Bulb and Kaplan turbines and their efficiency envelope.

The free vortex approach to radial equilibrium introduced in section 4.5 is often used in relating velocity triangles and rotor blade sections, and once the stream surfaces have been established the approach outlined already in this chapter may be used to determine the necessary geometry and blade profiles. The stress considerations are discussed in section 7.4. Turbine blades tend to be of long chord and low camber at the tip and relatively thick and cambered at the root section, as sketched in Fig. 7.7. The root

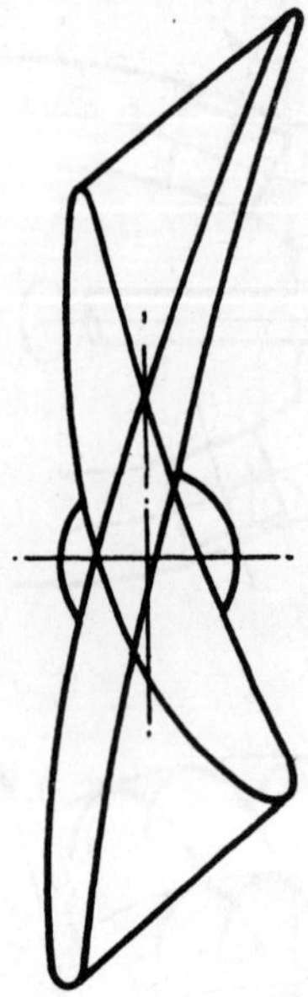

**Figure 7.7** Hub and tip profiles for a water turbine.

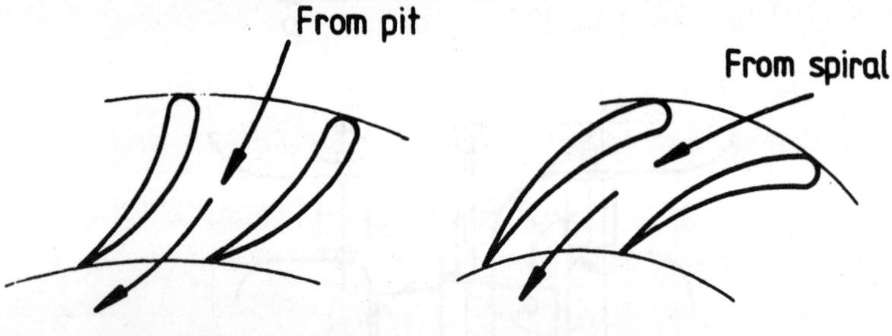


**Figure 7.8** Typical guide vane profiles for a water turbine.

section is dictated by stress considerations and by the need to give a good blend of section into the hub attachment; a typical layout is illustrated in Fig. 7.7 for a Kaplan design, in which the rotor blades must be rotated over a range of angles to give the control illustrated in Fig. 7.6.

Whether the machine has fixed rotor blades or moving Kaplan runner blades, the main control of output is achieved by the provision of moving guide vanes or wicket blades. These have alternative shapes as indicated in Fig. 7.8 where pit vanes are the more normal shapes used when water enters from the spiral casing are shown. The profile must be chosen to give good passage shapes for all openings as well as to control the outlet direction into the runner system. The nose shape must be such that it will tolerate the wide range of incidence it meets as the vane is moved over the control

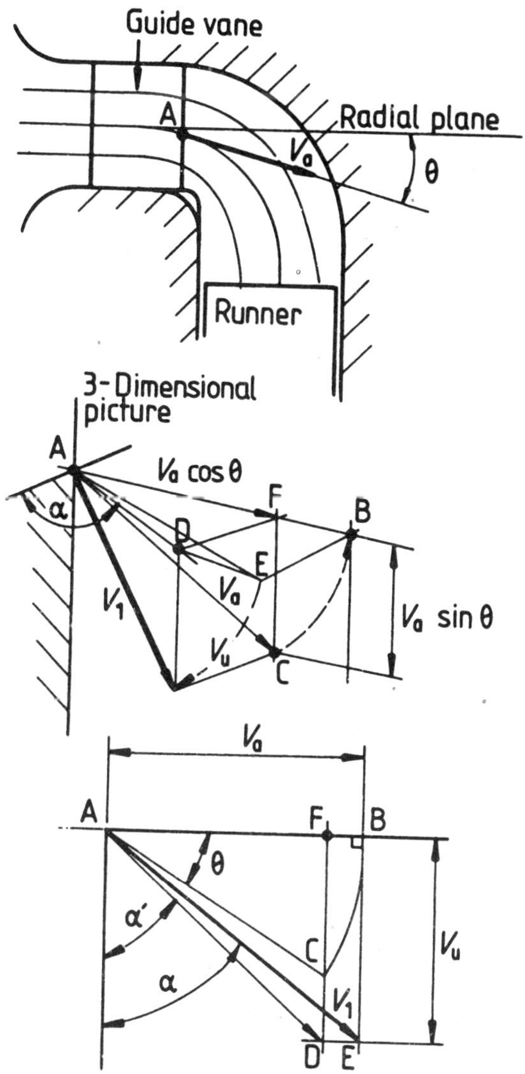


**Figure 7.9** Construction for relating guide vane setting to velocity triangle (as outlined by Nechleba, 1957).

range, and in most cases is a 'blunt' large radius which blends well into the main profile, as seen in the sketches in Fig. 7.8. When the machine is truly axial, with the guide vanes coaxial with the runner blades, there is very little difficulty in relating the blade settings. However, when the layout is that usually used in the large output Kaplan designs there is a change in direction between the radial plane in which the guide vanes are placed and the axial direction which is referenced for the runner blade sections. Figure 7.9

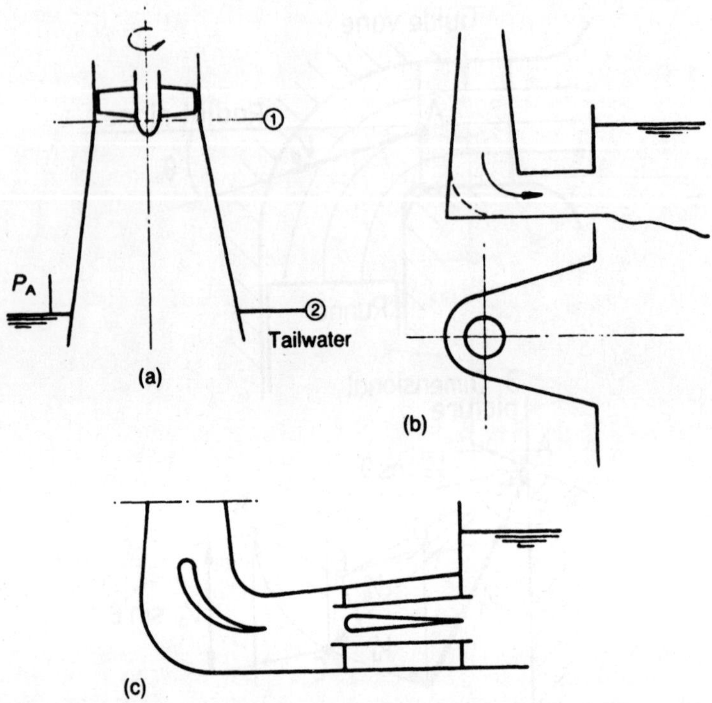


**Figure 7.10** Some draft tubes used in water turbines.

illustrates this three-dimensional problem, and demonstrates a simple correction technique that allows for the stream surface effect on the flow direction, and thus allows the setting angle measured in the radial plane for the guide vanes $\alpha$ to be related to the Euler triangle angle $\alpha'$.

The draft tube or outlet diffuser, fitted after the runner to recover some of the kinetic energy leaving the rotor blades, may follow any of the forms sketched in Fig. 7.10, variation (c) being often used with vertical axis machines. If the cone option (a) is considered to establish the principle, applying the equation of energy to the two planes 1 and 2 gives

$$\frac{p_1}{\rho} + \frac{V_1^2}{2} + gZ_1 = \frac{p_2}{\rho} + \frac{V_2^2}{2} + gZ_2 + (\text{losses between 1 and 2}) \quad (7.5)$$

If plane 2 is datum, $gZ_2$ is zero, and the draft tube efficiency is defined as

$$\eta_{DT} = \text{pressure regain}/(V_1^2 - V_2^2)/2$$

equation (7.5) becomes

$$\frac{p_2 - p_1}{\rho} = gZ_1 + \eta_{DT}\,(V_1^2 - V_2^2)/2 \quad (7.6)$$

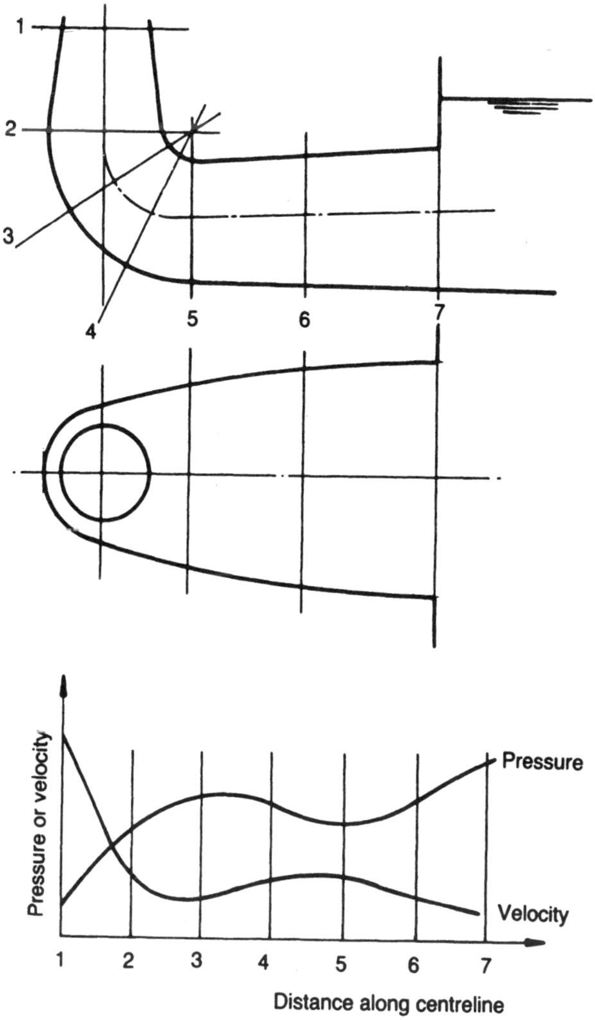


**Figure 7.11** Typical draft tube and the pressure and velocity distributions.

Kaplan performed tests on simple draft tube details like cones, bends and combinations of simple elements, and later developments have resulted in the shape illustrated in Fig. 7.11; typical pressure and velocity variations are also shown. In many cases the outlet from the runner is very low, below the tail water level in the case of the reversible pump turbine units, where cavitation limitations when pumping fix the runner level. The text by Nechleba (1957) outlines some of the history of water turbines and draft tubes, and recent work reported at the International Association for Hydraulic Research (IAHR) for hydraulic machinery symposia suggests a

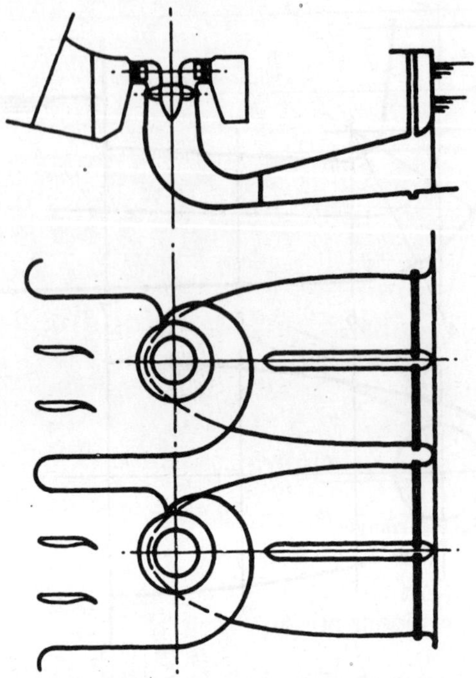

**Figure 7.12** Spiral and draft tube shapes in a barrage installation.

contribution to the overall losses of about 8% from the draft tube, falling as the head drop over a Kaplan machine increases.

The spiral casing acts as the device that directs the flow into the guide vanes and thence the runner, so its function is the exact reverse of the volute in a centrifugal pump. In many machines the spiral shape is used, but in the run of river barrage machines the shapes may be as in Fig. 7.12, the profiles being formed in the concrete mass of the dam.

To illustrate typical values of velocity and the shape of the velocity triangles a short solution to a student problem follows.

Each Kaplan turbine in a large river barrage scheme rotates at 65.2 rev $\text{min}^{-1}$, has a tip diameter of 8 m and a hub : tip ratio of 0.4. During proving trials, with a river level difference of 11 m, the gauged flow rate was 500 $\text{m}^3\,\text{s}^{-1}$ and the electrical output was 45 MW. If it is given that the alternator efficiency was 96% and the mechanical efficiency 97%, determine the plant efficiency, the hydraulic efficiency and the velocity triangles for the rotor root and tip sections, assuming that the blading followed the free vortex principle. If the draft tube has an area ratio of 1.5 : 1, has its exit centre line 6 m below the tailwater and its entry section 2 m above tailwater, estimate the turbine outlet pressure, and comment upon this figure. It may

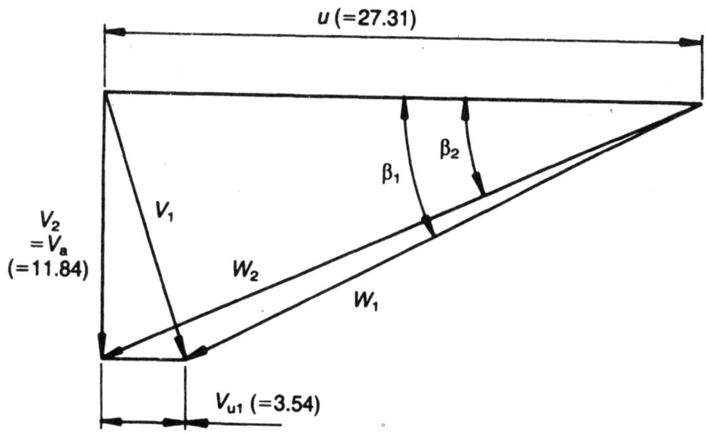


**Figure 7.13** Tip velocity triangle.

be assumed that the vapour pressure of water is 2500 N m$^{-2}$, and the draft tube efficiency is 80%.

The hydraulic power is given by

$$g \times 11 \times 500 \times 10^3 = 54\,\text{MW}$$

$$\eta_0 = \frac{45}{54} = 0.834$$

Thus

$$\eta_H = \frac{0.834}{0.96 \times 0.97} = 0.896$$

Also

$$gH_{\text{Euler}} = g \times 11 \times 0.896 = 96.69\,\text{J Kg}^{-1}$$
$$= u_1 V_{u1} \qquad \text{(since } V_{u2} \text{ assumed zero)}$$

The velocity at the runner tip is $u_t = 27.31\,\text{m s}^{-1}$, and at the runner hub is $u_h = 10.92\,\text{m s}^{-1}$. Therefore at the tip section $V_{u1} = 3.54\,\text{m s}^{-1}$. Using the annulus dimensions and neglecting blade blockage, $V_a = 11.84\,\text{m s}^{-1}$. The tip velocity triangle resulting is Fig. 7.13 and $\beta_1 = 26.48°$, $\beta_2 = 23.43°$.

Assuming a free vortex blade flow system, at the hub $V_{u1} = 8.85\,\text{m s}^{-1}$. Figure 7.14 results for the hub velocity triangle, and $\beta_1 = 80.1°$, $\beta_2 = 47.3°$. This is based on zero outlet whirl, so $V_2 = V_a = 11.84\,\text{m s}^{-1}$.

In Fig. 7.15 the area in plane A is given by $(\pi/4)\,(8)^2 = 50.2\,\text{m}^2$, and that at plane B is $75.4\,\text{m}^2$. Applying the energy equation between A and B,

$$gH_A + \frac{V_A^2}{2} + gZ_A = gH_2 + \frac{V_B^2}{2} + gZ_B + \text{losses}$$

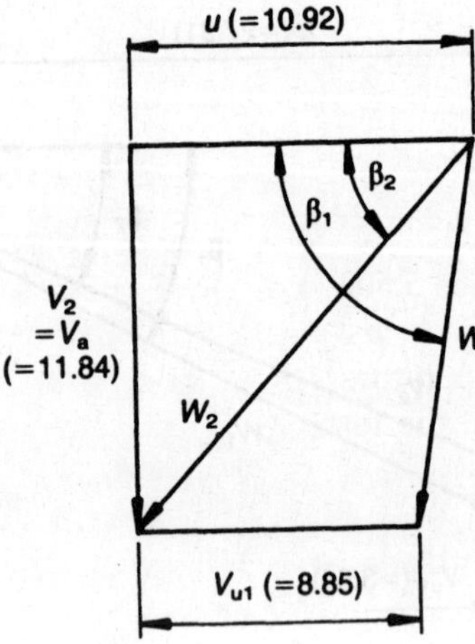


**Figure 7.14** Hub velocity triangle.

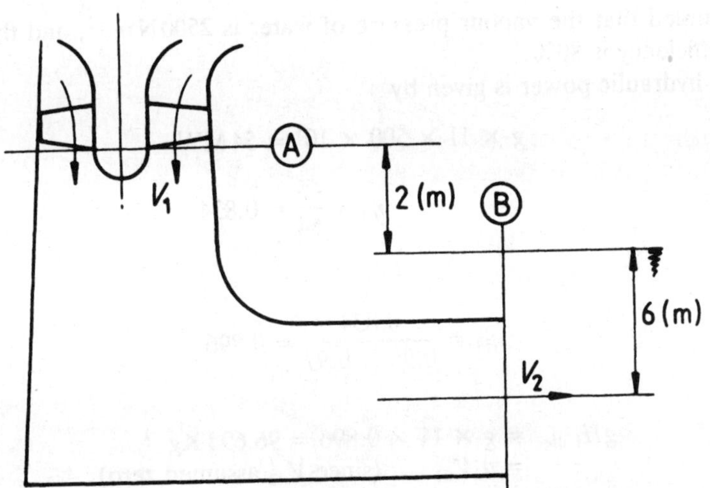


**Figure 7.15** Area profile.

At B,

$$gH_2 = \frac{10^5}{10^3} + (g6) = 158.86\,\mathrm{J\,kg^{-1}}$$

and $V_B = 6.63\,\mathrm{m\,s^{-1}}$. At A,

$$V^A = \frac{500}{50.27} = 9.95\,\mathrm{m\,s^{-1}}$$

The theoretical energy regain in the draft tube is

$$\frac{9.95^2 - 6.63^2}{2} = 27.52\,\mathrm{J\,kg^{-1}}$$

The actual regain is

$$0.8 \times 27.52 = 22.018\,\mathrm{J\,kg^{-1}}$$

Thus the draft-tube loss is $5.502\,\mathrm{J\,kg^{-1}}$. Substitution in the energy equation yields $gH_A = 58.36\,\mathrm{J\,kg^{-1}}$ or $p_A = 0.584 \times 10^5\,\mathrm{N\,m^{-2}}$.

Since at normal temperatures the vapour pressure for water is about $2500\,\mathrm{N\,m^{-2}}$, cavitation at the turbine runner outlet is unlikely.

## 7.4 Forces on blades and their implications for design

### 7.4.1 Static blades

Clearly, guide vanes are subject to varying loads as the angles change during the control cycle. When closed, for example as in Fig. 7.16, if $\Delta p$ is the pressure difference across the row of blades, the force exerted $F$ is given by

$$F = \Delta pth \tag{7.7}$$

$\Delta p$ depends upon the inlet pressure and the discharge pressure levels in a pump, and upon the difference between the spiral casing pressure and that in the turbine space for water turbines, which is a function of the tailwater level and whether the space is full of water or empty. The torque needed to keep the vanes closed is

$$T = FX \tag{7.8}$$

This, with an allowance for dynamic or shock effects which is based on experience, is used to decide the blade fulcrum pin sizes and profile thicknesses when using acceptable stress levels.

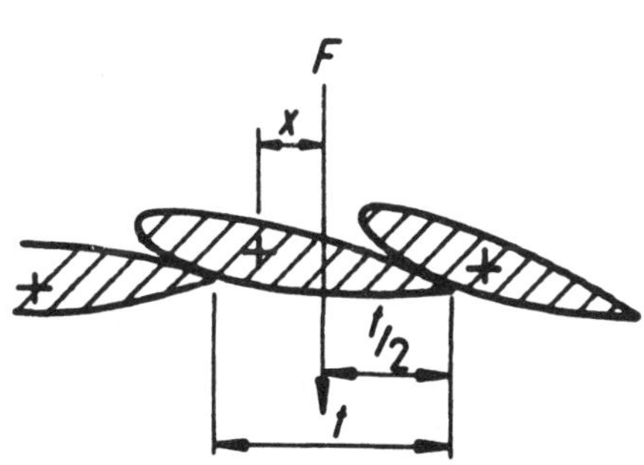


**Figure 7.16** Force on a guide vane.

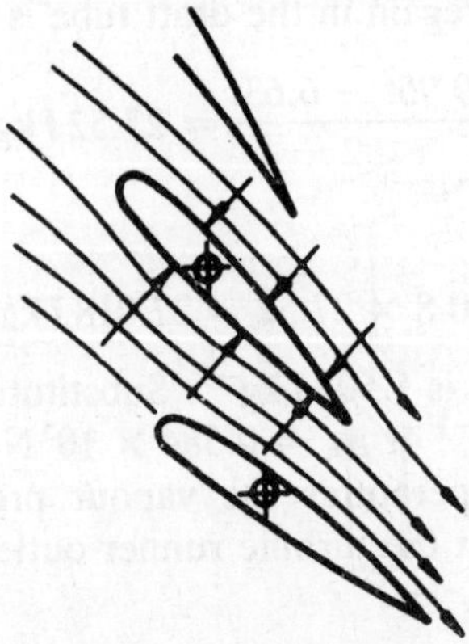

**Figure 7.17** Streamlines for open guide vanes.

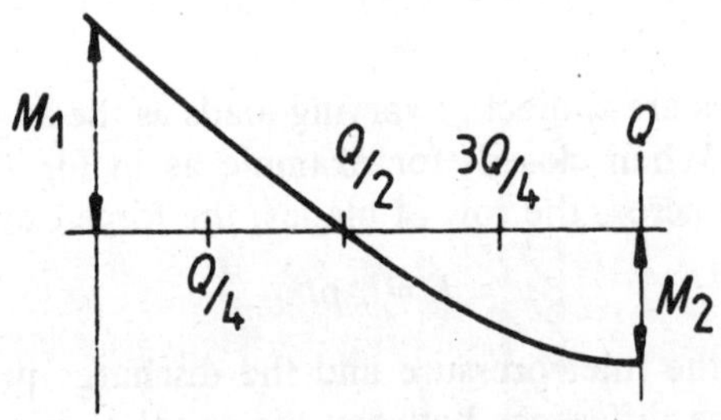


**Figure 7.18** Variation of guide vane moment with flow rate.

If the vanes are open, as sketched in Fig. 7.17 for turbine guide vanes, it is possible by plotting velocity changes round the profiles to estimate the pressure levels and thus to determine the net fluid torque applied to the vane about any chosen pivot point. Figure 7.18 illustrates the way this is likely to vary, the pivot point being chosen to balance maximum moments $M_1$ and $M_2$. These values are used to determine pivot and blade thickness sizes, and also to design the linkages used to transmit the moving effect from actuator to vanes.

### 7.4.2 Rotating blades

In Fig. 7.19 water turbine blades are used again to illustrate the pressure distribution round a blade profile and also how the hydraulic load varies along the blade length. As well as these loads there are the mechanical loads, both centrifugal and tangential, due to the mass and rotation of the blade, so that considering the blade as a cantilever will lead to the stresses at

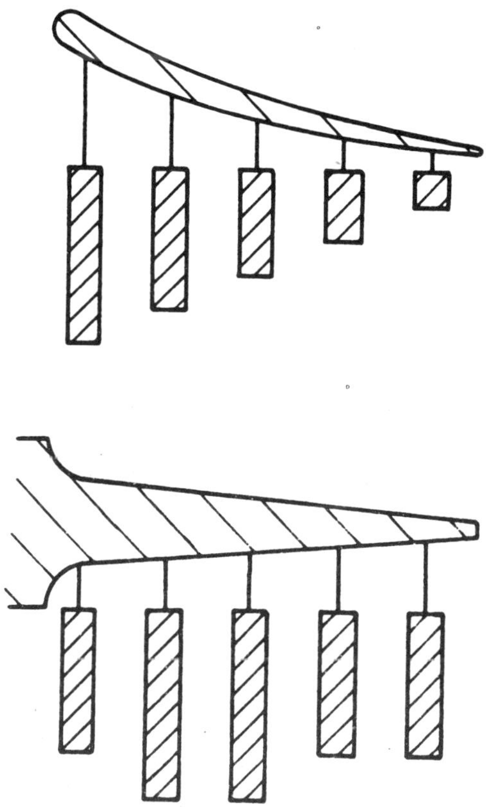

**Figure 7.19** Diagrammatic representation of fluid forces on a rotor blade.

all sections to be established. Clearly the root section will be most highly stressed, and the profile must be adequate mechanically as well as aerodynamically. The stress situation is extremely complex as is the root fixing loading and stressing, since in variable geometry blades the loads change considerably, resulting in the need for a complete study of the variations in stress during the design stage. The article by Barp, Schweizer and Flury (1973) illustrates the empirical nature of the problem in water turbine design.

## 7.5 Concluding remarks

As in the discussion of centrifugal machines, it is necessary to consider the thrust loads that occur due to the interaction of the blades with the fluid. Clearly the main effect will be axial thrust. This may be calculated using the equation given as the preamble in section 4.2, or by assuming that the pressure change acts over the annulus swept by the blades. In single-stage machines the thrust bearing is sized to take this calculated value plus an operating margin, or in the case of very large machines a balance disc or piston is fitted, as described in section 6.5. It will be noted that if the axis of the machine is vertical, the whole weight of the rotating assembly must be taken by the thrust bearing in addition to the hydraulic load. In this simple approach, complications are met during start-up, where the water column in pumps, for example, can give a very large component that may differ in direction from the usual load. Stepannof (1957b) and other hydraulic texts deal with this problem.

Since this is not intended to be a treatise on the complete design of machine elements, the texts and papers cited may be consulted for further detail. For a very useful exposition of the application of computer techniques to water turbine design, Chacour and Graybill (1977) may be consulted, as can articles in such journals as the *Escher Wyss News* and *Water Power and Construction*.

## 7.6 Exercises

7.1 An axial flow water turbine is to generate 750 kW when installed in a small 'run of river' station where the level difference is usually 2.5 m. Propose a suitable outline design if the angular velocity of the runner is to be 7.15 rad s$^{-1}$, and the tip diameter is limited to 5 m. It may be assumed that $\eta_m = 95.5\%$, $\eta_{\text{altemator}} = 97\%$ and $\eta_H = 92\%$.

7.2 An axial flow fan is to deliver air to a ventilation system at the rate of 4 m$^3$ s$^{-1}$ with a pressure rate of 175 Pa. The ventilation system is designed with a mean velocity of 15 m s$^{-1}$ and the fan intake is at $10^5$ Pa and 290 K. Perform an outline design of a suitable fan, assuming a 'hydraulic' efficiency of 75% and a rotational speed of 1450 rpm. Discuss any assumptions you may make. Can the factory use a constant section to produce the fan?

7.3 A small axial water turbine is fitted to a 'run of river' station. The level difference is 2.5 m, the rotational speed 7.15 rad s$^{-1}$, the tip diameter is 5 m, the hub diameter 2.5 m. When rotating at 7.15 rad s$^{-1}$, the estimated flow rate was 50 m s$^{-1}$. If 12 Clark Y-section blades with a chord at the hub section of 500 mm are fitted, estimate the runner blade setting angle at the hub (noting that this angle is 90° – the stagger angle), and the power developed at the shaft, assuming that hydraulic efficiency is 92% and ideal flow conditions apply.

7.4 Each Kaplan turbine in a large river barrage scheme rotates at 65.2 revolutions per minute, has a tip diameter of 8 m and a hub tip ratio of 0.35. When on proving trials with a river level difference of 11 m the gauged river flow through each machine was $500\,m^3\,s^{-1}$ and the electrical output was 45 MW. Assuming that the mechanical efficiency was 96% and the alternator efficiency was 97%, determine the hydraulic efficiency and the velocity triangles at the tip and root sections assuming free vortex principles and zero outlet whirl. If the draft tube has an area ratio of 1.5:1, has its exit centreline 6 m below the tailwater level, and has an efficiency of 80%, estimate the outlet pressure from the turbine and comment on this level. The water vapour pressure is $2500\,N\,m^{-2}$.

# 8 Axial turbines and compressors for compressible flow

## 8.1 Introduction

The principal difference between axial turbines passing gases and their incompressible counterparts is the compressibility of the fluid being moved. The basic fluid mechanics is the same, and the profile behaviour information has already been covered in Chapter 4. The same can be said of compressors and blowers.

Gas and steam turbines are in wide use as the main driver in electricity generation, in gas pipeline booster stations, and as prime movers. Gas turbines are also in use in such industries as the production of industrial process gases, and in the utilization of surplus energy in blast furnace plant as an aid to fuel economy. Sizes of machine vary from one or two megawatts to 660 MW in the base load stations supplying the grid system in the UK. Gas turbines are at the lower end of that range, typical of the size being the Tornado machine producing 6 MW currently being produced by Ruston Gas Turbines.

Compressors of the axial type are used in aircraft engines, blast furnaces, petrochemical plant, nitric acid production, natural gas liquifaction systems, and in the process industries. Machines have been supplied to give pressure rises in excess of 25 bar, and for flow ranges from about $20\,m^3\,s^{-1}$ to over $350\,m^3\,s^{-1}$. The largest power input is of the order of 88 MW to a single unit, though a string of machines may be supplied to allow for smaller units and for the provision of intermediate pressures.

Figure 8.1 illustrates a simple axial compressor, indicating the way the annulus changes, and Fig. 8.2 shows a gas turbine and the annulus area's change through the flow path. The simple steam turbine shown in Fig. 8.3 illustrates the greater area change needed by the large specific volume increase through the flow path. The principles underlying the layout of these machines are introduced in this chapter, but for detailed examination of the problems the references cited must be consulted.

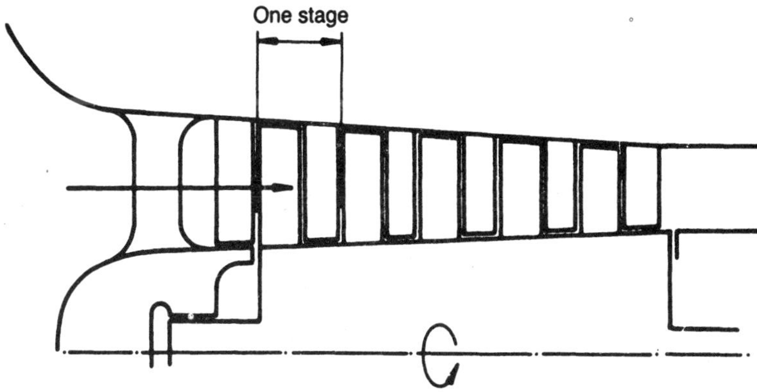


**Figure 8.1** Simple compressor.

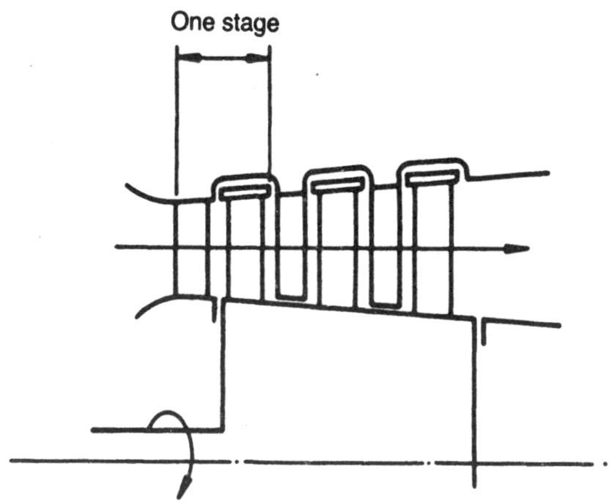


**Figure 8.2** Simple gas turbine.

## 8.2 Approach to axial compressor principles

In multistage machines like axial compressors the total pressure rise is shared between the stages, so that an important consideration is the stage pressure rise. Other matters of fundamental interest are the blade profiles used and their behaviour, the numbers in successive rows, the axial spacing between sets of blades and the interaction of the flow leaving one cascade on the pattern of flow at inlet to the next. Radial equilibrium, and the effects of boundary layer development on the inner and outer annulus on blade work distribution, are other factors to be considered in this treatment. The principles have already been introduced in Chapter 4, and these will be seen applied in the worked example which is included in the discussion. The

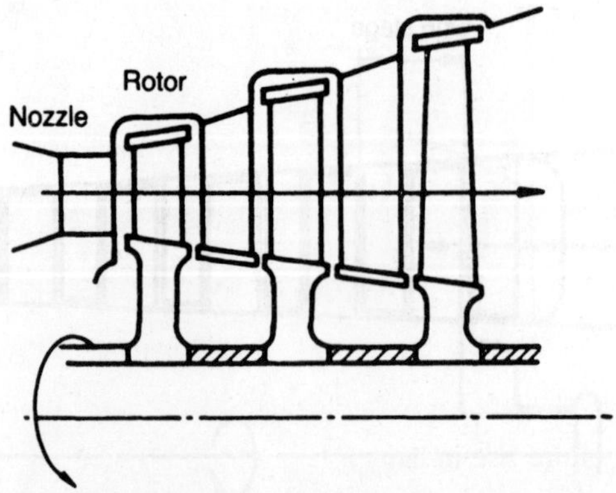


**Figure 8.3** Simple steam turbine (in wet steam units the increase in blade length is more marked).

problem of designing profiles, or of applying known information to a given situation, has not yet been covered, so that the classical English approaches due to Howell (1945) and Carter (1948) will be introduced, and Howell's correlation applied to a sample compressor problem.

Howell approached the problem by basing his correlation on nominal conditions, which he defined as those pertaining to a cascade deflection which is 80% of its maximum 'stalling' deflection. His view was that the nominal deflection $\varepsilon^*$ for a number of cascades studied was a function of the nominal gas angle $\alpha_2^*$, the space:chord ratio and the Reynolds number. He therefore proposed a correlation of these factors with the stalling deflection (defined as $\varepsilon_s = 1.25\varepsilon^*$) and claimed that the camber was apparently not a dependent variable. The use of his approach will now be explained by partly solving a compressor design problem with explanatory comments where necessary. It will be seen that in following this line of solution, the choice of the camber angle is arbitrary.

A compressor is required to deliver air at the rate of $50\,\text{kg}\,\text{s}^{-1}$ and provide a pressure ratio of 5:1, the inlet stagnation conditions being 288 K and $10^5\,\text{N}\,\text{m}^{-2}$. The target efficiency is 86%.

$$\text{pressure ratio} = 5 = (1 + 0.86\Delta T/T_{01})^{3.5}$$

Thus $\Delta T = 195.5$ and $T_{02} = 483.5\,\text{K}$.

The number of stages must now be chosen. In aircraft compressors, to limit length and weight, there is a tendency to go for as low a number as possible consistent with surge effects due to early stage stall, but in industrial machines there is a tendency to settle for lower ratios. Choosing here 10 stages gives 19.6° rise per stage. As Horlock (1958) discusses, the early stages may be unloaded to reduce surge risk at low speeds and thus increase the temperature rise in later stages, as illustrated in Fig. 8.4; however, for the purpose of the example this will not be implemented.

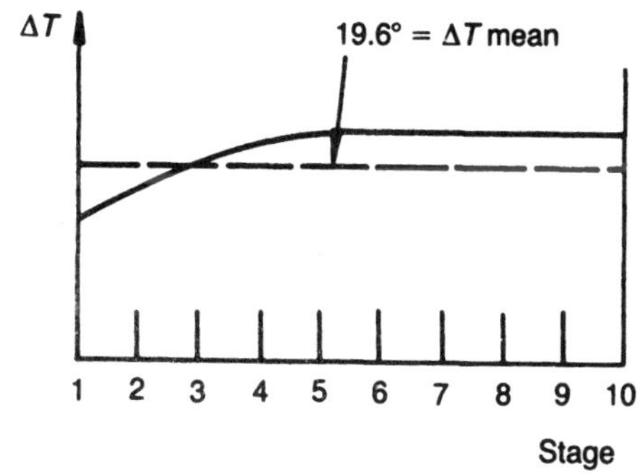


**Figure 8.4** Temperature rise in late stages.

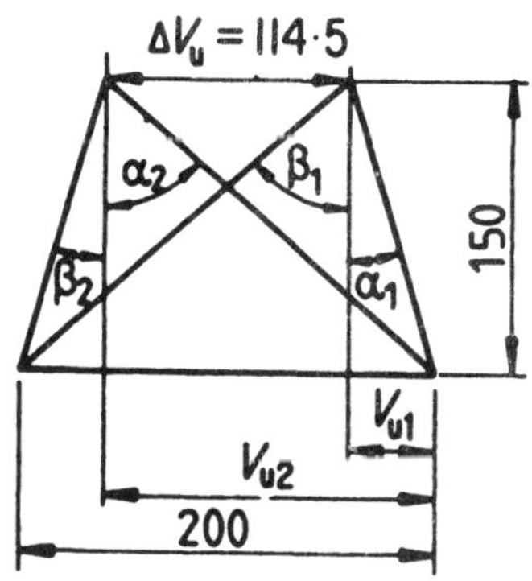


**Figure 8.5** Velocity triangle for axial compressor.

Good practice suggests an axial velocity of $150\,\text{m}\,\text{s}^{-1}$ and a mean peripheral speed of $200\,\text{m}\,\text{s}^{-1}$ (the latter is related to noise generation, which is not within the scope of this treatment).

The application of these values to the first stage, and the assumption of 50% reaction at mean blade height, allows the velocity triangles to be constructed.

$$C_p\Delta T = \Omega U \Delta V_u$$

Howell's work done factor (see Fig. 4.20) is $\Omega = 0.86$. Therefore, $V_{u\ \text{mean}} = 114.5\,\text{m}\,\text{s}^{-1}$. The velocity triangle of Fig. 8.5 results, and

$$\alpha_1 = \beta_2 = 15.9°$$
$$\alpha_2 = \beta_1 = 46.35°$$
$$\varepsilon = 30.45$$
$$V_1 = 156.05\,\text{m}\,\text{s}^{-1}$$
$$W_1 = 217.31\,\text{m}\,\text{s}^{-1}$$

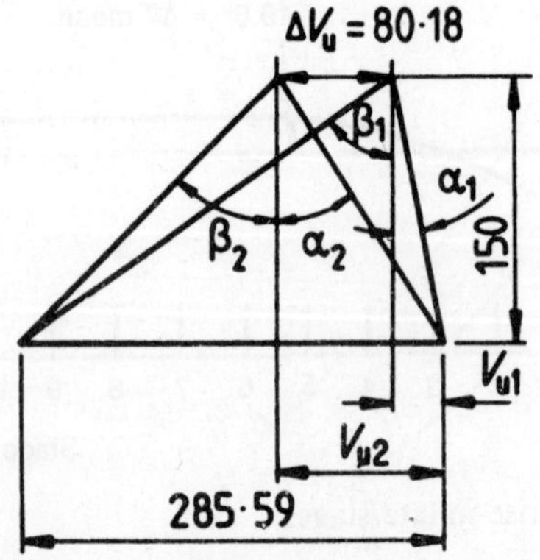


**Figure 8.6** Tip velocity triangle for axial compressor.

Since stagnation conditions are quoted, the static conditions at the inlet must be calculated to allow density and flow area to be found.

$$p_1 = 0.86 \times 10^5\,\mathrm{N\,m^{-2}}$$
$$T_1 = 275.88\,\mathrm{K}$$

thus $\rho_1 = 1.087\,\mathrm{kg\,m^{-3}}$, and the flow area is given by the equation

$$A = 50/1.087 \times 150$$

Therefore the annulus dimensions are tip diameter 0.674 m, mean diameter 0.472 m and hub diameter 0.27 m. Thus, since the mean peripheral velocity is 200 m s$^{-1}$, the rotational speed is 8093 rev min$^{-1}$.

If these sizes and speed are not acceptable, owing to space or driver choice, another trial must be made and new values determined.

The mean height triangles were found; now the tip section will be examined. Figure 8.6 results from applying the free vortex principle, from which the maximum fluid velocity at the tip appears to be 322.59 m s$^{-1}$ for the first stage blades. The acoustic velocity is

$$a = \sqrt{(1.4 \times 287 \times 275.88)} = 332.94\,\mathrm{m\,s^{-1}}$$

The corresponding Mach number is 0.97. This is rather high, suggesting that measures to adjust the velocities need to be taken to reduce this value to possibly 0.85 or lower.

We now return to the mean height section. Howell's correlation will be applied to the blade section design. The deflection $\varepsilon^* = 30.45$, and $\beta_2 = 15.9°$, so from Fig. 8.7 the space : chord ratio is approximately 0.7 ($\beta_2$ is the outlet angle from the blades in the triangles, and is $\alpha_2$ in Fig. 8.7). Since the blade height is 202 mm, then for an aspect ratio of about 3–4 a chord of 50 mm may be acceptable, yielding a value for spacing of 35.4 mm; thus the

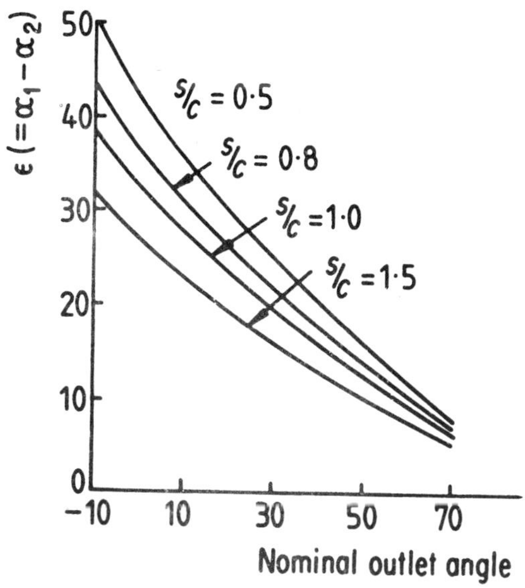


**Figure 8.7** Deflection against outlet angle with varying *s/c* ratio (Howell (1945) courtesy of the Institution of Mechanical Engineers).

number of blades is approximately 42. Selection of a prime number to avoid vibration interaction problems suggests 43 blades, resulting in a chord of 49.3 mm and an aspect ratio of about 4.1. If a C4 profile is used (as in Howell's work) the gas and blade angles may be related with the camber line, with assumptions of incidence; using Fig. 8.8 the optimum incidence appears to be 2.5°, so the blade inlet angle becomes 46.35 − 2.5 = 43.85°. To determine the deviation, equation (4.22) may be used

$$\delta = m\theta\sqrt{(s/c)}$$

where

$$m = 0.23(2a/c)^2 + 0.1(\alpha_2^*/50)$$

as proposed by Howell ($a$ is the distance of the point of maximum camber from the leading edge, and $\alpha_2^*$ is the nominal outlet angle).

Assuming a circular arc camber line,

$$m = 0.23(1)^2 + 0.1(15.9/50)$$

to give $\delta = 0.22\theta$ and $\beta_2' = \beta_2 - 0.22\theta$. Thus

$$\theta = \beta_1' - \beta_2' = 43.85 - 15.9 + 0.22\theta$$

and $\theta = 35.8°$. The stagger angle is

$$\gamma = 43.85 - 35.8/2 = 26°$$

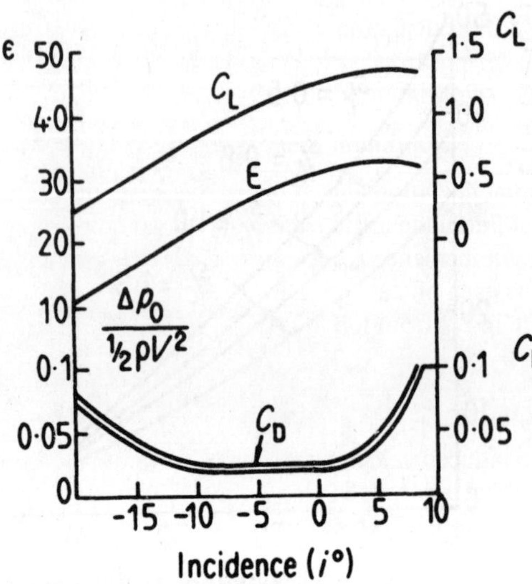


**Figure 8.8** Performance of a cascade (Howell (1945) courtesy of the Institution of Mechanical Engineers).

Following Howell (1945) and using his data (Fig. 8.8), then assuming

$$C_L = 0.85$$

$$C_D = 0.025$$

and using equation (4.20), the probable maximum row efficiency is 94.1.

The camber radius and other geometric functions may be calculated using simple trigonometry, and similar calculations performed for other sections. The sections may be related radially, as discussed in section 7.4, to give the required stress level calculations.

Carter's (1948) correlation proceeded to relate the optimum lift:drag ratio to the deflection, gas incidence angle and outlet angle. A design following his approach would be as in Table 8.1. It should be noted that Carter considered variations in outlet angle to be unimportant over the range $0 < \beta_2 < 40°$.

Both the correlations and others like them are effectively two dimensional, based on static idealized cascade rig test data, and the only corrections for three-dimensional flow are to use the work done factor and radial equilibrium approaches to relate the blade sections. One school of thought is to use the actuator disc technique pioneered by Hawthorne and developed by Horlock (1958). This lends itself to computer-based design, as is well illustrated by Railly (1961) and Railly and Howard (1962). The alternative approach is to use data obtained from actual tests on machines. McKenzie (1980) used a low speed four-stage compressor with a hub:tip

**Table 8.1** Steps in the calculation of optimum performance, following Carter (1948)

- Estimate a value of optimum incidence $i_{opt}$; thus

$$\text{optimum inlet angle} = \text{gas angle} + i_{opt}$$

- Find the optimum deviation angle using the equation for deviation and Fig. 3.21.
- Calculate the outlet blade angle from gas angle and deviation.
- Find the optimum incidence $i_{opt}$ for an isolated aerofoil, Carter (1948, 1961).
- Calculate $C_L$ for each blade.
- Find the change in $i_{opt}$ from foil in cascade to isolated foil:

$$i = i_{opt} - i_{opt(original)} = \frac{C_n C_L}{2s/c}$$

  where $C_n$ is the induced velocity perpendicular to the mean velocity due to unit circulation and pitching, found using Betz rules as $f(s/c)(\gamma)$.
- Find $i_{opt} = i_{opt(original)} + i$, and compare with the value estimated. Iterate if necessary.

ratio of 0.8, provided with constant section blades with C5 profiles on circular arc camber lines. He showed that for constant stagger angle the maximum efficiency occurred at a constant flow coefficient irrespective of the camber and space:chord ratio, from which he derived a correlation of stagger angle to mean gas angle:

$$\tan\gamma = \tan\beta_m - 0.213 \tag{8.1}$$

He showed from exhaustive traverse tests that the deviation followed the equation

$$\delta = (2 + \theta/3)(s/c)^{1/3} \tag{8.2}$$

He compared actual data available for low hub:tip ratios giving alternative information, and also proposed two design charts which could be used. He also made a comparison with the fan approach of Hay, Metcalfe and Reizes (1978) discussed in Chapter 7, and found fairly close agreement. He also considered the difficulty of accurately predicting stall, which the other approaches share, and demonstrated considerable scatter for the performance plots covered; he remarked that low speed performance prediction is reasonable but is of doubtful validity when applied to high speed compressors. The approach is as valid as the Howell and Carter correlations, however, as it is based upon actual machine data, and may be applied to compressors in the same speed, flow and pressure range.

The first stage has been examined, and the design of the rest of the blade rows would proceed in the same way, with allowance being made for density change – the blade heights must reduce progressively. Some designs use an annulus based on constant hub diameter, used in some aircraft designs, where a 'waisting' effect accommodates auxiliaries in the area of the outlet casing. Many others keep the outer diameter constant to simplify stator blade fixings; blade sections are constant over a number of stages, the length being changed to suit annulus height requirements. This reduces costs with very little reduction in efficiency.

When the blade profile dimensions have been determined, the axial spacing of the blades, the root fixings, and the stress levels are found. The axial spacing has to allow for good fluid coupling between the blade rows and at the same time provide for relative expansion between the rotor assembly and the stator; a typical industrial axial spacing appears to be up to 25% of the blade chord, but one company has stated that, on low pressure rise stages, varying the spacing from 14% of chord to as much as 50% gave changes in efficiency that were considered to be commercially acceptable. Tip clearance too must be chosen to minimize flow loss but still allow for radial growth of the rotating assembly into the stator. Blade stressing was covered in Chapter 7 in outline, and root fixings using pins or fir tree root types are matters of company policy. Carchedi and Wood (1982) illustrate how a commercial design may be carried out with the working restrictions on blade profile choice, and how general mechanical design decisions are reached.

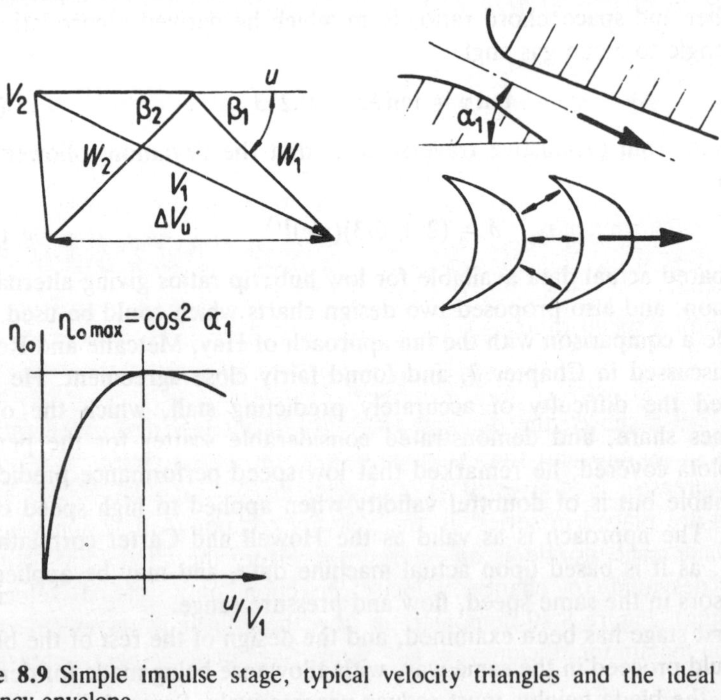


**Figure 8.9** Simple impulse stage, typical velocity triangles and the ideal overall efficiency envelope.

## 8.3 Axial turbine principles

### 8.3.1 General principles

The basic principles were introduced in sections 1.5.2 and 1.7.3, as were the usual parameters used. Before proceeding to discuss the correlations used for loss and performance prediction, it is necessary to cover the simple relations governing the enthalpy change in a stage with the velocity triangles.

Consider first the 'impulse' stage, illustrated in Fig. 8.9. The nozzle discharges gas at an angle of $\alpha_1$ and this impinges on the rotor in such a way that the gas angle is the blade inlet angle (deviation and incidence are ignored). All the enthalpy drop is assumed to take place in the nozzle. The diagram work is

$$u\Delta V_u = 2u(V_1 \cos \alpha_1 - u) \tag{8.3}$$

The energy available is $V_1^2/2$ (assuming no loss in the nozzle; this is related to the enthalpy drop as shown in section 1.7, and $V_1$ is then the isentropic velocity).

The diagram efficiency is

$$\eta_D = \frac{2u(V_1 \cos \alpha_1 - u)}{V_1^2/2}$$

$$\eta_D = \frac{4u}{V_1}\left(\cos \alpha_1 - \frac{u}{V_1}\right)$$

Differentiating with respect to $u/V_1$ and equating to zero suggests that maximum $\eta_D$ occurs when the blade:speed ratio is

$$u/V_1 = \frac{\cos \alpha_1}{2} \tag{8.4}$$

the maximum efficiency is

$$\eta_D = \cos^2 \alpha_1 \tag{8.5}$$

and at optimum speed ratio

$$\text{work output} = 2u^2 \tag{8.6}$$

As Fig. 8.9 also shows, the diagram efficiency varies with the speed ratio, reaching a maximum and then falling away again as the speed ratio continues to increase.

If stages are placed in series, in the velocity compounded design where only the first stage nozzle is an energy convertor, the succeeding stator rows are designed to change velocity direction only (called velocity compounding). Figure 8.10 demonstrates how the efficiency curves change, and also the maximum efficiency, as the number of impulse stages increases.

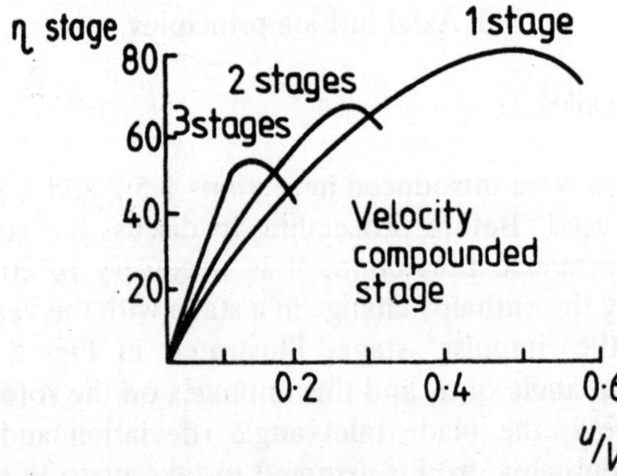


**Figure 8.10** Effect of the number of stages on the efficiency envelope.

Now consider the 50% reaction case. By similar reasoning, and following the velocity triangle shapes shown in Fig. 1.11, the optimum speed ratio is given by the relation:

$$\frac{u}{V_1} = \cos \alpha_1 \tag{8.7}$$

The maximum diagram efficiency is given by:

$$\eta_D = \frac{2\cos^2 \alpha_1}{1 + \cos^2 \alpha_1} \tag{8.8}$$

and

$$\text{work output at optimum speed ratio} = u^2 \tag{8.9}$$

Figure 8.11, following Kearton (1958) compares the efficiency envelopes for multistage reaction and multistage impulse machines based upon a nozzle angle of 20° and a nozzle efficiency of 94%.

This discussion, together with those in Chapter 1, allows the construction of velocity triangles. The next stage in analysis or design consists in relating the gas directions to blade angles, and giving an approach to the estimation of losses and efficiencies, and allowances for three-dimensional flow and other effects.

The main correlations are based on data obtained using perfect gases such as air; all gas turbines, and many steam turbines, pass such fluids, because superheated steam follows the same laws as air. The only exception to this rule is the turbine passing wet steam, as in the low pressure cylinders of a large main load machine and in machines such as those used in nuclear generators; the particular problems associated with these machines will be discussed.

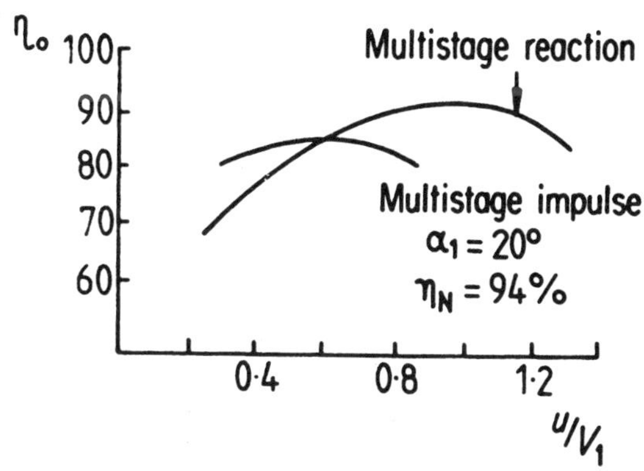


**Figure 8.11** Comparison of multistage impulse and reaction machine efficiency envelopes.

Horlock (1966) in a useful background study, brought together two main streams of information in comparing the correlations from steam turbine and gas turbine technology. He compared the early steam turbine data reported by Kearton (1958), Stodola (1945) and Guy (1939), for example, with the gas turbine studies published by Zweifel (1945), Ainley and Matheison (1957) and Soderberg (1949) among others. Horlock demonstrated the wide differences between the approaches reviewed, and a private study comparing them with some company data, indicated an almost 10% band in efficiency prediction for low reaction and 'impulse' blading designs.

Zweifel (1945) used data obtained for both accelerating and decelerating cascades with typical large deflections, and argued that for typical zero reaction blading a tangential blade loading coefficient $\psi_T$ should be used rather than the more usual pressure coefficient or lift coefficient. He also advanced a correlation based on profiles with a value of $\psi_T = 0.8$, and used it to predict the number of blades that would give effective performance for axial turbocharger turbines.

Ainley and Matheison (1957) suggested that the total loss of a cascade of blades consisted of profile loss $Y_F$, secondary loss $Y_S$ and tip clearance loss $Y_C$ (Fig. 8.12). They correlated $Y_P$ for zero-incidence flow conditions with blades having a thickness:chord ratio of 20%, and applied corrections for incidence and thickness. They proposed a combined formula for $Y_S$ and $Y_C$:

$$Y_S + Y_C = \left[\lambda + B\frac{\delta}{H}\right]\left(\frac{C_L}{s/c}\right)^2 \frac{\cos^2 \alpha_2}{\cos^3 \alpha_m} \qquad (8.10)$$

$B$ and $\delta$ are defined in Fig. 8.13, and $\lambda$ is also shown. The data were presented for low reaction blading, and in correspondence Ainley suggested the corrected line as being a better correlation for higher reaction turbines.

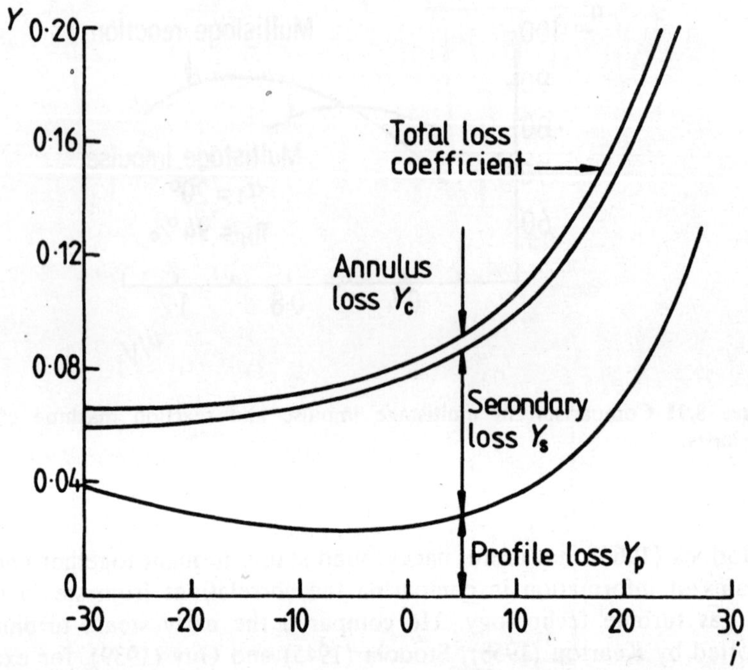


**Figure 8.12** Loss coefficient proposed by Ainley and Matheison (after Ainley and Matheison, 1957).

Soderberg, using Zweifel's information and other data, related losses on the basis of space: chord ratio, Reynolds number, aspect ratio, thickness: chord ratio and profile geometry, assuming zero incidence and correcting for flow conditions with some incidence and for other deviations from the basis on which the data was obtained. Horlock (1966) may be consulted for a thorough comparison of this and other approaches, and demonstrates the wide differences between them and the essential complexity of the problem.

An alternative approach is to use information obtained from machines to complete the correlation. One example of this approach is that of Craig and Cox (1970) for steam turbines. They identified two loss groupings. Group one comprised losses due to profiles, secondary flows, clearance effects, annulus and cavities for both stator and rotor rows; group two losses were due to gland leakage, windage, wetness effects, and partial admission. The group one effects are effectively corrected for, as in the approach of Ainley and Matheison (1957), with company experience fed into the databank. The discussion of the group two losses highlights the problems facing the steam turbine manufacturer, for example in the exhaust casing systems. These must carry the wet steam leaving the low pressure turbines into the condensers with the maximum effectiveness in the minimum axial distance,

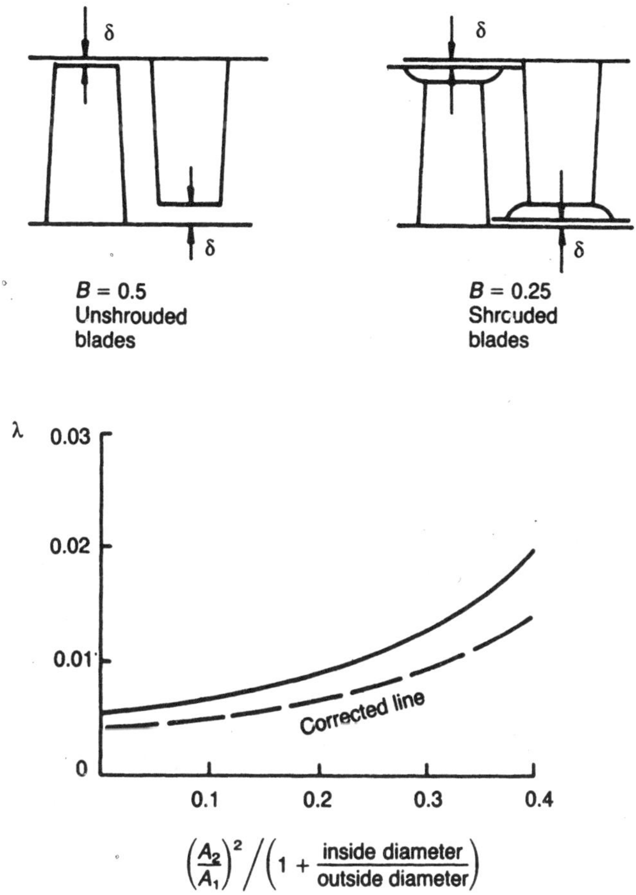


**Figure 8.13** $\lambda$ factor proposed by Ainley and Matheison (after Ainley and Matheison, 1957).

owing to the need to keep the total length of the turbine to an acceptable value. Flow is strongly three dimensional, and steam turbine manufacturers have used extensive air model testing to improve their designs.

The comparisons of Horlock referred to above, and the later study by Dunham (1970), related to the profile data then available, which was obtained at Reynolds numbers based on chord of about $10^6$ and at Mach numbers well below sonic velocities. As power output has increased, annulus dimensions and turbine cylinders have also become larger, with low pressure turbine blade tip velocities attaining supersonic values. This, and the cost pressure causing designers to reduce machine sizes and optimize the number of stages, has led to higher velocities, and the work of designers has gone hand-in-hand with analytical and experimental studies.

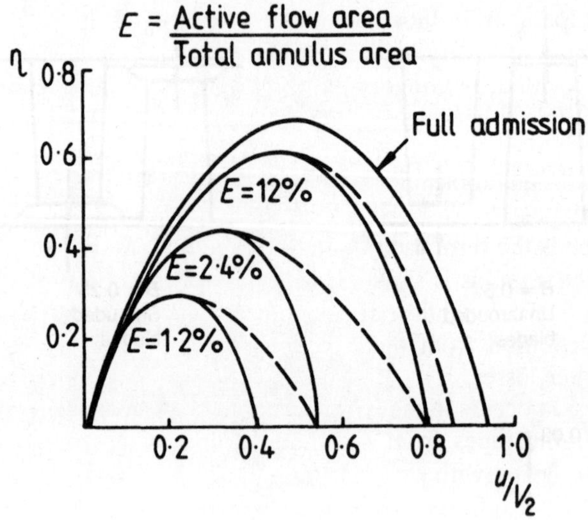


**Figure 8.14** Effect of partial admission.

Wilson (1987) approached the preliminary design of axial flow gas turbines with the Craig and Cox correlation in mind. He identified twin objectives: to correlate efficiency with reaction, loading coefficient (defined as $u\Delta V_u/u^2$) and flow coefficient; and to provide an informed choice of vortex flow distribution across the turbine annulus and allow an optimum annulus profile to be chosen for turbines working at various expansion ratios. He had as a further objective the improvement of the Craig and Cox loss estimation method, already discussed, along with other published approaches. The paper draws extensively from the earlier work already discussed in this chapter and proposes some simple guidelines for reaction choice, hub tip ratio, stage number, and loading coefficient. He also suggests an equation for swirl variation to give an almost constant nozzle outlet angle across the turbine annulus. The paper ends with loss correlations in terms of pressure loss ratio statements based on Craig and Cox and upon Brown (1972).

The economic and engineering aspects of steam and turbine design are well illustrated by the contributions of Burn (1978), Parsons (1972) and Smith (1975), the latter being an extremely well illustrated treatment of wet steam problems. Hesketh and Muscroft (1990) detail the engineering of the steam turbine generators for the Sizewell B nuclear station giving an insight into the problems to be solved and the reasoning behind the solutions selected in constructing the plant. Recent studies of wet steam problems are discussed by Laali (1991) and Tanuma and Sakamoto (1991).

For further reading refer to Fielding (1981) which is a look at losses from a different standpoint from those quoted above; Hill and Lewis (1974) is an experimental study of a very high divergence turbine; a study of secondary flow losses is provided by Gregory-Smith (1982) and in later papers by the same author and his colleagues.

### 8.3.2 Partial admission problem

Partial admission is the term used to describe the flow situation where only a proportion of the annulus is supplied with gas, so that the rotor blades are moving from a live throughflow sector into a 'dead' zone where the gas is carried by the rotation from the live flow and its energy is dissipated in windage and other losses.

Partial admission occurs in many turbocharger turbines, owing to the need to keep discharging pipes from individual cylinders of the engines being blown from interfering with one another. In this case a different sector of the annulus is live at different times in the engine cycle, so that the flow is very three dimensional and complex, giving rise to large losses. In small steam turbines a complete sector may be blanked off permanently to adjust steam flows to give a set output, so carry-over fluid flow losses from the live to dead sectors account for a large part of the losses. Evidence of this is given in Fig. 8.14, based on work published by Stenning (1953) and by Suter and Traupel (1959). Clearly seen are the effects of greatly reduced active flow annulus area.

## 8.4 Other problems

Since the temperatures are high there are large differential expansions, so that clearances, which are cold during erection, have to allow for the closing up due to expansion. This is particularly important with tip clearances, where adjustment is needed to prevent tips touching, particularly for compressor blades made from titanium where local fires may be caused. In gas turbines the problem is one of simple calculation owing to the light construction of the casings and blades, though arrangements have to be made to allow blades to take up their own attitude under load and expansion movement to reduce induced stresses. In steam turbines the high pressures require very heavy pressure vessel construction; the rotor heats up more quickly than the casing, so that the axial clearances during warm-up are more critical than those when cold or at running temperature.

Aircraft turbine temperatures are now exceeding 1600 °C so that elaborate cooling methods using hollow blades are now standard practice. Industrial machines like the Ruston Tornado are now utilizing the technique, as higher efficiencies are sought and extended life is a high priority.

Axial thrust loading is clearly a problem, and in industrial compressors with many stages is usually compensated for by introducing a balance piston or disc as described in section 6.5.2. In aircraft gas turbines and industrial gas turbines the thrust produced by the compressor is partly balanced by that from the turbine so that a thrust bearing is only needed for small thrust loads and for axial location. Where a turbine is used as a drive other measures are needed. For example, in multicylinder steam turbines the high pressure and middle pressure turbines are arranged so that steam inlet ends and the axial thrusts from the two units are opposed; the low pressure cylinders have their steam inlet belt in the centre with expansion in both axial directions, thus ensuring thrust balance. Gas or air turbines driving compressors use balance as in the gas turbine, but need thrust bearings in most other cases. As in the machines described in Chapter 7, radial thrusts are low and well within the capacity of the bearings conventionally used.

## 8.5 Computer-aided solutions

There are a number of CFD and CAD programmes now in use. Some are simply the provision of computer power using existing empirical data and the approaches noted above, but the reader is referred to a number of very recent papers and articles that indicate the power and capability of the software being developed. Typical are the papers by Scrivener *et al.* (1991), Hart *et al.* (1991) and Kobayashi *et al.* (1991) who were all contributors to a recent Institution of Mechanical Engineers conference.

## 8.6 Illustrative examples

### 8.6.1 Axial compressor example

An axial flow compressor is to deliver helium and has six stages equally loaded, with stage temperature rise 25 K. If the overall efficiency is to be 87%, determine the overall pressure ratio if $T_{01}$ is 288 K and $p_{01}$ is $10^5\,\mathrm{N\,m^{-2}}$, and the stage pressure ratio for the last and first stages. Calculate the blade height at entry to the last stage and the rotational speed if $V_1$ is $165\,\mathrm{m\,s^{-1}}$ and makes an angle of 20° to the axial direction, the mass flow is $13\,\mathrm{kg\,s^{-1}}$ and $D_{\mathrm{mean}}$ is 680 mm. The Howell work done factoı is 0.83 and the mean section has been designed to be 50% reaction. What is the maximum Mach number in the last stage?

From the data given

$$T_{06} = 288 + 6 \times 25 = 438\,\mathrm{K}$$
$$C_p = 5.193 \times 10^3\,\mathrm{J\,kg^{-1}\,K^{-1}}$$
$$M = 4.003$$

thus

$$R = \frac{8.314 \times 10^3}{4.003} = 2.077 \times 10^3\,\mathrm{J\,kg^{-1}\,K^{-1}}$$

and

$$C_v = 3.116\,\mathrm{J\,kg^{-1}\,K^{-1}}$$

$$n = \frac{C_p}{C_v} = 1.67$$

$$\frac{n}{n-1} = 2.493$$

Therefore

$$\left(\frac{p_{02}}{p_{01}}\right)_{\text{overall}} - \left(1 + 0.87 \times \frac{150}{288}\right)^{2.493} = 2.54:1$$

$$p_{02} = 2.54 \times 10\,\mathrm{N^5\,m^{-2}}$$

At the first stage

$$\left(\frac{p_{02}}{p_{01}}\right)_{\text{first stage}} = \left(1 + 0.88 \times \frac{25}{288}\right)^{2.493} = 1.2$$

and, from Fig. 1.17,

$$\eta_p = 0.88$$

At the last stage $T_0$, at inlet to this stage, is 438 − 25 = 413 °C and

$$\frac{p_{02}}{p_{01}} = 1.138$$

The stagnation pressure at inlet to the stage is

$$2.54 \times 10^5/1.138 = 2.232 \times 10^5\,\mathrm{N\,m^{-2}}$$

To construct the velocity triangles at the sixth stage:

$$25 \times 5.193 \times 10^3 = 0.83 \times u\Delta V_u$$

Turning to Fig. 8.15

$$V_{u1} = 60.055$$

$$V_A = 165 \sin 20 = 56.43\,\mathrm{m\,s^{-1}}$$

$$V_{u2} = (u - 60.055)$$

$$\Delta V_u = (u - 60.055) - 60.055$$

thus, $V_u = u - 120.11$, and

$$25 \times 5.193 \times 10^3 = 0.83 \times u(u - 120.11)$$

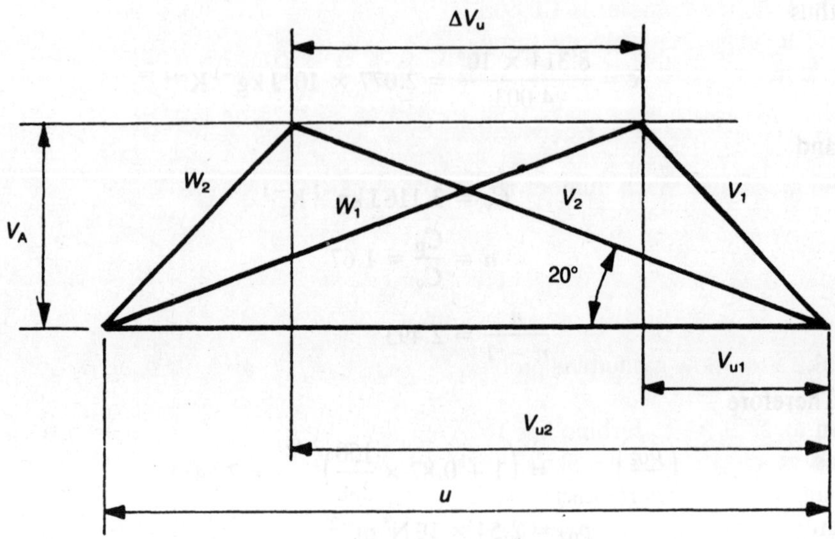


**Figure 8.15** Axial compressor example – velocity triangles.

Solving this equation gives $u = 460.055$ and

$$N = \frac{460.055 \times 2 \times 30}{0.68 \times \pi} = 12\,921 \text{ rpm}$$

$$V_{u2} = 400 \text{ m s}^{-1}$$

$$V_2 = 432.7 \text{ m s}^{-1} = W_1$$

The static temperature and pressure at inlet to the last stage are now needed:

$$413 = T_5 + \frac{175.6^2}{2 \times 5.193 \times 10^3}, \qquad T_5 = 410 \text{ K}$$

$$p_5/p_{05} = (410/413)^{2.493} = 0.98, \qquad p_5 = 2.192 \times 10^5 \text{ N m}^{-2}$$

thus

$$\rho_5 = \frac{2.192 \times 10^5}{2.077 \times 10^3 \times 410} = 0.257 \text{ kg m}^{-3}$$

and

$$Q = \frac{13}{0.257} = 50.5 \text{ m}^3 \text{ s}^{-1}$$

Since the annulus height is small compared with the mean diameter

$$\frac{50.5}{56.43} = \pi \times 0.68h, \qquad h = 0.419 \text{ m}$$

Thus the tip diameter is 1.099 m and the hub diameter is 0.261 m.

The maximum velocity through the stage is $W_1$ and, since the acoustic velocity $a$ is

$$a = 1.67 \times 2.077 \times 10^3 \times 410 = 1192.52$$

the maximum Mach number is

$$\frac{432.7}{1192.52} = 0.363$$

### 8.6.2 Axial flow gas turbine problem

An axial flow gas turbine has 16 stages, equally loaded and is supplied with gas at the rate of 12.5 kg s$^{-1}$. Determine the shaft power, the inlct tip diameter and the gas angles at the root and tip sections of the first-stage rotor. It may be assumed that the rotational speed is 5000 rpm, the inlet static conditions are 1150 K and $7 \times 10^5$ N m$^{-2}$, the outlet static conditions are $1.05 \times 10^5$ N m$^{-2}$ and 750 K, the turbine rotor inner diameter is 1 m for all stages, the reheat factor is 1.02, $\eta_m$ is 90%, the axial velocity is 100 m s$^{-1}$ in all stages, and all root sections are designed to the 50% reaction principle.

For the combustion gas products $n = 1.333$ and $C_p = 1.145 \times 10^3$ J kg$^{-1}$K$^{-1}$. For isentropic expansion across the turbine the gas laws give

$$T_{\text{out}} = 703.36\,\text{K}$$

from the 16th stage, thus

$$\begin{aligned}\Delta h_{\text{isentropic}} &= (1150 - 703.36) \times 1.145 \times 10^3\\ &= 511.4\,\text{kJ kg}^{-1}\\ \Delta h_{\text{actual}} &= (1150 - 750) \times 1.145 \times 10^3\\ &= 458\,\text{kJ kg}^{-1}\end{aligned}$$

The static to static efficiency is

$$\eta_{ss} = \frac{458}{511.4} = 0.896$$

Since

$$R = \frac{\eta_{\text{overall}}}{\eta_{\text{stage}}}$$

$$\eta_{\text{stage}} = \frac{0.896}{1.02} = 0.878$$

Since

$$\text{reheat factor} = \text{number of stages} \times \frac{\Delta h_{\text{isentropic}} \text{ per stage}}{\text{overall } \Delta h_{\text{isentropic}}}$$

$$1.02 = 16 \frac{\Delta h_{\text{isentropic}} \text{ per stage}}{511.4}$$

$$\Delta h_{\text{isentropic}} \text{ per stage} = 32.6 \times 10^3 \text{ J kg}^{-1}$$

and

$$\Delta h_{\text{actual}} \text{ per stage} = 32.6 \times 0.878 = 28.62 \times 10^3 \text{ J kg}^{-1}$$

At inlet to the stage

$$\rho = \frac{7.5 \times 10^5}{287 \times 1150} = 2.27 \text{ kg m}^{-3}$$

the annulus area is

$$\dot{m}/\rho V_a = 0.055 \text{ m}^2$$

and, since $D_h = 1$ m, $D_{tip} = 1.034$. Also, $u_h = 261.8 \text{ m s}^{-1}$ and $u_t = 270.7 \text{ m s}^{-1}$.

At the hub section reaction = 50%, which gives

$$28.62 \times 10^3 = 261.8 \times \Delta V_u$$

$$\Delta V_u = 105.73 \text{ m s}^{-1}$$

The velocity triangles are shown in Fig. 8.16, from which

$$V_{u1} = 183.77, \qquad V_{u2} = 78.05$$

$$\beta_1 = 37.97°, \qquad \beta_2 = 61.45$$

At the tip section, since $V_u \times R$ = constant,

$$V_{u1} = 177.73, \qquad V_{u2} = 75.48$$

$$\beta_1 = 37.05°, \qquad \beta_2 = 60.64$$

The shaft power is given by

$$\text{shaft power} = 1.145 \times 10^3 \times \Delta h_{\text{actual}} \times \eta_m$$

$$= 1.145 \times 10^3 \times 28.62 \times 10^3 \times 0.9$$

$$= 29.49 \text{ MW}$$

## 8.7 Exercises

8.1 A reaction steam turbine rotates at 1500 rpm and consumes steam at the rate of 20000 kg h$^{-1}$. At a certain stage in the machine the steam is at

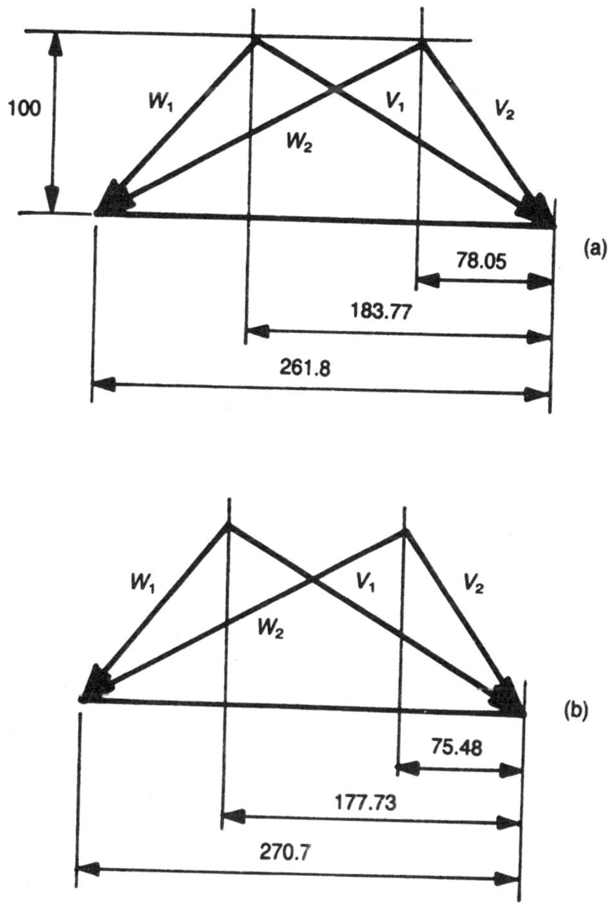


**Figure 8.16** Axial flow gas turbine problem – velocity triangles.

a pressure of 0.4 bar with a dryness fraction of 0.93, and the stage develops 0.25 MW. The axial velocity is 0.7 of the mean blade section peripheral velocity and is constant through the stage. The reaction at the mean radius is 50%, and the blade profiles are the same for both rotor and stator blades with the outlet angles being at 20° to the peripheral direction. Determine the inlet annulus dimensions for this stage, if the total losses are estimated to be 30% of the developed power.

8.2 A four-stage axial steam turbine is designed on the assumption of equal loading per stage and a constant axial velocity through the machine, with a mean diameter for all stages of 1.5 m. A reaction of 50% was used for each stage at the mean section with the nozzle angle at this diameter of 20° and a rotor blade inlet angle of 55° to the peripheral direction. The design reheat factor was 1.04, the rotational speed 3000 rpm. When tested under steam inlet conditions of 220 bar and 973 K, the exhaust conditions were 0.5 bar and 373 K. For these conditions calculate the steam mass flow rate, the diagram work, and the overall efficiency, if the blade height at the last stage is 35 mm.

8.3 A 10-stage axial flow gas turbine is to run at 3000 rpm when passing gas at a mass flow rate of 70 kg s$^{-1}$. The gas is that normally produced by hydrocarbon combustion. It is assumed that the blading is designed to the 50% reaction concept at the mean diameter of 1.5 m, the diagram output is to be 16.5 MW, the axial velocity of 100 m s$^{-1}$ is constant through the machine, and all stages are equally loaded. If the inlet stagnation conditions are $7.5 \times 10^5$ N m$^{-2}$ and 1000 K for the first stage, determine the annulus dimensions and the mean blade height velocity diagrams. Assuming free vortex conditions across the leading and trailing edges determine the tip and hub velocity triangles and discuss whether a constant section could be used, specifying which section would be the most suitable.

8.4 The first stage of an axial flow compressor passing air has a hub:tip ratio of 0.6, and $V_1 = V_A = 140$ m s$^{-1}$. The rotational speed is 6000 rpm, the temperature rise is 20 °C, and the Howell work done factor is 0.925. The inlet stagnation conditions are $1.01 \times 10^5$ N m$^{-2}$ and 288 K and the stage efficiency is 89%. Find for the flow condition that the maximum value of Mach number (relative) is 0.95, the tip diameter, the blade angles at the tip section, the mass flow rate, the stage pressure ratio, and the blade angles at the root section.

8.5 A 12-stage axial flow compressor is to compress helium gas at the rate of 350 kg s$^{-1}$. The overall pressure ratio is to be 3.75, the stages are equally loaded, the design efficiency is to be 85%, and the inlet stagnation conditions are 5 bar and 350 K. Determine for the fifth stage the hub diameter, the tip and hub velocity triangles, and the blade angles and stagger angle at the tip section if the tip chord is 150 mm and the incidence angle is 2°. It may be assumed that the rotational speed is 3000 rpm, the tip diameter is 3.5 m and constant through the machine, the blading follows the free vortex principle and is 50% reaction in design at the tip section, the axial velocity is 120 m s$^{-1}$ and constant through the machine, and circular arc camber lines have been used with a radius of 500 mm at the tip section.

# 9 Radial flow turbines

## 9.1 Introduction

The flow path in radial turbines may be centrifugal as in the Ljungström design or, more commonly, centripetal as in most of the machines found in both water power installations and in gas power generation plant. The two types of flow path are illustrated in Figs 9.1–9.3.

The water turbine in Fig. 9.1 is a typical Francis design, and may be installed vertically as shown or horizontally depending on the station. The head or level range applied to these machines at present is from 10 m to over 500 m, the power developed ranging from 10 kW to over 450 MW, with runners varying in diameter to well over 6 m. Some of these machines are used in pumped storage schemes as will be described in Chapter 10, but do not differ in essentials from those used purely for power generation.

The inward flow gas turbine in Fig. 9.2 may be used in turbochargers, in small gas turbines, or as expanders in the cryogenic industry. They are rugged in design and, particularly when fitted in turbochargers, may rotate at speeds over 100 000 rev min$^{-1}$. Their flexibility in terms of flow range and power range control is much less than the axial turbine designs, so they tend only to be used for small power installations.

The outward flow machine is less common, but has been used to some effect with steam as the large increase in specific volume is more easily allowed for. The Ljungström turbine (Fig. 9.3) is one example of this design, and has been supplied to generate powers of up to 100 MW, but is offered for small inhouse generating plant delivering 10–35 MW typically, and for pass-out turbine applications.

In the following sections these machines will be discussed in the order of this introduction, with a worked example to assist understanding.

## 9.2 Water turbines

The early designs proposed by Francis were purely radial, with both guide vanes and runner blades disposed in the radial plane, and with the rotor

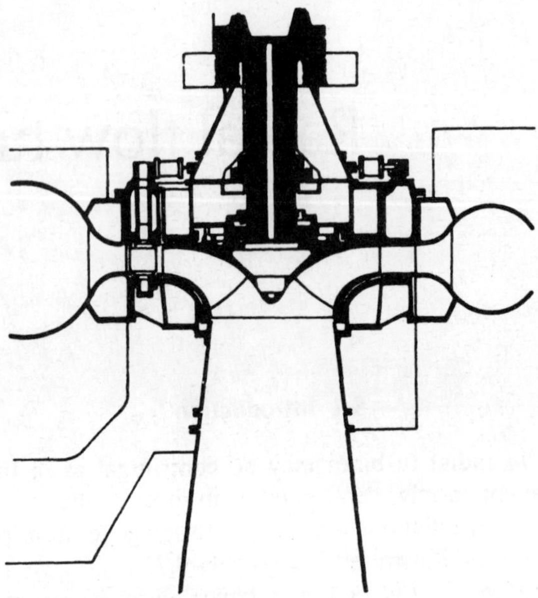

**Figure 9.1** Typical Francis turbine.

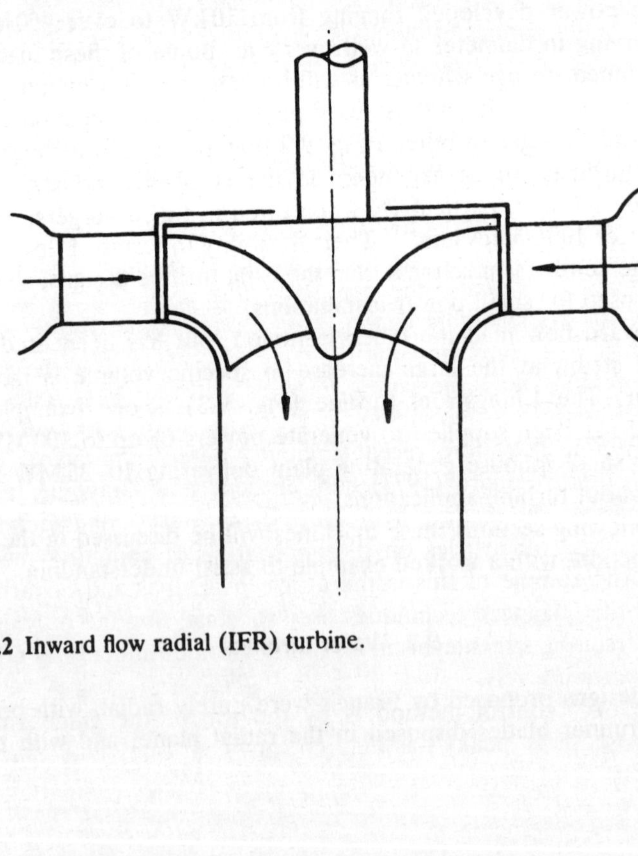

**Figure 9.2** Inward flow radial (IFR) turbine.

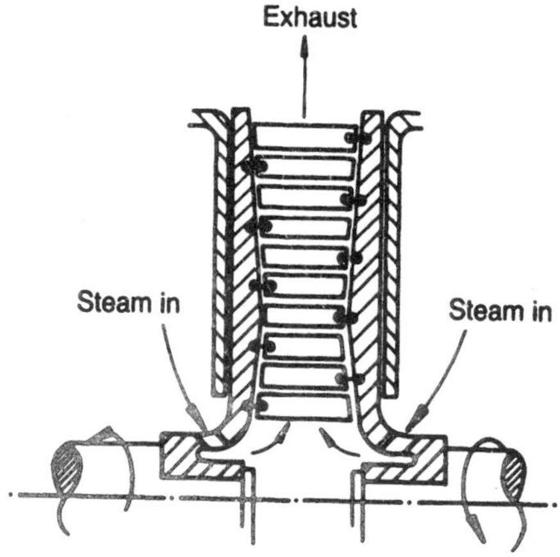


**Figure 9.3** Outward flow radial turbine.

having no shroud as in the Kaplan and propellor turbines. As heads and powers increased, the flow path became completely three dimensional, with the runner passages being mixed flow as Fig. 9.1 indicates. The machine is still known as the Francis turbine.

Francis machines are used for a wide range of level differences from 50 m to well over 500 m, which is into the Pelton impulse turbine range. Figure 9.4 shows how the elevational profile changes with specific speed, the lower value being associated with the highest level drop through the turbine. Since the flow path is three dimensional, the stream surfaces may be established in any of the ways outlined in Chapter 5. Some method like that discussed in section 7.2 for the Kaplan is used to relate guide vane setting angles to the runner inlet angles, and also the inlet and outlet angles of the rotor blades, since there is a considerable curvature of the stream surfaces and of the leading and trailing edges shown in Fig. 9.4. The shapes of both guide vanes and runner blades follow the same principles as those in the axial turbines, and the routine step by step method that may be used is well illustrated by Nechleba (1957). The more recent computer-based approach is discussed by, among others, Pollard (1973) and Chacour and Graybill (1977). The house journals of the leading hydroelectric machinery manufacturers may also be consulted for articles on the application of computer methods to machines. An example of this is the article by Dubas and Schuch (1987). This details the advanced techniques used for analysing the pressure distribution and resulting stress levels in a Francis turbine runner, and predict the probable vibration levels.

Since the only control method is by the use of the guide vanes, the efficiency envelope is 'peaky', as Fig. 9.5 shows for one guide vane setting.

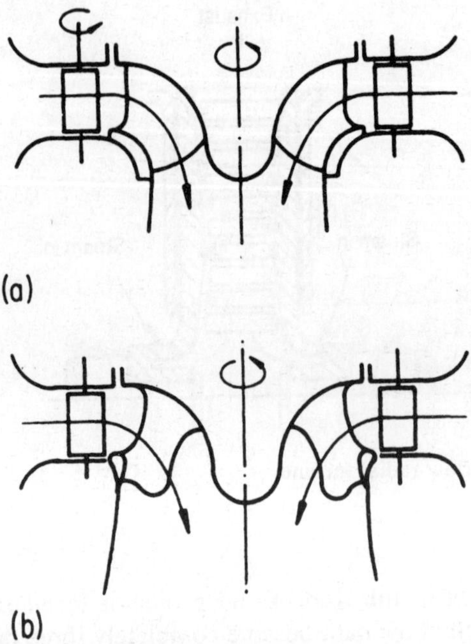


**Figure 9.4** Illustration of the typical Francis turbine runner shape: (a) for low specific speeds; (b) for high specific speeds.

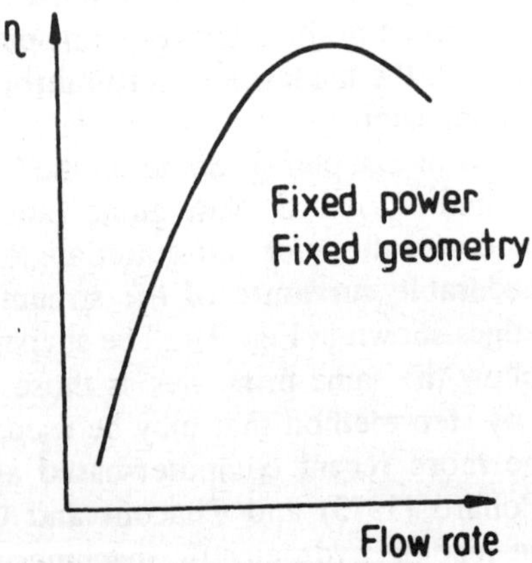


**Figure 9.5** Efficiency envelope.

The envelope is not so wide as that of a Kaplan machine, but the maximum efficiency compares very well, as Fig. 9.6 demonstrates.

Operating problems are discussed in the house journals and in such journals as *Water Power and Dam Construction*. Typical of the former are

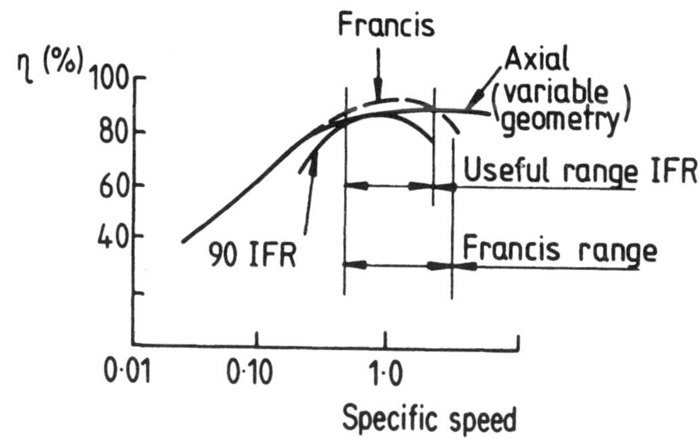


**Figure 9.6** Comparison of the Francis and IFR efficiency envelopes.

papers by Grein and Staehle (1978), Anghern, Holler and Barp (1977) and Grein and Bachmann (1976) who detail the development problems associated with these machines.

### 9.2.1 Francis turbine problem

Figure 9.7 illustrates a Francis turbine. When the angular velocity was $29.3\,\text{rad}\,\text{s}^{-1}$ and the flow rate was $12\,\text{m}^3\,\text{s}^{-1}$ the total energy at inlet to the guide vanes was $500\,\text{J}\,\text{kg}^{-1}$, the pressure loss through the guide vanes was estimated to be $0.3 \times 10^5\,\text{N}\,\text{m}^{-2}$, the pressure loss through the runner passages was estimated to be $0.15 \times 10^5\,\text{N}\,\text{m}^{-2}$, and the pressure at inlet to the runner was found to be $3.4 \times 10^5\,\text{N}\,\text{m}^{-2}$. The pressure at inlet to the draft tube was measured to be $0.86 \times 10^5\,\text{N}\,\text{m}^{-2}$. Estimate the hydraulic power and the necessary setting angle for the guide vanes referred to the tangential direction, if the hydraulic efficiency may be assumed to be 94%.

Draft tube inlet area $= 1.25^2/4 = 1.227\,\text{m}^2$. Therefore,

$$V_3 = 12/1.227 = 9.78\,\text{m}\,\text{s}^{-1}$$

Energy levels through the flow path:

$$\begin{aligned}
\text{total energy at inlet to guide vanes} &= 500\,\text{J}\,\text{kg}^{-1}\\
\text{static energy at inlet to runner} &= 340\,\text{J}\,\text{kg}^{-1}\\
\text{static energy at inlet to draft tube} &= 86\,\text{J}\,\text{kg}^{-1}
\end{aligned}$$

Applying the energy equation to the guide vane passages

$$\begin{aligned}
500 &= 340 + V_2^2/2 + 0.3 \times 10^5/10^3 + 1.5g\\
V_2 &= 15.185\,\text{m}\,\text{s}^{-1}
\end{aligned}$$

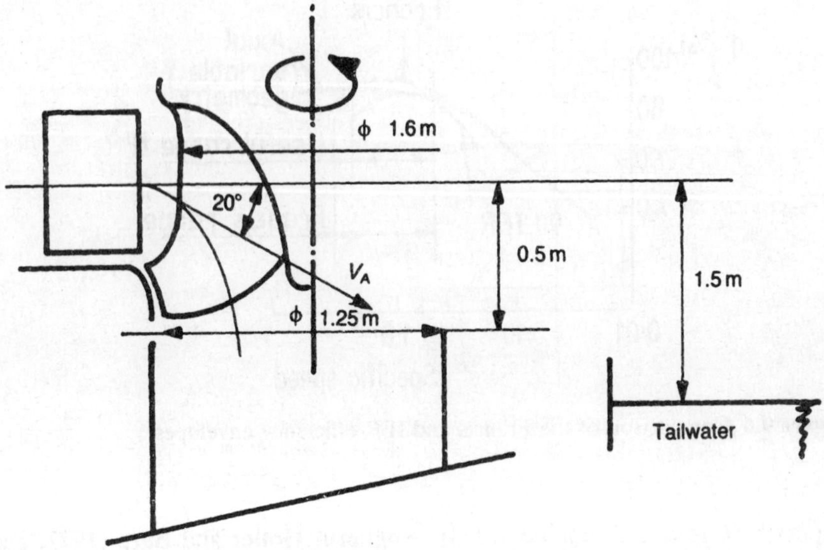


**Figure 9.7** Francis turbine problem.

Similarly, across the runner,

$$340 + \frac{15.185^2}{2} + 1.5g = 86 + \frac{9.78^2}{2} + 1g + \frac{0.15 \times 10^5}{10^3} + \text{work done}$$

thus

$$\text{work done} = 311\,\text{J kg}^{-1}$$

The hydraulic power = work done × mass flow = 3.74 MW. Since $\eta_{\text{hyd}}$ = 0.94:

$$\text{shaft power} = 3.51\,\text{MW}$$

and

$$gH_e = \frac{311}{0.94} = 330.85 = u_2 V_{u2} - u_3 V_{u3}$$

Taking zero outlet whirl conditions, since $u_2 = 23.44\,\text{m s}^{-1}$

$$V_{u2} = 14.11\,\text{m s}^{-1}$$

The inlet velocity triangle becomes as shown in Fig. 9.8, and

$$\alpha_2 = 21.69°, \qquad \beta_2 = 31.02°$$

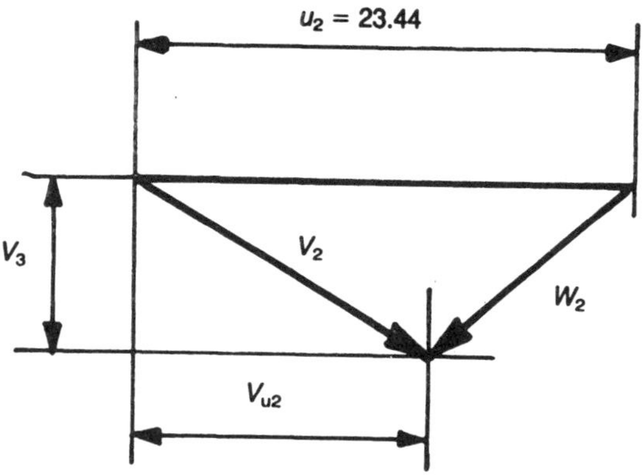


**Figure 9.8** Velocity triangles for Francis turbine problem.

Using the setting angle correction shown in Fig. 7.7,

$$\alpha' = 20.49°$$

## 9.3 Radial inflow gas turbine

This machine, used extensively in turbochargers and in cryogenic applications, is illustrated in Fig. 9.2. The flow path is centripetal, and unlike the Francis design the majority of turbines have fixed geometry nozzles and runners, the output being varied by altering gas flow rate in cryogenic systems and by controlling fuel rate in gas turbines. The machines are, like the Francis turbine, typically single stage, having a similar efficiency envelope, and are much used in low flow applications where their overall efficiency is better than the equivalent axial machine.

Typical idealized velocity triangles and $h$–$s$ charts were discussed in section 1.7, as were the common efficiency definitions. The conventional 90° inlet angle for the rotor blades gives rise, when zero outlet whirl is assumed, to large outlet angles in the exducer section (Fig. 9.9). The velocities also tend to be large, as the worked example that ends this section shows, and this leads to one of the limiting considerations discussed in section 9.3.2.

Approaches to loss correlation and design criteria follow those outlined in Chapter 8 for water turbines, and studies by Benson (1970) and by Rohlik (1968) supplement the working formulae proposed by Balje (1981).

### 9.3.1 Nozzle systems

Two nozzle systems are available, guide vanes and a vaneless space. In the former case the guide vane angle seems to be in the range 10–30°, but there

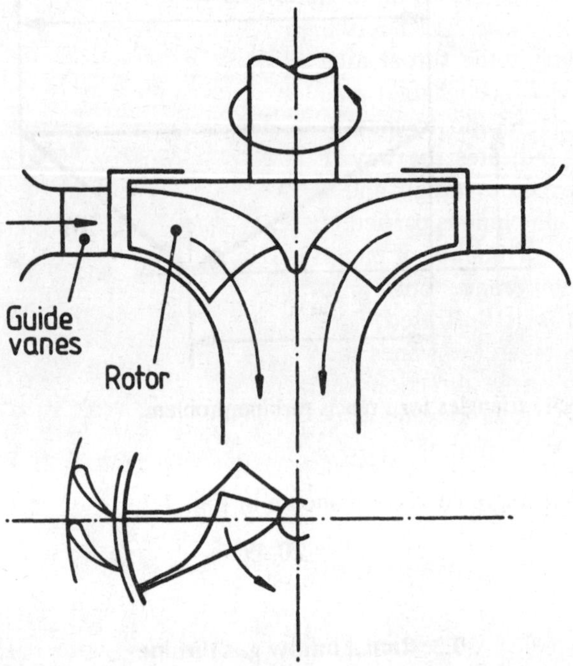


**Figure 9.9** Sketch of an IFR turbine showing the twist in the exducer.

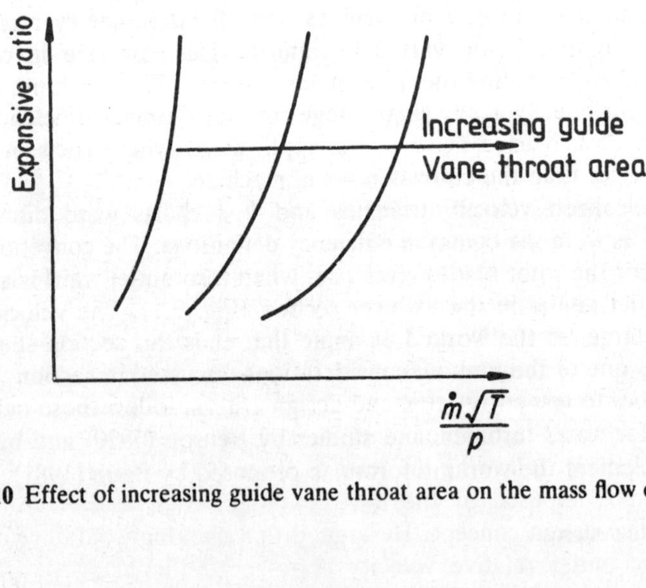


**Figure 9.10** Effect of increasing guide vane throat area on the mass flow coefficient.

is need for correction to allow for flow in the vaneless space between the vanes and the rotor, as Knoerschild (1961) has discussed. The most important parameter is the throat area (the minimum area), and this is the approach used by turbocharger manufacturers. For example, Fig. 9.10 shows a plot of pressure ratio to mass flow parameter for four nozzle areas; it is typical, and indicates the way that the mass flow range of the machine is affected by nozzle design and geometry. The way in which turbine and compressor matching is carried out using changes in geometry in the nozzles is discussed in a number of papers presented at an Institution of Mechanical Engineers conference on *Turbocharging* held in 1978, and the various methods are well illustrated.

Where there are no vanes the flow follows the Archimedian spiral path approximately, the direction being modified by real fluid effects, but the height of the nozzle passage is used to control the area imposed on the gas in the same way that the nozzle ring is manipulated. The smaller turbochargers use the vaneless designs as they are cheaper to produce, and give a wider range of flows even though the efficiency is somewhat reduced.

### 9.3.2 Rotor geometry

Apart from the flow direction, the flow problems met are the same as those found in centrifugal compressors. The blade angle at inlet is normally 90°, but secondary flow is a problem as in the compressor (Fig. 9.11) so that there is a balance needed between control and friction loss due to too many vanes. Jamieson (1955) Wallace (1958) and Hiett and Johnston (1964), among others, have considered the choice of vane numbers. Jamieson suggested the minimum number to avoid flow reversal in the passages, and Fig. 9.12 is based on his correlation, suggesting the minimum number of vanes against nozzle angle. Wallace used a more rigorous approach, and comments that more vanes are needed than Jamieson's minimum. Both approaches suggest a large number of vanes at the inlet, but flow blockage considerations mean that only half the blades continue through the exducer, the others stopping short as illustrated in radial compressors. Hiett and Johnston (1964) reported tests on a rotor with a nozzle angle of 77° and a 12-blade rotor that gave a total to static efficiency of 84% measured at optimum flow conditions. They found that doubling the number of blades gave an improvement in efficiency of 1%, which suggests that the data of Fig. 9.12 form a reasonable compromise.

Wood (1963), following the lead of Balje (1981), discussed the limits on performance imposed by the flow conditions that resulted from the zero outlet whirl design concept. He argued that choking would occur when the maximum outlet relative velocity approached the local speed of sound. Wood produced the plot (Fig. 9.13) of stagnation pressure ratio against a ratio $(V_3/V_0)^2$, showing two lines, one for a Mach number of unity and one

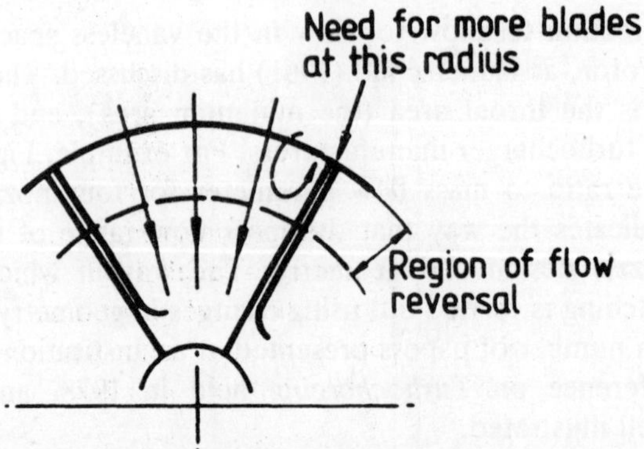


**Figure 9.11** Need for flow control at inlet to an IFR rotor.

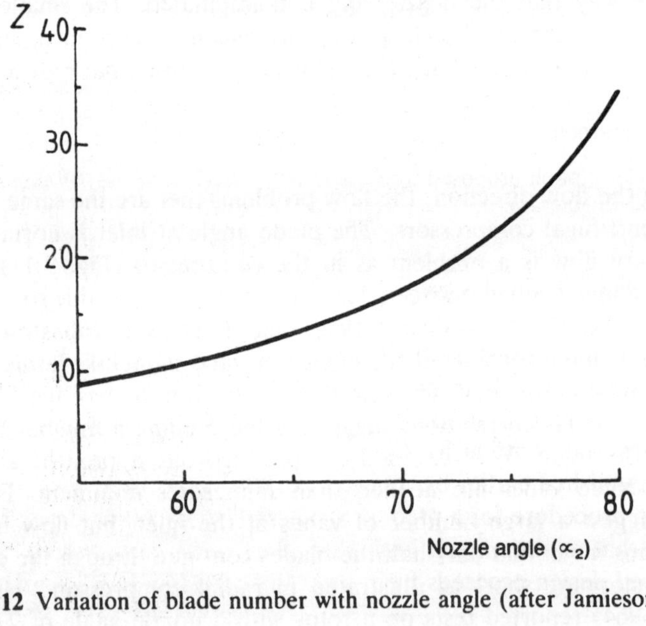


**Figure 9.12** Variation of blade number with nozzle angle (after Jamieson, 1955).

for a more typical value of 0.7 which is better for use in machine design. The velocity $V_0$ is the so-called spouting velocity, defined as

$$V_0 = \sqrt{[2C_p(T_{01} - T_{03ss})]} \tag{9.1}$$

(the two temperatures are defined in Chapter 1). The plot is for air as the working fluid, and an efficiency of 90% was assumed, but the principle is well illustrated. It may be noted that this attempt at correlation is based on the design condition of zero outlet whirl, and that there is very little in-

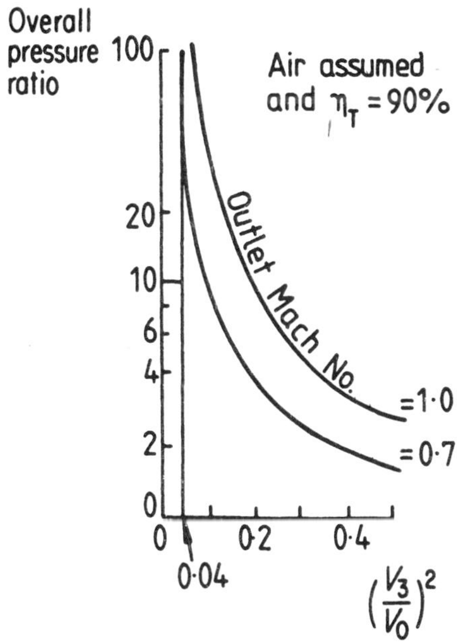


**Figure 9.13** Correlation proposed by Wood (1963; courtesy of the American Society of Mechanical Engineers).

formation available on the probable effects for other flow conditions that may occur.

It is not possible in this discussion to do anything but introduce the basic theory of machines which are the essential component of modern high speed turbochargers. A recent contribution to an analytical approach to optimizing such turbines was written by Chen and Baines (1992). Using data and techniques published earlier by their co-workers and others, they present an optimization procedure for a preliminary design study which depends on the proposition that minimum exit loss is associated with zero outlet whirl from the rotor, and design proceeds from a specified loading coefficient.

### 9.3.3 Worked example

An inward flow radial gas turbine is provided with a nozzle ring designed to give a nozzle angle of 25° to the tangent. The impeller diameter is 160 mm, the exducer maximum diameter is 115 mm and the hub : tip ratio at outlet is 0.2. The rotor vanes are radial at inlet, and the design was based on the 50% reaction concept. When running at 50 000 rev min$^{-1}$ the inlet stagnation conditions were found to be $3.6 \times 10^5$ N m$^{-2}$ and 1100 K. Assuming a nozzle efficiency of 93% a total to static efficiency of 82.5%, suggest the probable mass flow rate, the exducer blade angles at root and tip, and the maximum

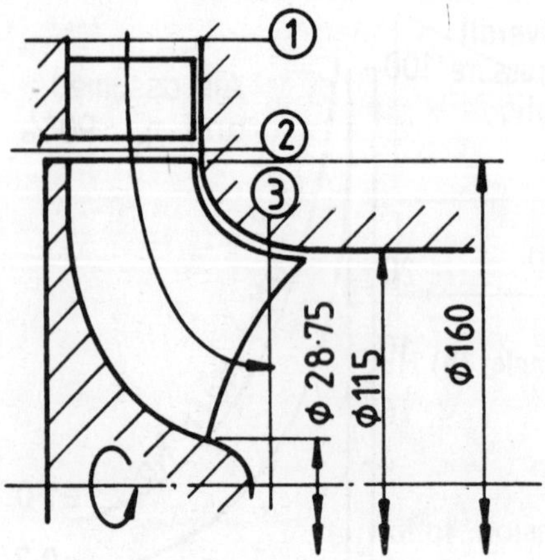


**Figure 9.14** Worked example – flow path.

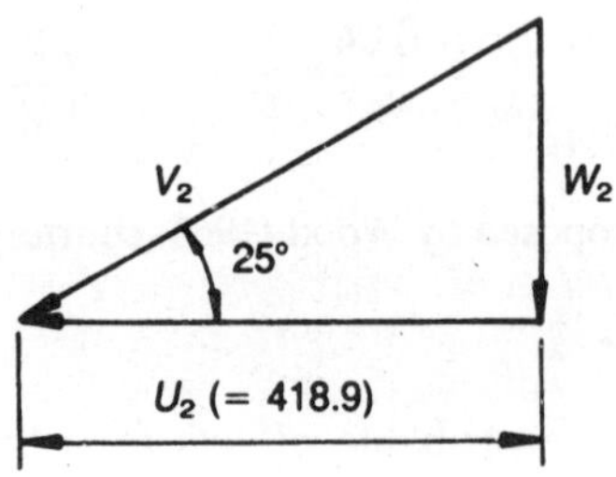


**Figure 9.15** Worked example – velocity triangle.

Mach number at outlet. It may be assumed that the axial velocity in the outlet is $150\,\mathrm{m\,s^{-1}}$ ($C_p = 1.145 \times 10^3\,\mathrm{J\,kg^{-1}}$).

The flow path is shown in Fig. 9.14, and it is assumed that there is zero outlet whirl: $u_2 = 418.9\,\mathrm{m\,s^{-1}}$; $u_{3\ \mathrm{tip}} = 301.1\,\mathrm{m\,s^{-1}}$; $u_{3\ \mathrm{hub}} = 60.22\,\mathrm{m\,s^{-1}}$.

Following the velocity triangle in Fig. 9.15 and assuming that there is no deviation of flow from the radial direction, the velocity $V_2$ from the nozzle is found from the equation

$$V_2 = \sqrt{(\eta_n 2 C_p \Delta T_{\mathrm{actual}})}$$

From the triangle, $V_2 = 462.2\,\mathrm{m\,s^{-1}}$ and $W_2 = 195.34\,\mathrm{m\,s^{-1}}$. Thus $\Delta T_{\mathrm{actual}} = 100.31\,\mathrm{K}$, and since 50% reaction applies the total temperature drop over the turbine is 200.62 K. Therefore, since $T_{01} = 1100\,\mathrm{K}$,

$$\eta_{\mathrm{TS}} = 0.8 = 200.62/1100 - T_{3\mathrm{s}}$$

$$T_{3\mathrm{s}} = 849.23\,\mathrm{K}$$

$$T_{03} = 899.38\,\mathrm{K}$$

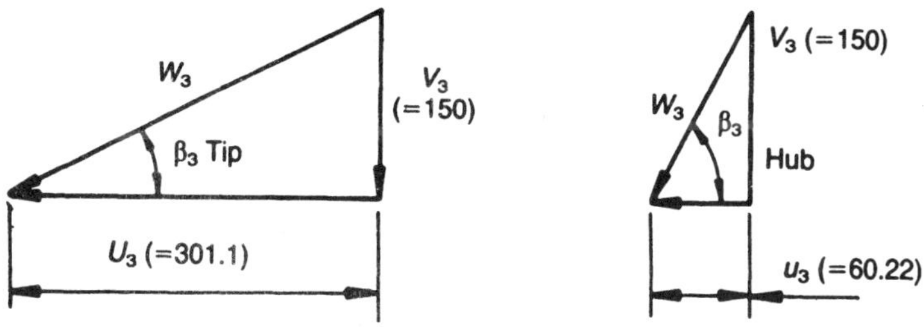


**Figure 9.16** (a) Tip triangle; (b) Hub triangle.

With isentropic expansion, following Fig. 1.26, $(k/k - 1 = 4)$

$$\frac{p_{01}}{p_3} = \left(\frac{T_{01}}{T_{3s}}\right)^4 = \left(\frac{1100}{849.23}\right)^4$$

and

$$p_3 = 3.6 \times 10^5 \left(\frac{849.23}{1100}\right)^4 = 1.28 \times 10^5 \,\mathrm{N\,m^{-2}}$$

Since $V_3 = 150\,\mathrm{m\,s^{-1}}$,

$$T_{03} = T_3 + 150^2/2C_p$$

$$T_3 = 889.56\,\mathrm{K}$$

$$\rho_3 = \frac{1.28 \times 10^5}{287 \times 889.56} = 0.501\,\mathrm{kg\,m^3}$$

Thus

$$\dot{m} = 0.051 \times 150 \times \pi/4(0.115^2 - 0.02875^2)$$
$$= 0.732\,\mathrm{kg\,s^{-1}}$$

For the exducer section, the velocity triangles are as in Fig. 9.16. Consider the maximum diameter first: $\beta_3 = 26.48°$ and $W_3 = 336.39\,\mathrm{m\,s^{-1}}$. Since the acoustic velocity is

$$a = 1.33 \times 287 \times 889.56 = 582.71\,\mathrm{m\,s^{-1}}$$

the maximum Mach number is 0.577.

For the hub diameter, $\beta_3 = 68°$ and $W_3 = 161.64\,\mathrm{m\,s^{-1}}$.

The change in $\beta_3$ along the exducer trailing edge is typical of the usual twist in the vanes needed at the rotor outlet.

## 9.4 Ljungström or radial outflow turbine

The working fluid in this machine is usually steam, which enters the centre and then flows outward through rows of blades that are arranged so that alternate rows rotate in opposite directions (Fig. 9.3). The two discs each drive an alternator. The arrangement offers a compact design of short length, but because of overhang problems has not been developed to provide the large powers that the axial machines currently deliver. The principles and practice of the design are covered by Kearton (1958) and a simplified discussion only will be given here.

If it is assumed that all blades have the same profile, camber and spacing, apart from the first and last rows, that the velocity ratio is the same for all rows, and that the clearance radially between the rows is neglected, a simple outline of theory may be attempted. Consider the three rows shown in Fig. 9.17, and the velocity triangles that are also drawn there. The inlet triangle is based upon the assumption that there is zero inlet whirl, and the others have been drawn using the assumptions just stated. The triangles have been drawn with the blades in the manner shown to emphasize the interrelations.

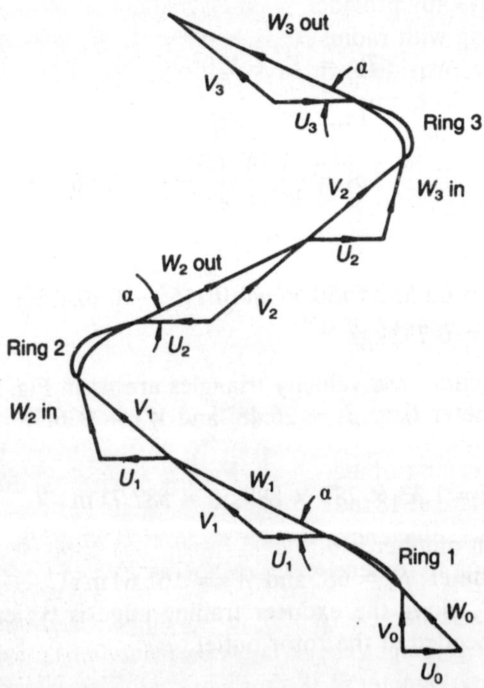


**Figure 9.17** Build-up of velocity triangles for three rings of moving blades in a Ljungström-type machine.

Appyling the Euler equation for the first row, the energy change, since $V_{u0}$ is zero, is

$$gH_1 = U_1V_{u1}$$

For the second row the energy change is

$$gH_2 = U_2V_{u2} - (-U_1V_{u1}) = U_2V_{u2} + U_1V_{u1}$$

Similarly, for the third row,

$$gH_3 = U_3V_{u3} + U_2V_{u2}$$

and for each succeeding row up to row $n$,

$$gH_n = U_nV_{un} + U_{n-1}V_{un-1}$$

The sum of all the energy transfers is the total energy change from inlet to outlet. If the energy transfer for a stage for maximum utilization is considered, it can be shown that this condition is met for a radial stage when $U/W = \cos\alpha/2$ and $gH$ (maximum) $= 2U^2$, which is the same as for an impulse machine of the axial type. This demonstrates that, whereas two rows of blades are needed in an impulse axial turbine, only one is needed in the Ljungström machine.

Early designs were only provided with a radial path, with the height of the blade rows increasing with radius. However, with high superheat available, modern designs are provided with one or two stages of axially disposed blades at the maximum radius of the discs. A number of machines of up to 60 MW were installed by the UK Central Electricity Generating Board, but present designs offered by a number of companies appear to be of a size tailored to in-house generating plant for large factory sites; typical powers having been outlined in the introduction to this chapter. Kearton describes the ingenious techniques used to attach blade rows and allow for lateral and radial relative expansion and for the very long blades needed at outlet with radial stress and movement.

## 9.5 Exercises

9.1 A mixed flow water turbine of the Francis type installed with its axis vertical, is to rotate at 18 rad s$^{-1}$. Estimate the power developed and the overall hydraulic efficiency of the machine given the following information relating to dimensions and flow losses.

- Guide vanes: discharge angle 20° to the tangential direction elevation of mean passage height above tailwater level 4 m.
- Runner: inlet diameter 3300 mm, mean outlet diameter 2150 mm, blade inlet and outlet angles referred to the tangential direction are 85° and 25°, respectively, inlet flow area 3 m$^2$ outlet flow area 3.1 m$^2$.

- Draft tube: inlet 3.25 m above tailwater level, outlet area 11.5 $m^2$.
- Tests at the duty flow rate indicate that losses related to velocity were: guide vane loss, 0.05 × outlet absolute velocity energy; runner loss, 0.2 × outlet relative velocity energy; draft tube loss, 0.5 × inlet absolute velocity energy.

9.2 An inward radial flow gas turbine has been designed on the 50% reaction principle to give 100 kW at 50 000 rpm when the inlet gas stagnation pressure was $3 \times 10^5 \, N \, m^{-2}$ and the stagnation temperature at inlet was 850 K. When on test at the specified power the stagnation outlet temperature was found to be 670 K and the nozzle efficiency and total to static efficiencies were calculated to be 97% and 86%, respectively, and there was zero inlet whirl. The rotor outer diameter was 165 mm, the rotor blades were radial at inlet, the mean diameter of the nozzle throats was 175 mm, the passages 30 mm high and the blockage due to the blades was 15%.

Determine for the conditions stated (a) the gas mass flow rate, and (b) if the axial velocity at outlet is measured as $300 \, m \, s^{-1}$ find the outlet tip Mach number based on the relative velocity.

9.3 An inward flow gas turbine designed for the 50% reaction case is provided with a 150 mm rotor and is to rotate at 50 000 rpm. The nozzle guide vane outlet angle is 35° referred to the tangential direction and the rotor blades are radial at inlet.

When operating at the design point flow rate of $0.56 \, kg \, s^{-1}$ the inlet stagnation conditions were $3 \times 10^5 \, N \, m^{-2}$ and 850 K, and the nozzle and total to static efficiencies were calculated to be 97% and 83%, respectively.

For this operating condition, assuming zero outlet whirl, estimate the diagram power, the outlet pressure and temperature and the necessary nozzle guide vane peripheral throat area.

9.4 An inward flow gas turbine is provided with a vaneless casing and a rotor that is 150 mm in diameter. The outlet section has an outer diameter of 105 mm and a hub : tip ratio of 0.25. The design approach is to use 50% reaction, and the blades on the rotor are radial at inlet. When delivering 120 kW at 50 000 rpm the casing inlet stagnation conditions were $3.5 \times 10^5 \, N \, m^{-2}$ and 1000 K. If the inlet casing 'nozzle' efficiency is assumed to be 93%, the total to static efficiency is 83% and the shaft mechanical efficiency is 80%, suggest the exducer angles at the tip and hub sections and the maximum Mach number at outlet. Assume that that the inlet casing delivers the gas to the rotor at an average angle of 35° referred to the tangential direction. If a zero reaction design approach is used with the geometry nozzle velocity and power output

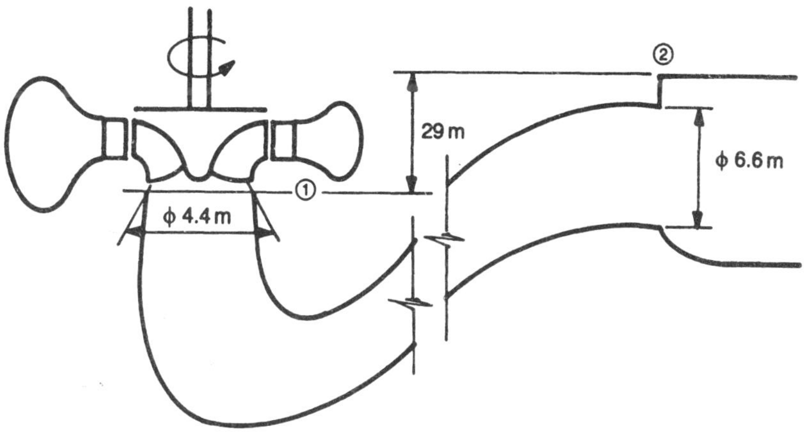


**Figure 9.18** Exercise 9.5.

quoted above, suggest the effect on the exducer section angles and gas flow through the rotor.

9.5 Figurc 9.18 shows part of the longitudinal section of a reversible pump–turbine in a hydroelectric station. When pumping, the unit delivers $74\,m^3\,s^{-1}$ under an entry head of 318 m when rotating at 333 rpm with an efficiency of 88.5%.

When generating, the flow is to be $76.5\,m^3\,s^{-1}$ with an efficiency of 93.5%. The draft tube efficiency when turbining is 80% and when pumping the friction factor $f$ in the formula $gH = 4\,fLV^2/2D$ may be assumed to be 0.005, $V$ being the mean between the inlet and outlet draft velocities, and the diameter $D$ the mean of the inlet and outlet diameters.

Determine the minimum pressure at inlet to the draft tube in $N\,m^{-2}$ and suggest the value of the Thoma cavitation parameter for this condition.

If a tenth-scale model is used to study performance suggest, when operated in the pumping mode, the flow rate, head rise, and power required, if the model speed is limited to 3000 rpm.

9.6 Figure 9.19 illustrates a Francis turbine installation. When commissioned the machine was supplied with a total head above atmospheric level of 100 m in the spiral casing, the runner speed was $29.3\,rad\,s^{-1}$ and the generated power was 50 MW. The computed 'wire to water' efficiency was 90%. Determine the flow rate, and the pressure at inlet to the runner and the draft tube.

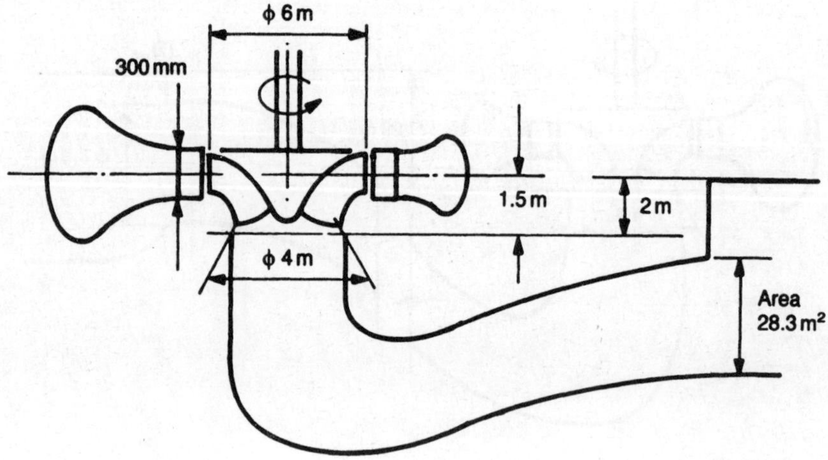


**Figure 9.19** Exercise 9.6.

It may be assumed that the draft tube efficiency is 83% and there is zero inlet whirl, the guide vane setting relative to the tangential direction is 25°, and the loss through the guide vanes is estimated to be 0.06 × absolute kinetic energy at inlet to the runner.

# 10 Special machine applications

## 10.1 Introduction

This concluding chapter discusses the effects of viscosity change, of fluid properties, and of gas content on the operation of pumps, the ways in which output control of machines can be achieved, the steady state interaction between pumps and systems, and the hydraulic problems to be solved in pumped-storage systems.

## 10.2 Problems involved in special pumping applications

### 10.2.1 Gas suspension problems

Pumps find difficulty in coping with gas in suspensions and tend to vapour lock and cease to pump when gas fills up the suction area. Merry (1976) illustrates the effect on pump performance of air in measured quantities, and the dramatic effect on the curves is clearly demonstrated. A general rule often followed is that if air is present in a proportion above about 10% the pump performance is unacceptable. In the process industries, problems arise in plant when the pump is shut down for a period and heat soaks from the pump case into the fluid. The local temperature rises to a level that causes vaporization and a local pocket of gas forms; this causes the effect known as vapour locking, and the pump will not pump. Cooling passages in the pump casing are usually provided to avoid the problem with those chemicals susceptible to this problem. In other cases provision is made to ensure that fluid is not retained in the casing.

Where, as part of the system operation there is a risk of air ingestion, as in site water-level maintenance pumping equipment, for instance, special provision is made to ensure that air is evacuated from the pump, continuously if necessary. Arrangements which do this are known as priming systems, and are of various types. The simplest consists of a tank in the suction line, and the pump draws from its bottom, so if flow is stopped enough liquid is retained in the tank to allow the pump to attain its normal suction pressure and thus stimulate flow from the suction source. More sophisticated designs have a special pump casing, with a large chamber in the suction zone which retains liquid. This liquid is drawn into the impeller on start-up and is delivered to the volute from where it recirculates, after giving up entrained gas, to the suction to entrain gas from the empty suction line, and then passes through the impeller. This process goes on until the suction line is full of liquid (the pump is primed) and the pump gives a normal delivery flow rate. There are a number of patented recirculation systems, described by Rachmann (1967), and reference to this contribution and to pump handbooks may be made for more detail. These types are called self-priming as they rely on the dynamic action of the impeller.

A special pump design which is inherently self-priming is the regenerative pump (Fig. 10.1). The rotor has radial channels machined in opposing faces, and the casing has channels provided as can be seen. Flow spirals in a manner similar to the fluid flywheel. The pump as applied to bilge pumping aboard ship is an example of self-priming action. Papers by Burton (1966), Yamasaki and Tomita (1971) and Tomita, Yamasaki and Sasahara (1973) may be consulted for further reading.

For many years contractors' pumps have been provided with external priming systems. Typically the diesel engine driving the pump provides air extraction through its exhaust manifold. A cut-off valve operates when the

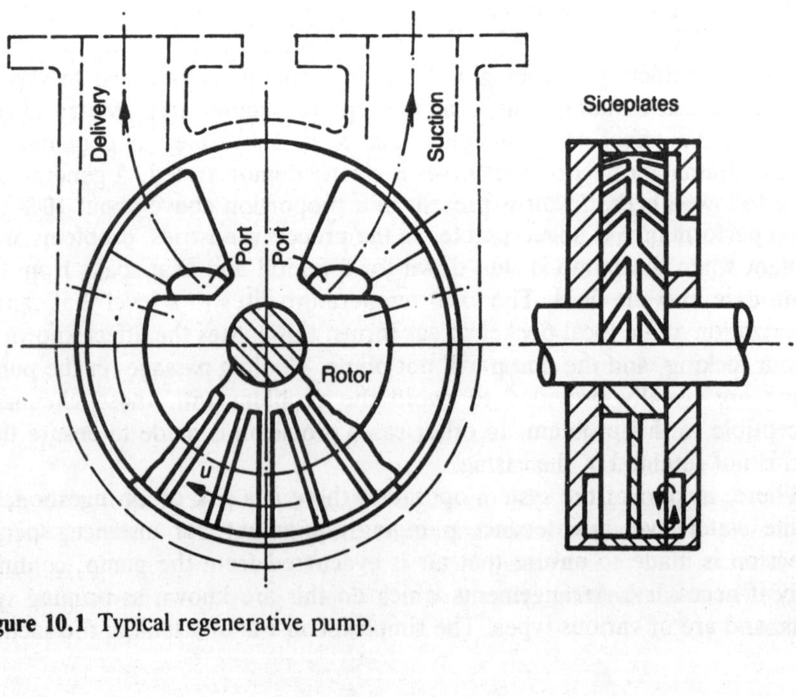


**Figure 10.1** Typical regenerative pump.

pump is primed, preventing water ingress to the engine. Jet pump extraction systems are used and, in vehicle-mounted fire pumps, liquid ring pumps (Fig. 10.2) are used to provide prime.

### 10.2.2 Liquid–solid suspension pumping

Gravel, limestone, coal and power station ash are now pumped as a suspension in water over considerable distances. In addition, sewage and chemical suspensions need to be pumped. Some of the solids are aggressive and some of the liquids are corrosive, so that pump parts need to be able to resist abrasion, corrosion and erosion. The pump designer has therefore to provide for wear, to give a suitable service life and an additional problem is to design to avoid blockage.

Figure 10.3 illustrates typical solids handling impellers designed to avoid blockage; if a solid mass can get into the impeller it will pass through. Typical centrifugal machines are similar to clean-water pumps but, with the special impellers mentioned, the provision of easy access (split casings) and easily replaced casing components in areas of high wear. A special type of machine (the 'Egger' Turo design) with a recessed impeller is shown in Fig. 10.4.

Pumps may be constructed of 'Ni-hard' or of chromium–molybdenum cast iron, which can be hardened after machining for longer life. Linings of rubber, nylon, polyurethane and more exotic materials have been used, but the consensus appears to be that they are not often cost effective, so that metallic materials tend to be favoured. Ceramic sleeves are used in some designs to protect the shaft from abrasion under the seals, and give good service; the material is not suitable in moving parts owing to its brittle nature.

Reference may be made to contributions by Burgess and Reizes (1976), Odrowas–Pieniazek (1979), Ahmad, Goulas and Baker (1981), Willis and Truscott (1976) and Bain and Bonnington (1970) for background information. For current developments the continuing series of hydro-transport conferences may be consulted.

### 10.2.3 Effect of viscosity change

The discussions in previous chapters have been devoted to fluids that have viscosities similar to that of water. Figure 10.5 illustrates the fall-off in efficiency and pressure rise as viscosities increase. The rapid fall-off in efficiency begins when kinematic viscosities exceed 100 centistokes. This effect arises due to increased friction, so that as well as the loss in performance there is a heating effect that can have undesirable effects on the seal system.

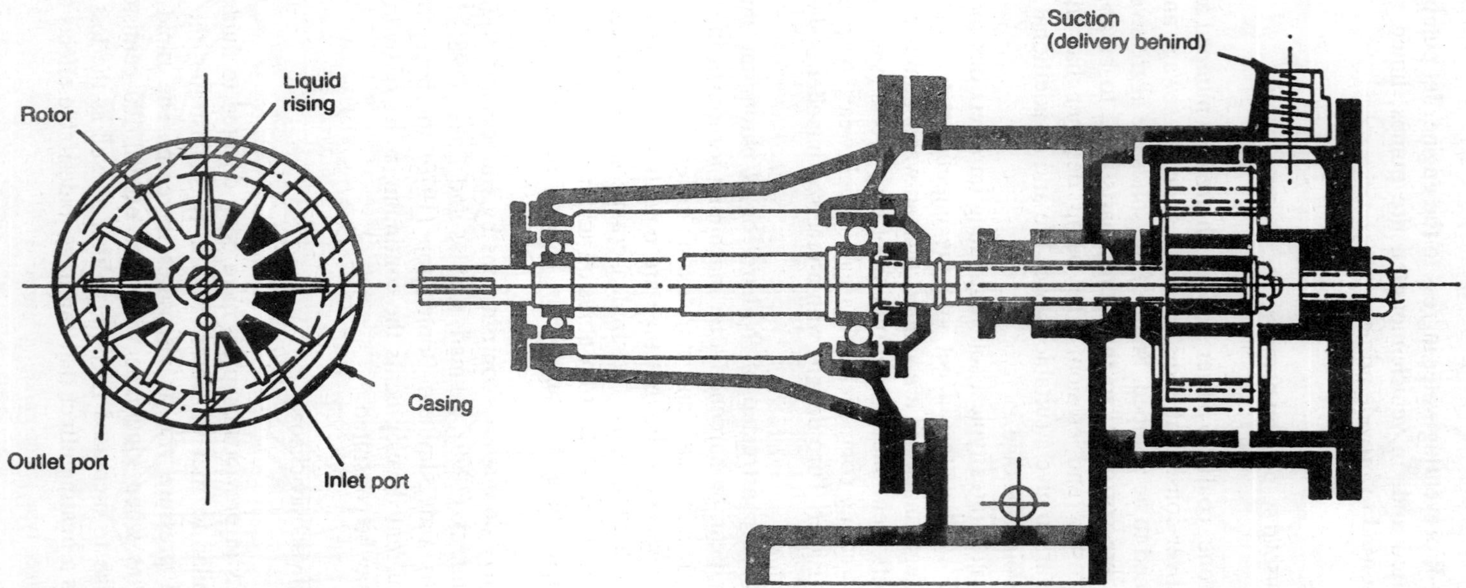


**Figure 10.2** Liquid ring pump.

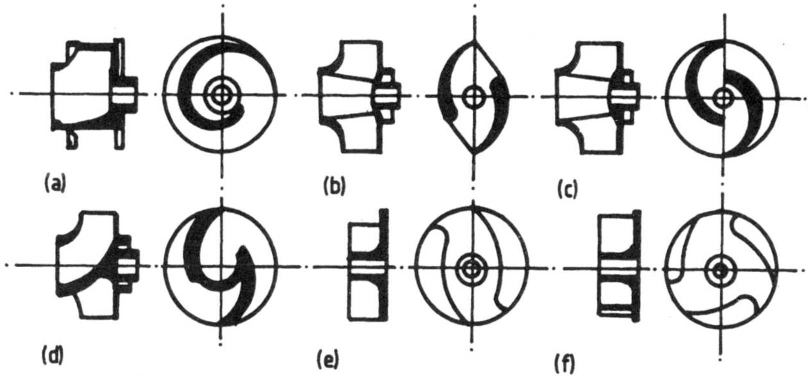


**Figure 10.3** Range of solids-handling impellers.

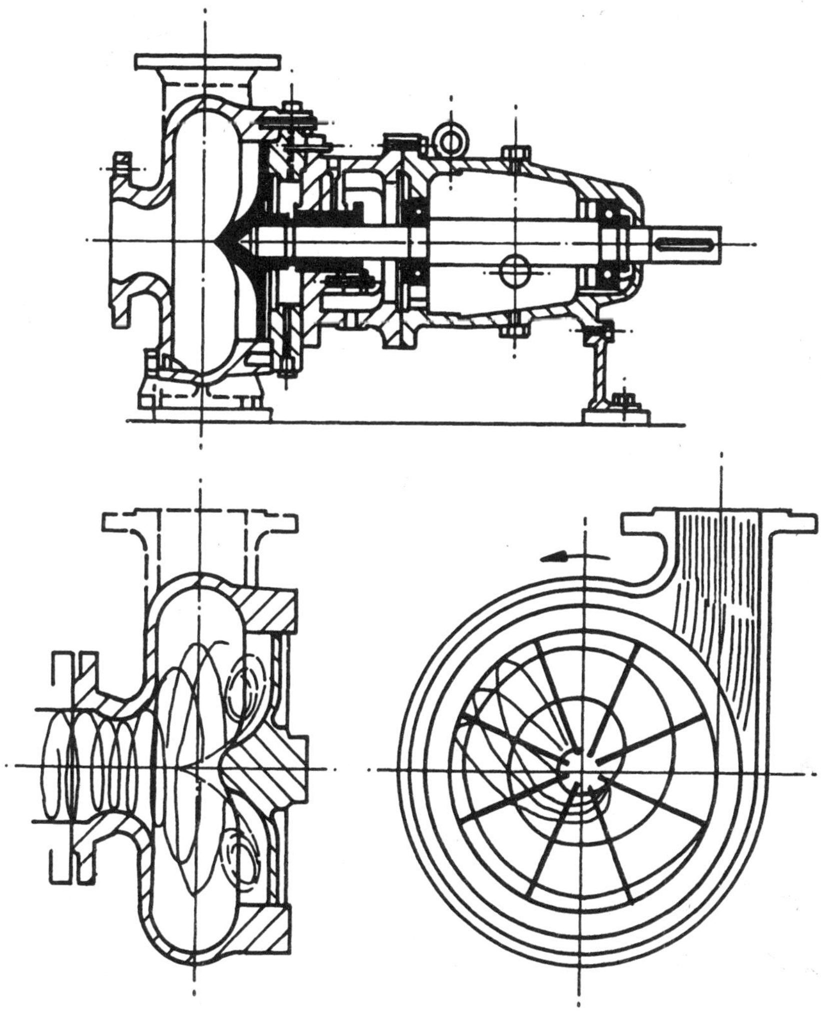

**Figure 10.4** Egger type solids-handling pump.

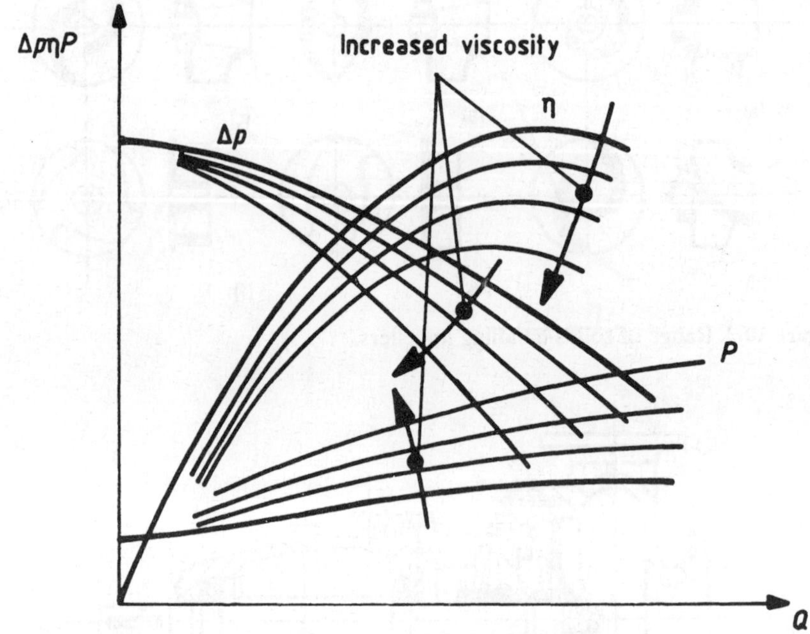


**Figure 10.5** Effect of viscosity change on constant speed centrifugal pump characteristic behaviour.

## 10.3 Pumped storage systems

The size of the generating units supplying electricity grid systems has steadily increased in the pursuit of efficiency, so that the base load is now being supplied in the UK by a few large stations, some of which are nuclear. The efficiency of these units is high, but the fuel burn of a thermal station is higher at half load than at full load, so that it is not economical to run at anything but full load. This poses problems for the grid control and planning engineers, for as Fig. 10.6 shows, the grid demand varies throughout the day from very low levels during the night period to two peak periods during the normal day. There is an additional problem posed by sudden changes in demand, such as those resulting from sudden drops in ambient temperature or from surges in consumption at the end of popular TV programmes. Keeping plant in spinning reserve to respond is expensive, and response is slow due to the need to avoid thermal shock. One solution to the problem is the installation of pumped storage hydroelectric plant, where power from the grid at times of low demand is used to pump water into high-level storage reservoirs, which is then allowed to flow through turbines at time of peak loading on the grid. Two advantages are offered – fast response to demand, and the use of off-peak electricity to provide storage capacity.

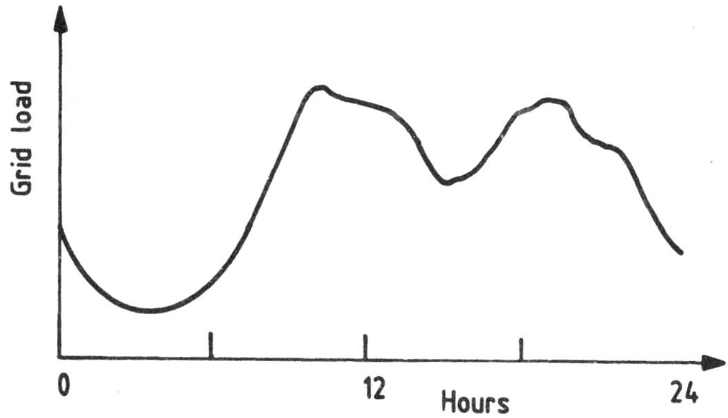


**Figure 10.6** Typical load variation on the National Grid over 24 hours.

Early installations used the four-machine layout, that is, a turbine, an alternator, a pump and a motor; both hydraulic machines were arranged to run at the optimum efficiency. Many plants have been used with very high level differences, so that multistage pumps provide storage, and Francis or Pelton turbines absorb the water power available. More recently three-machine layouts have been used, both vertical and horizontal installations being employed, with a single-electrical machine doubling as alternator and motor, and using separate turbine and pump machines. Typical of these is the station at Blaenau Ffestinniog in North Wales, provided with a Francis turbine designed to give about 79 MW and a multistage pump to lift water to a storage reservoir about 300 m above the lower lake. This machine string is vertical, with the pump below the turbine for cavitation submergence reasons; a coupling is installed that allows the turbine to generate without having the drag of the pump, which is dewatered and filled with air during the power generation cycle. In other sets the pump is always run in air during the generation phase, but in some not disconnected. Other sets are horizontal. For example, the machines at Vianden, which form phase one of the station, are placed with the electrical machine between the turbine and the pump, and a small Pelton turbine is provided to reduce electrical loading during the start-up period. The operating sequence is shown in Fig. 10.7.

Alternative machine layouts are illustrated in Fig. 10.8. As can be seen, the two-machine arrangement permits lower excavation costs and less expensive machinery because only one hydraulic unit is used. Extra electrical equipment is needed; a phase changer is required to change over the direction of the turbine–pump when moving from the pumping mode to the generating mode. There is a compromise required between efficiency when pumping or turbining, and a number of studies have been published on the problem. The pump–turbine, when generating, has a lower efficiency than a

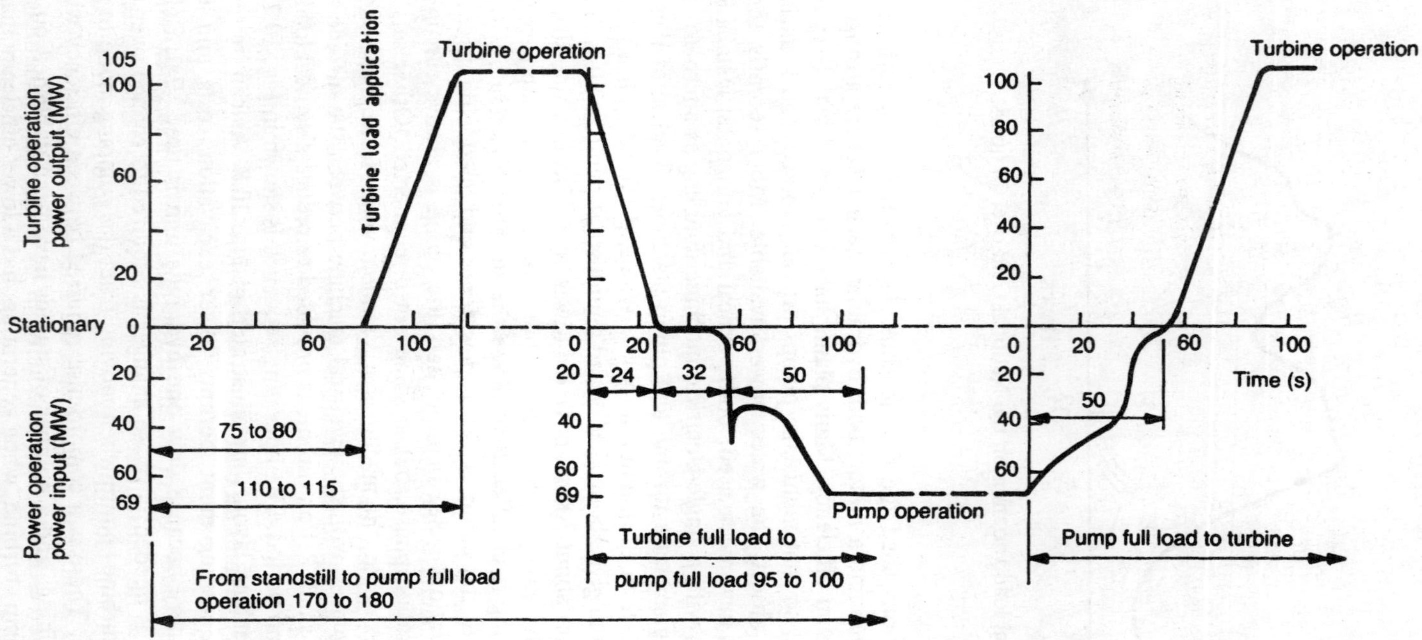


**Figure 10.7** Operating schedule, showing the times for changeover from one operating mode to another (based on the Vianden Station: Grein and Bachmann, 1976).

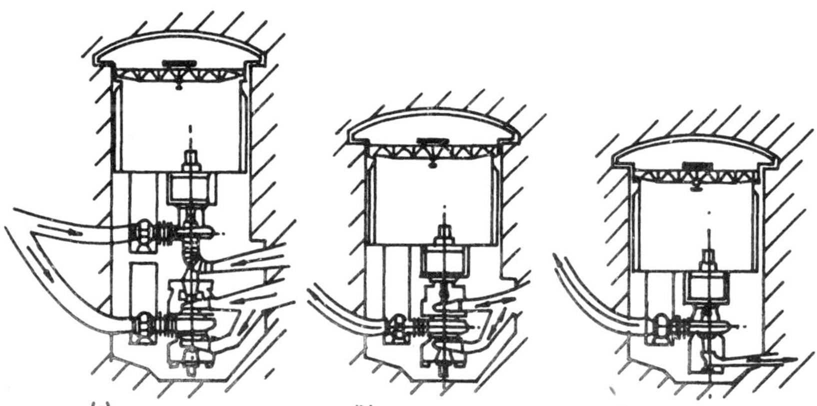


**Figure 10.8** A comparison of the machine layouts for pumped storage stations: (a) two-machine system – pump plus turbine; (b) pump which operates in reverse as a turbine; (c) turbine which operates as a pump – the usual single hydraulic unit system used.

Francis machine under the same level drop and peak efficiency occurs at a lower value of pressure coefficient but the efficiency ranges being comparable. The condition for this is that the machine was designed as a turbine but capable of being run as a pump. A machine designed as a pump but run as a turbine has about the same turbine efficiency as the reversible machine just discussed but a very restricted range of effective operation, and tends only to be used for high level change applications where several pump stages are required. In practice there is a choice of the ratio of pumping and turbining flow rates to optimize the operation in both modes, and this is linked with grid loading patterns to accommodate economically the periods of peak and low demand. A good discussion of this and related problems will be found in Meier (1966) Meier *et al.* (1971) and Muhlemann (1972). The Biannual proceedings of the International Association for Hydraulic Research (IAHR) section for hydraulic machines, cavitation and equipment may be consulted for up-to-date information.

## 10.4 Comments on output control of rotating machines

Turbines, whether water, gas or steam driven, are used to a large extent to generate power at a constant speed because of the need for frequency control. Marine propulsion units, and those used for some forms of traction drives, work over a range of running speeds. The control of steam and gas turbines is achieved either by fuel scheduling or by suitable combinations of valve settings. Water turbines use guide vane control, and in Kaplan

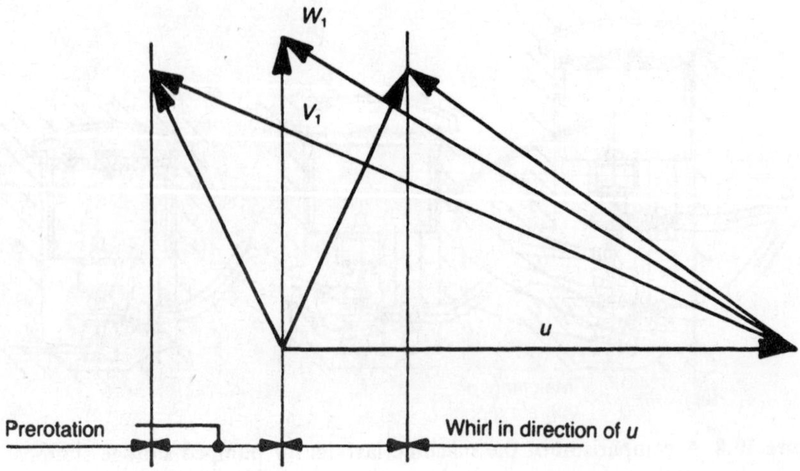


**Figure 10.9** Effect of inlet whirl on the inlet triangle for a pump.

machines finer control is provided by the use of variable settings of the runner blades, as has already been outlined in Chapters 8 and 9.

Output control of compressors and blowers can be provided by speed control or variable stator or rotor geometry. Speed control is the simplest, particularly if the machine is large and driven by a steam turbine, but the drive is expensive if a thyristor-controlled AC motor is used or if a Ward Leonard DC set is fitted. The alternative systems, however, make the compressor more expensive, though stator control can give good control without a large penalty in efficiency loss over a range of flows down to about 60% of the design value. Variable rotor-blade geometry can give good control over a wider range of flows, down to perhaps 30% of design, but at much greater expense and complication since the control system needs to provide a connection between the static actuator and the rotating assembly.

Pumps and fans may be simply controlled by using valves in the outlet system, but at the expense of power since this method creates control by increasing pressure loss. Stepannof (1957a) surveyed the problem and demonstrated that such machines may effectively be controlled, with significant power savings, by using variable geometry inlet guide vanes. Figure 10.9 illustrates how varying the vane angles affects the inlet velocity triangles, the important vector being the inlet whirl component of the absolute velocity. The Euler equation (1.4) states that

$$gH = u_2V_{u2} - u_1V_{u1}$$

The zero inlet whirl 'design' condition suppresses the second term on the right-hand side of the equation, and if the whirl component is in the opposite direction to rotation (prewhirl) the result in the equation is to in-

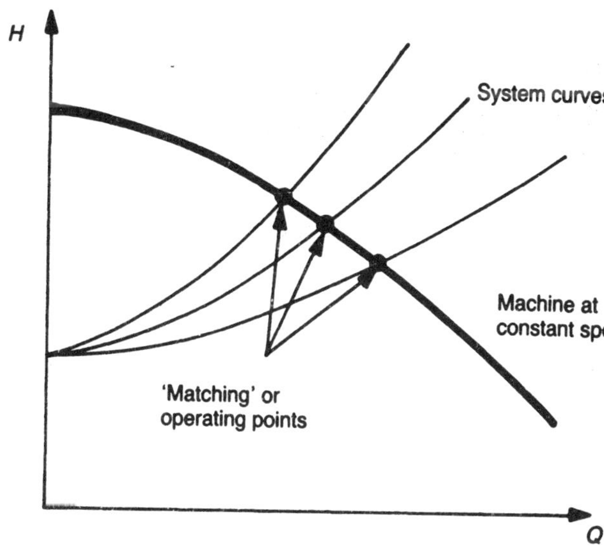


**Figure 10.10** Sketch of pumped system curves superimposed on the constant speed curve for a centrifugal pump.

crease the energy risc, since the $u_1V_{u1}$ term changes its sign and adds to the other part of the equation. There is also a small change in axial velocity, as can be seen from the figure. If, on the other hand, the whirl component is in the same direction as the peripheral velocity the effect is to reduce the energy rise with a small reduction in flow rate. Stepannof demonstrated for a fan that a vane movement of 53° from the neutral position, producing whirl in the direction of rotation, reduced flow to 70% of the design value and the power to about 70% of the rated level.

Flow control in fan-supplied systems leading to energy savings was considered by Woods-Ballard (1982). He showed that a fixed-output fan may cost £1200 and an inverter motor drive for speed control would cost £4500 involving a payback time of 2.6 years in terms of the saved energy costs at 1982 prices. The cost of an inlet guide vane system could be of the order of £1750 with a payback time of about 1.3 years. The application was a variable air-volume type of air conditioning system, and indicates the order of relative costs involved when choosing the best method of providing variable flow rate and energy input.

Similar considerations apply for pumps, though most are driven at constant speed, so that changes in system resistance illustrated in Fig. 10.10 are achieved by valving in the discharge line to bring the crossover point to the level needed. Larger pumps, for example boiler feed pumps, tend to be driven by steam turbines, and output control is provided with variable-geometry rotor blading where the economics allow for the extra first costs.

# Appendix – Solutions to exercises

1.1 See Fig. A.1.

$$\omega = \frac{1500}{30}, \pi = 157.1\,\text{rad}\,\text{s}^{-1},\ u_1 = 5.89\,\text{m}\,\text{s}^{-1},\ u_2 = 23.96\,\text{m}\,\text{s}^{-1}.$$

From inlet Δ, $V_{R1} = V_1 = 5.89\,\text{m}\,\text{s}^{-1}$, therefore

$$Q = 5.89 \times 0.049 \times \pi \times 0.075 = 0.068\,\text{m}^3\,\text{s}^{-1}$$

$gH_E = 23.96\ V_{u2}$, $V_{u2}$ from outlet Δ $= 21.45\,\text{m}\,\text{s}^{-1}$ (since $V_{R2} = 1.45\,\text{m}\,\text{s}^{-1}$) and $gH_E = 513.91\,\text{J}\,\text{kg}^{-1}$. Therefore

$$\Delta p = 513.91 \times 10^5 = 5.139 \times 10^5\,\text{N}\,\text{m}^{-2}$$

Hydraulic power $= \rho Q \times gH_E = 34.95\,\text{kW}$.

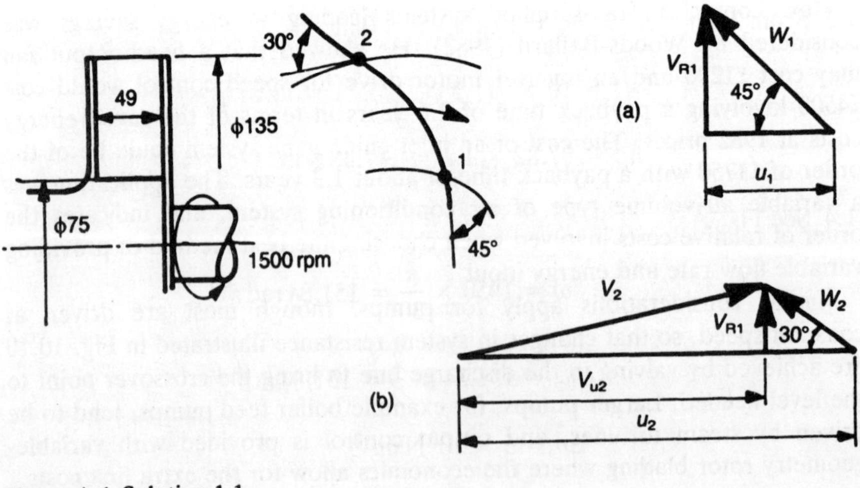


**Figure A.1** Solution 1.1.

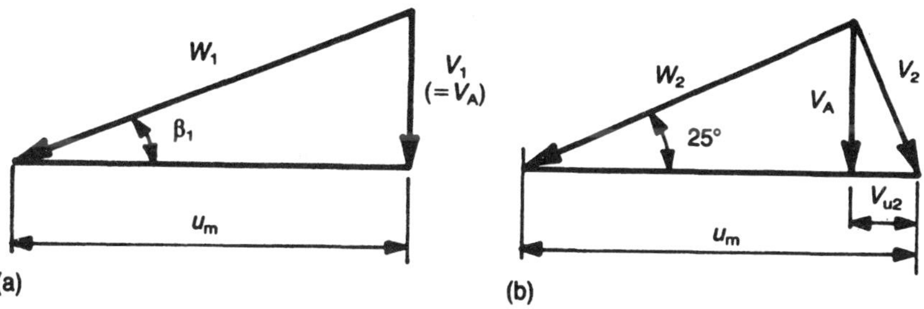


**Figure A.2** Solution 1.2.

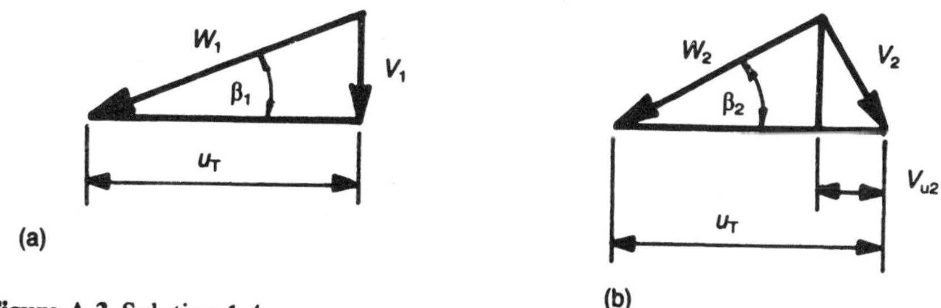


**Figure A.3** Solution 1.4.

1.2 See Fig. A.2.

Mean diameter = 1.125 m, therefore $u_m = 25.31\,\mathrm{m\,s^{-1}}$
From the outlet triangles, $V_{u2} = 3.865\,\mathrm{m\,s^{-1}}$, therefore

$$gH_E = 25.31 \times 3.865 = 97.82\,\mathrm{J\,kg^{-1}}$$

1.3 From Exercise 1.2 $V_A = 10\,\mathrm{m\,s^{-1}}$, so that

$$Q = 10 \times \frac{\pi}{4}(1.5^2 - 0.75^2) = 13.25\,\mathrm{m^3\,s^{-1}}$$

$$\Delta p = 97.82 \times 1.2 = 117.38\,\mathrm{N\,m^{-2}}$$

1.4 See Fig. A.3.

$$\omega = 1450 \times \frac{\pi}{30} = 151.84\,\mathrm{rad\,s^{-1}}$$

$$U_{tip} = 151.84 \times \frac{2}{2} = 151.84\,\mathrm{m\,s^{-1}}$$

$$\Delta p = 5 \times 10^3\,\mathrm{N\,m^{-2}}$$

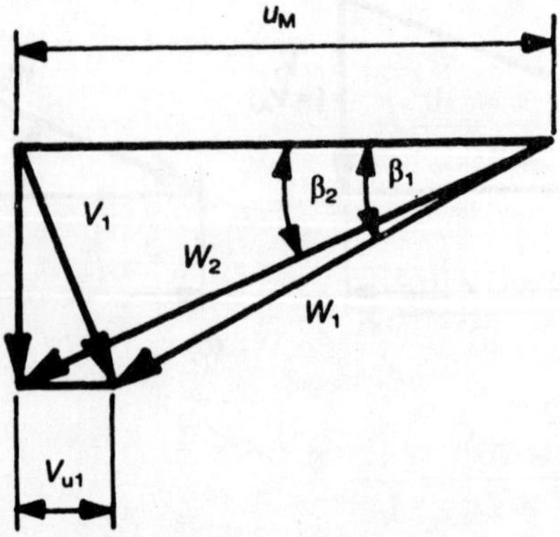


**Figure A.4** Solution 1.5.

therefore

$$gH_E = 5 \times 10^3/1.2 = 4167\,\mathrm{J\,kg^{-1}} = 151.84\ V_{u2}$$
$$V_{u2} = 27.44\,\mathrm{m\,s^{-1}}$$

$\beta_1 = 19.91°$, $\beta_2 = 23.85°$ and

$$Q = 55 \times \frac{\pi}{4}(2^2 - 0.8^2) = 145.15\,\mathrm{m^3\,s^{-1}}$$

therefore

$$\text{power} = 4167 \times 145.14 \times 1.2 = 0.726\,\mathrm{MW}$$

1.5 See Fig. A.4.
$\omega = 45\,\mathrm{rad\,s^{-1}}$, $U_{tip} = 33.75\,\mathrm{m\,s^{-1}}$, $U_{hub} = 13.95\,\mathrm{m\,s^{-1}}$, $U_{mean} = 23.85\,\mathrm{m\,s^{-1}}$ and $gH = 8 \times g$. Therefore

$$gH_E = 0.92 \times 8 \times g = 72.2\,\mathrm{J\,kg^{-1}}$$

$72.2 = 23.85\ V_{u1}$, therefore $V_{u1} = 3.03\,\mathrm{m\,s^{-1}}$, $\beta_1 = 25.66°$, $\beta_2 = 22.75°$, flow rate $= 14.65\,\mathrm{m^3\,s^{-1}}$.

$$\text{Electrical power} = 89 \times 14.65 \times 10^3 \times 0.95$$
$$= 1.092\,\mathrm{MW}$$

2.1 80% of design speed $= 185 \times 0.8 = 148\,\mathrm{rad\,s^{-1}}$.

$$\frac{4.5 \times 10^5}{185^2 \times D^2} = \frac{\Delta P}{148^2 \times D^2}, \text{ therefore } \Delta p = 2.88 \times 10^5\,\mathrm{N\,m^{-2}}$$

$$\frac{0.28}{185D^3} = \frac{Q}{148D^3}, \quad \text{therefore } Q = 0.224\,\mathrm{m^3\,s^{-1}}$$

At 185 rad s$^{-1}$

$$P = \frac{4.5 \times 10^5}{10^3} \times \frac{0.28 \times 10^3}{0.85} = 0.107\,\text{MW}$$

At 148 rad s$^{-1}$

$$P = \frac{2.88 \times 10^5}{10^3} \times \frac{0.224 \times 10^3}{0.85} = 0.076\,\text{MW}$$

2.2

$$\frac{4.5 \times 10^5}{185^2 \times D_1^2} = \frac{4.5 \times 10^5}{148^2 \times D_2^2}, \quad \frac{D_2}{D_1} = 1.25:1$$

$$\frac{0.28}{185 D_1^3} = \frac{Q}{148 D_2^3}, \quad Q = 0.28\,\text{m}^3\,\text{s}^{-1}$$

$$P = \frac{0.28 \times 10^3}{0.85} \times \frac{4.5 \times 10^5}{10^3} = 0.108\,\text{MW}$$

2.3

$$20 = \frac{2900\sqrt{0.03}}{(H)^{3/4}}$$

$$H = \left(\frac{2900}{20}\sqrt{0.03}\right)^{4/3} = 73.54\,\text{m per stage}$$

$$\text{number of stages} = \frac{820}{73.54} = 11.15 \text{ (i.e. 12 stages)}$$

2.4

$$\frac{37.5 \times 10^6}{0.93} = 18g \times Q \times 10^3$$

$$Q = 228.35\,\text{m}^3\,\text{s}^{-1}$$

$$\frac{18}{90^2 \times Dp^2} = \frac{6}{Nm^2 Dm^2} \quad \text{and} \quad \frac{37.5 \times 10^6}{10^3 20^3 Dp^5} = \frac{45 \times 10^3}{10^3 Nm^3 Dm^5}$$

$P_1 = 37.5\,\text{MW}$, $H_1 = 18\,\text{m}$, $\eta_1 = 0.93$, $N_1 = 90\,\text{rpm}$, $H_2 = 6\,\text{m}$, $P_2 = 45\,\text{kW}$ and $\eta_2 = 0.93$.

$$\frac{H_1}{N_1^2 D_1^2} = \frac{H_2}{N_2^2 D_2^2}, \quad \frac{P_1}{\rho N_1^3 D_1^5} = \frac{P_2}{\rho N_2^3 D_2^5}, \quad \frac{Q_1}{N_1 D_1^3} = \frac{Q_2}{N_2 D_2^3}$$

and

$$ks = \frac{\omega\sqrt{P}}{(gH)^{3/4}} = \text{constant}$$

are all valid if losses are neglected (implied by $\eta_1 = \eta_2$). Then,

$$\frac{18}{90^2 D_1^2} = \frac{6}{N_2^2 D_2^2}$$

$$\frac{90 \times \sqrt{(37.5 \times 10^6)}}{(18g)^{5/4}} = \frac{N_2 \sqrt{45 \times 10^3}}{(6g)^{5/4}}$$

$$N_2 = 658\,\text{rpm}$$

Therefore

$$\frac{18}{90^2 D_1^2} = \frac{6}{658^2 D_2^2}$$

$$\frac{D_1}{D_2} = \frac{658}{90}\sqrt{\left(\frac{18}{6}\right)} = 12.66{:}1$$

$$\frac{45 \times 10^3}{0.93} = 10^3 \times Q_2 \times 6g,$$

$$Q_2 = 0.822\,\text{m}^3\,\text{s}^{-1}$$

2.5 $D_1/D_2 = 3{:}1$, $N_1 = 1450\,\text{rpm}$, $Q_1 = 5\,\text{m}^3\,\text{s}^{-1}$, $\Delta p_1 = 450$ and $\eta_1 = 0.78$.

$$\rho_1 = \frac{9.89 \times 10^5}{287 \times 283} = 1.218\,\text{kg}\,\text{m}^{-3}$$

$$\rho_2 = \frac{10^5}{287 \times 298} = 1.169\,\text{kg}\,\text{m}^{-3}$$

From tables, $\nu_1 = 14.5\,\text{mm}^2\,\text{s}^{-1}$ and $\nu_2 = 16\,\text{mm}^2\text{s}^{-1}$. Since $Re$ must be the same (for dynamic similarity), then

$$\frac{1450 \times 3^2}{14.5} = \frac{N_2 \times 1^2}{16}, \qquad N_2 = 14\,400\,\text{rpm}$$

Since

$$\frac{Q}{ND^3} = \text{const}, \qquad \frac{\Delta p}{N^2 D^2} = \text{const}$$

$$Q_2 = 1.839\,\text{m}^3\,\text{s}^{-1}, \qquad \Delta p_2 = 4.38\,\text{N}\,\text{m}^{-2}$$

To find $\eta_2$, use equation (2.10):

$$\frac{1 - \eta_1}{1 - \eta_2} = 0.3 + 0.7\left(\frac{Re_2}{Re_1}\right)^{0.2}$$

therefore $\eta_2 = 78.5\%$ and

$$P_2 = \frac{1.839 \times 1.169}{0.785} \times \frac{4.38}{1.169}\left(= \dot{m}\frac{\Delta p}{\rho}\right)$$
$$= 10.26\,\text{W}$$

2.6 $P_1 = 100\,\text{MW}$ output, therefore hydraulic power full size = 108.696 MW. The full size flow rate is given by

$$108.696 \times 10^6 = 20g \times Q \times 10^3, \qquad Q = 554\,\text{m}^3\,\text{s}^{-1}$$

Using

$$\frac{gH}{N^2D^2} = \text{const}, \qquad \frac{Q}{ND^3} = \text{const}$$

$$\frac{554}{93.7 \times 7^3} = \frac{0.9}{N^2D_2^3}, \qquad \frac{20g}{93.7^2 \times 7^2} = \frac{5g}{N_2^2D_2^2}$$

Solving gives $D_2 = 0.399\,\text{m}$ and $N_2 = 821.92\,\text{rpm}$.
Using the Hutton formula for $\eta_2$

$$\frac{1-\eta_1}{1-\eta_2} = 0.3 + 0.7\left(\frac{Re_2}{Re_1}\right)^{0.2}$$

Since $R_e = \omega D^2/\nu$ and $\nu$ same for both machines, $\eta_2 = 87.57\%$. Therefore

$$P_2 = 0.9 \times 10^3 \times 5g \times 0.8757 = 38.66\,\text{kW}$$

If air is used as the flow medium in modelling

$$\nu_{\text{water}} = 1\,\text{mm}^2\,\text{s}^{-1} \text{ at } 20\,°\text{C}$$
$$\nu_{\text{air}} = 17\,\text{mm}^2\,\text{s}^{-1} \text{ at } 20\,°\text{C and } 1\,\text{bar}$$

If $Re$ is to be the same, then

$$\frac{\omega_{\text{air}}}{17} = \frac{\omega_{\text{water}}}{1}, \qquad \omega_{\text{air}} = 13\,973\,\text{rpm}$$

Clearly this will pose stress problems, so one solution could be to follow other similarity laws, including characteristic number and to correct for efficiency variation.

2.7 For dynamic similarity

$$\frac{P_{02}}{P_{01}} = f\left(\frac{\dot{m}\sqrt{(RT_{01})}}{D^2p_{01}} \times \frac{ND}{\sqrt{(RT_{01})}} \times R_E \times M_n \times \ldots\right)$$

For helium

$$R = R_0/M = 8.3143/4.003 = 2.077\,\text{kJ}\,\text{kg}^{-1}\,\text{K}^{-1}$$

From tables, $\mu_{\text{helium}} = 20.8 \times 10^{-6}\,\text{kg}\,\text{m}^{-1}\text{s}^{-1}$, $\rho_{\text{helium}} = 0.1626\,\text{kg}\,\text{m}^{-3}$, $\nu_{\text{air}} = 1.568 \times 10^{-5}\,\text{m}^2\,\text{s}^{-1}$ and $\rho_{\text{air}} = 1.16\,\text{kg}\,\text{m}^{-3}$.

$$\left|\frac{P_{02}}{P_{01}}\right|_{\text{helium}} = \left|\frac{P_{02}}{P_{01}}\right|_{\text{air}} = 12$$

$$\left|\frac{\dot{m}\sqrt{RT_0}}{D^2 P_{01}}\right|_{\text{helium}} = \left|\frac{\dot{m}\sqrt{RT_{01}}}{D^2 P_{01}}\right|_{\text{air}}$$

$$\frac{12000\sqrt{2.077 \times 300}}{D^2 \times 10 \times 10^5} = \frac{\dot{m}_A\sqrt{0.287 \times 300}}{D^2 1 \times 10^5}$$

Therefore $\dot{m}_{\text{air}} = 3228.2\,\text{kg}\,\text{h}^{-1}$

$$\left|\frac{ND}{\sqrt{RT_{01}}}\right|_{\text{helium}} = \left|\frac{ND}{\sqrt{RT_{01}}}\right|_{\text{air}}$$

$$\frac{3000D}{\sqrt{2.077 \times 300}} = \frac{ND}{\sqrt{0.287 \times 300}}$$

from which $N$ is 1115.2 rpm. But $R_E$ should also be constant since

$$R_E = \frac{\rho\omega D^2}{\mu}$$

$$\frac{0.1626 \times 3000 \times 2\pi \times D^2}{20.8 \times 10^{-6} \times 60} = \frac{N \times D^2}{1.568 \times 10^{-5}} \times \frac{2\pi}{60}$$

Clearly, both criteria for $N$ cannot be satisfied! One approach would be to test at 1115.2 rpm and correct power (efficiency) by using $R_E$ corrections. Clearly there are problems when modelling using air.

2.8 Again, as for Exercise 2.7

$$\frac{P_{02}}{P_{01}} = f\left(\frac{\dot{m}\sqrt{(RT_{01})}}{D^2 p_{01}} \times \frac{ND}{\sqrt{(RT_{01})}} \times R_E \times M_n \times \ldots\right)$$

$\dfrac{P_{02}}{P_{01}}$ will be the same in both cases

$$\left|\frac{ND}{\sqrt{(2.291 \times 660)}}\right|_{\text{gas}} = \frac{6000D}{\sqrt{(0.287 \times 288)}},$$

Therefore $N = 25\,662.5$ rpm and

$$\left|\frac{\dot{m}\sqrt{(2.291 \times 660)}}{1.4 \times 10^6 D^2}\right|_{\text{gas}} = \left|\frac{15\sqrt{(0.287 \times 288)}}{10 \times 10^4 D^2}\right|_{\text{air}}$$

$$\dot{m}_{\text{gas}} = 49.1\,\text{kg}\,\text{s}^{-1}.$$

3.1 Vapour pressure head from Fig. 3.8 = 0.345 m; therefore, vapour pressure energy = 3.385 J kg$^{-1}$.

Since atmospheric pressure = $10^5$ N m$^{-2}$ total energy at the pump is

$$-5.5g + \frac{10^5}{10^3} - 8.5 = 37.55\,\text{J kg}^{-1}$$

Now, $NPSE_A = 37.55 - 3.385 = 34.165$ J kg$^{-1}$ (or $NPSH_A = 3.48$ m of water). If atmospheric pressure falls to $0.85 \times 10^5$ N m$^{-2}$, then

$$NPSE_A = -5.5g + \frac{0.85 \times 10^5}{10^3} - 8.5 - 3.385$$

$$= 19.16\,\text{J kg}^{-1} \qquad (NPSH_A = 1.95\text{ m of water})$$

3.2 Vapour pressure energy = 3.385 J kg$^{-1}$

*Case (a).* Total energy at pump is

$$4g + \frac{10^5}{10^3} - 50 = 89.24\,\text{J kg}^{-1}$$

$$NPSE_A = 89.24 - 3.385 = 85.86\,\text{J kg}^{-1}$$

$$(NPSH_A = 8.75\text{ m of water})$$

*Case (b).*

$$NPSE_A = 4g + \frac{10^5}{10^3} \times 1.3 - 50 - 3.385 = 115.86\,\text{J kg}^{-1}$$

$$(NPSH_A = 11.81\text{ m of water})$$

3.3

$$NPSE_A = 100 = 4g + \frac{p}{10^3} - 100 - 3.385$$

$$P = 1.64 \times 10^5\,\text{N m}^{-2}$$

3.4 If the tank is at vapour pressure $NPSE_A = 5.5g - 50 = 3.955$ J kg$^{-1}$, and if free surface is at 3.5 bar gauge, then

$$NPSE_A = 5.5g + \frac{0.95 \times 10^5}{0.56 \times 10^3} + \frac{3.5 \times 10^5}{0.56 \times 10^3} - 50 - \frac{3.59 \times 10^5}{0.56 \times 10^3}$$

$$= 157.51\,\text{J kg}^{-1} \qquad (NPSH_A = 16.06\text{ m of butane})$$

3.5

$$NPSE_A = 4g + \frac{3.59 \times 10^5}{0.56 \times 10^3} - \text{loss} - \frac{3.59 \times 10^5}{0.56 \times 10^3} = 10\,\text{J kg}^{-1}$$

$$\text{loss} = 29.24\,\text{J kg}^{-1}$$

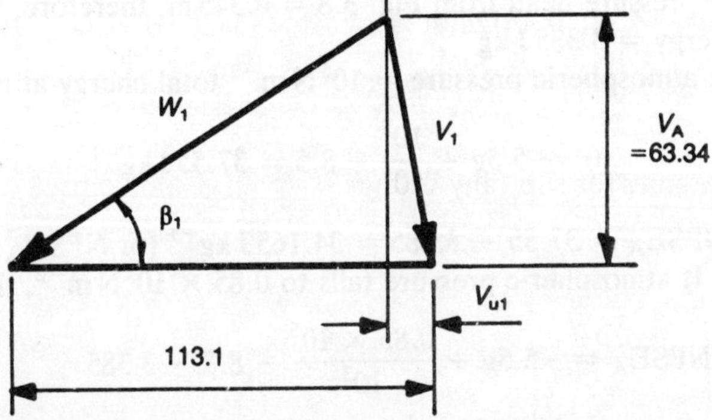


**Figure A.5** Solution 6.1.

3.6 For the pump

$$k_s = \frac{\frac{1450\pi}{30}\sqrt{\left(\frac{80}{3600}\right)}}{(g \times 50)^{3/4}} = 0.217$$

From Fig. 3.9, $\sigma = 0.07$, therefore

$$\text{NPSH}_R = 0.07 \times 50 = 3.5\,\text{m}, \qquad \text{NPSE}_R = 34.34\,\text{J kg}^{-1}$$

$$\text{NPSE}_A = -Xg + 72.65\,\text{J kg}^{-1} \text{ and, in the limit,}$$

$$\text{NPSE}_R = \text{NPSE}_A - 10 = 34.34, \qquad \text{NPSE}_A = 44.34$$

Therefore

$$44.34 = -Xg + 72.65$$
$$X = 2.9\,\text{m suction lift}$$

6.1 $C_p = 5.193\,\text{kJ kg}^{-1}\text{K}^{-1}$ and $R = 8.3143/4.003\ 2.077\,\text{kJ kg}^{-1}\text{K}^{-1}$. Therefore

$$C_v = 3.116\,\text{kJ kg}^{-1}\text{K}^{-1}, \qquad k = 1.67$$

$$u_A = \frac{0.09}{2} \times \frac{24\,000\pi}{30} = 113.1\,\text{m s}^{-1}$$

Figure A.5 shows the inlet velocity diagram and $V_{u1} = 3.32\,\text{m s}^{-1}$, $V_A = 63.34\,\text{m s}^{-1}$, $V_1 = 63.43\,\text{m s}^{-1}$ and $W_1 = 126.86\,\text{m s}^{-1}$. Therefore

$$\rho_A = 4 \times 10^5/2.077 \times 10^3 \times 300 = 0.642\,\text{kg m}^{-3}$$

$$\dot{m} = 0.642 \times 63.34 \times \frac{\pi}{4}(0.09^2 - 0.06^2) = 0.144\,\text{kg s}^{-1}$$

$$a = \sqrt{(kRT)} = \sqrt{(1.67 \times 2.077 \times 10^3 \times 300)} = 1020\,\text{m s}^{-1}$$

Therefore $M_{nmax} = 0.124$ at A, outside diameter $= 0.25$ and

$$u_2 = \frac{24\,000\pi}{30} \times 0.125 = 314.16\,\mathrm{m\,s^{-1}}$$

Using Stanitz for slip (any of those quoted could be used)

$$\frac{V_{u2}}{u_2} = 1 - \frac{0.63\pi}{12}, \qquad V_{u2} = 262.34\,\mathrm{m\,s^{-1}}$$

$$\frac{\Delta p}{\rho} = 8.242 \times 10^4\,\mathrm{J\,kg^{-1}} = u_2 V_{u2} (= h_{02} - h_{01})$$

Therefore

$$0.9 = 5.193 \times 10^3 \times T_{01}(T_{02}/T_{01} - 1)/8.242 \times 10^4$$

$$T_{01} = 300 + 63.43^2/2 \times 5.193 \times 10^3 = 300.39\,\mathrm{K}$$

and $T_{02}/T_{01} = 1.048/1$.

$$\frac{P_{02}}{P_{01}} = (1.048)^{\frac{1.67}{0.67}} = 1.124$$

Since

$$P_{01} = 4 \times 10^5 + \frac{63.43^2 \times 0.642}{2}$$

then

$$P_{02} = 4.013 \times 10^5\,\mathrm{N\,m^{-2}}$$

6.2 $w = 183.26\,\mathrm{rad\,s^{-1}}$, thus

$$u_2 = 183.76 \times 0.4 = 73.3\,\mathrm{m\,s^{-1}}$$

$$u_{1A} = 32.07\,\mathrm{m\,s^{-1}}, \qquad u_{1B} = 18.33\,\mathrm{m\,s^{-1}}$$

$$V_{R2} = 0.8/\pi \times 0.8 \times 0.075 = 4.24\,\mathrm{m\,s^{-1}}$$

Theoretical $V_{u2} = 64.2\,\mathrm{m\,s^{-1}}$. An estimate of 'slip' is needed for which a simple formula is

$$\frac{V_{u2}}{64.2} = 1 - \frac{\pi}{7}\sin 25, \qquad V_{u2} = 52.02\,\mathrm{m\,s^{-1}}$$

Assuming zero inlet whirl $V_{u1} = 0$

$$gH_E = 52.02 \times 73.3 = 4580\,\mathrm{J\,kg^{-1}}$$

considering the inlet annulus, flow through one inlet $= 0.4\,\mathrm{m^3\,s^{-1}}$, therefore

$$V_A = 0.4/\left(\frac{\pi}{4}\right)(0.35^2 - 0.2^2) = 6.17\,\mathrm{m\,s^{-1}}$$

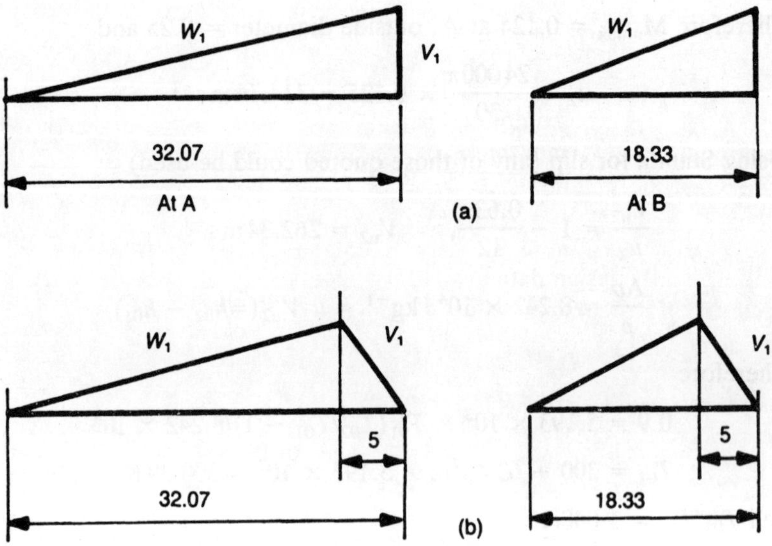


**Figure A.6** Solution 6.2.

Thus at A, with $V_{u1} = 0$, the velocity triangles are (Fig. A.6(a))

$$\beta_{1A} = 10.9°, \qquad \beta_{1B} = 18.6°$$

If $5\,m\,s^{-1}$ swirl is imposed in the direction of rotation, the velocity triangles are now (Fig. A.6(b))

$$\beta_{1A} = 12.84°, \qquad \beta_{1B} = 24.84°$$

(The incidence at A is tolerable, but that at B could lead to flow separation.)
If point A is considered

$$gH_E = 4580 - 5 \times 32.07 = 4424\,J\,kg^{-1}$$

Thus energy rise reduction $= 156\,J\,kg^{-1}$.

6.3 $\omega = 30000 \times \pi/30 = 3141.6\,rad\,s^{-1}$, therefore

$$u_{tip} = 0.27 \times 3141.6/2 = 424.1\,m\,s^{-1}$$

Using Stanitz

$$\frac{V_{u2}}{u_2} = 1 - 0.63\frac{\pi}{8}, \qquad V_{u2} = 319.2\,m\,s^{-1}$$

$$C_p\Delta T = 319.2 \times 424.1 = 1.354 \times 10^5\,J\,kg^{-1}$$

$$\Delta T = 134.7\,K$$

$$T_{02} = T_{01} + 134.7 = 424.7\,K$$

and

$$\frac{P_{02}}{P_{01}} = \left(1 + 0.79 \times \frac{134.7}{290}\right)^{3.5} = \frac{3.01}{1}$$

Using the simple assumption that $p_{03} = p_{02} = 3.76 \times 10^5\,\mathrm{N\,m^{-2}}$ and, since eight blades 3 mm thick are used, the impeller peripheral flow area is reduced to $0.021\,\mathrm{m}^2$

$$V_{R2} = 4.5/0.021\rho \qquad \text{(A.1)}$$

Since the only design data available is that $Mn$ is not to exceed 0.95, it is necessary to assume a value of $V_{R2} = 100\,\mathrm{m\,s^{-1}}$. Also, since $V_2 = 334.5\,\mathrm{m\,s^{-1}}$

$$\frac{334.5}{0.95} = a = \sqrt{(114 \times 287 \times T_2)}, \qquad T_2 = 308.56\,\mathrm{K}$$

Applying the gas laws,

$$\frac{p_{02}}{p_2} = \left(\frac{T_{02}}{T_2}\right)^{3.5}, \qquad p_2 = 1.23 \times 10^5\,\mathrm{N\,m^{-2}}$$

therefore

$$\rho_2 = 1.23 \times 10^5/287 \times 308.56 = 1.39\,\mathrm{kg\,m^{-3}}$$

This gives $V_{R2}$ from equation (A.1) as $154\,\mathrm{m\,s^{-1}}$. Assuming $V_{R2} = 175\,\mathrm{m\,s^{-1}}$, this yields $V_2 = 364\,\mathrm{m\,s^{-1}}$, $T_2 = 365.4\,\mathrm{K}$, $p_2 = 2.22 \times 10^5\,\mathrm{N\,m^{-2}}$ and $\rho_2 = 2.12\,\mathrm{kg\,m^{-3}}$, which compares with $\rho_2$ from equation (A.1) of $1.224\,\mathrm{kg\,m^{-3}}$. The iteration ceases, and $V_{R2}$ is $175\,\mathrm{m\,s^{-1}}$.

Considering the diffuser

$$V_{u3} = \frac{319.2 \times 270}{550} \text{ (free vortex law)}$$

$$V_{u3} = 156.5\,\mathrm{m\,s^{-1}}$$

$$C_p = 0.5 = \frac{\Delta p}{\tfrac{1}{2} \times 2.12 \times 364^2}$$

and

$$p_3 = 2.22 \times 10^5 + 7.354 \times 10^4$$

$$p_4 = 2.9554 \times 10^5\,\mathrm{N\,m^{-2}}$$

The $C_p$ approach does not reflect the true changes since both $\mu$ and $\rho$ vary with temperature.

6.4 At design, zero inlet whirl is assumed allowing for blade blockage, the flow area at outlet normal to flow $= 7.455 \times 10^{-3}\,\mathrm{m}^2$. Therefore

$$V_{R2} = 100 \times 10^3/3600 \times 7.4155 = 3.75\,\mathrm{m\,s^{-1}}$$

$$u_2 = 37.96\,\mathrm{m\,s^{-1}}$$

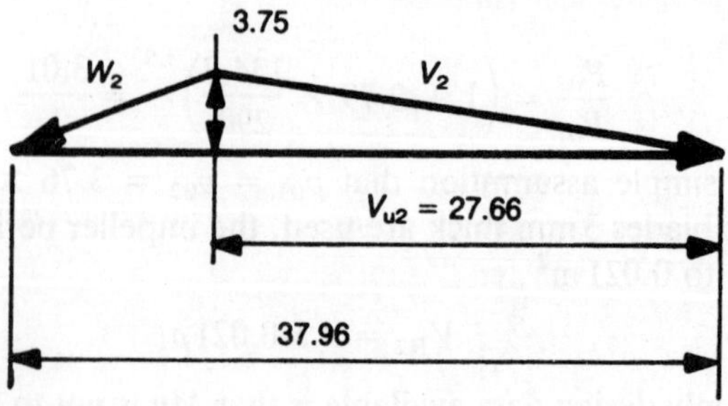


**Figure A.7** Solution 6.4.

From the velocity triangle (Fig. A.7)

$$V_{u2} = 27.66\,\text{m s}^{-1}$$

$$gH_E = 27.66 \times 37.96 = 1049.9\,\text{J kg}^{-1}$$

A slip allowance is needed. In this example Karassik's formula is used, which gives

$$V_{u2}/V_{u2\,\text{ideal}} = 0.65, \quad V_{u2} = 18.04$$

$$gH_E\ (\text{corrected}) = 648.9\,\text{J kg}^{-1}$$

Since the pump is of conventional design, $\eta_H$ will be used to get $gH_{\text{actual}}$. Using $gH = gH_{\text{corrected}}$ the characteristic number $k = 0.37$. From Fig. 2.6 $\eta = 0.73$ and, assuming a mechanical efficiency of 95% $\eta_{\text{hydraulic}} = 76\%$.

Recalculating $k$ using $520\,\text{J kg}^{-1}$ yields $k = 0.43$, giving a slightly lower $\eta_H$ of 75%, $\eta = 71.25\%$ and $gH = 514\,\text{J kg}^{-1}$. Therefore

$$\text{power} = \left(\frac{100}{3600} \times 10^3\right) \times \frac{514}{0.7125} = 20.04\,\text{kW}$$

Turning to the diffuser and using Fig. 6.23, $\eta_{\text{diffuser}} = 0.75$. Therefore,

$$\text{actual}\ \frac{\Delta p}{\rho} = \frac{0.75}{2}(V_2^2 - V_3^2)$$

Since $V_2 = 7.03$, $V_{u2} = 5.95$, $V_{u3} = 3.72$ and $V_3 = 4.33$ (using the free vortex principle). This gives

$$\frac{\Delta p}{\rho} = \frac{0.75}{2}(7.03^2 - 4.33^2) = 11.5\,\text{J kg}^{-1} = \text{extra generated pressure.}$$

6.5 $Q = 150/3600 = 0.0417\,\text{m}^3\,\text{s}^{-1}$.

The rotational speed is not specified, but since the pump is not of a special type, a typical rotational speed would be 1450 rpm on a 50 kHz supply. Thus,

$$0.75 = \frac{\frac{1450\pi}{30}\sqrt{0.0417}}{(gH)^{3/4}}$$

$$gH = 142.94\,\mathrm{J\,kg^{-1}}$$

$$\Delta p = 1.429 \times 10^5\,\mathrm{N\,m^{-2}}$$

Since the suction pressure is $0.85 \times 10^5\,\mathrm{N\,m^{-2}}$ the outlet pressure = $2.2794 \times 10^5\,\mathrm{N\,m^{-2}}$.

As the discussion in Section 6.6 indicates there are a number of solutions available, and here for illustration it will be assumed that the outlet pressure applies to the back plate and shroud down to the wear ring. Suction pressure applies to the area inside the wear ring diameter.

Area subjected to outlet pressure = $\pi/4(0.14^2 - 0.065^2) = 0.0121\,\mathrm{m^2}$.

Area subjected to suction pressure = $\pi/4(0.14)^2$, therefore

$$\begin{aligned}\text{axial thrust} &= 2.2794 \times 10^5 \times 0.0121 - 0.85 \times 10^5 \times 0.0154\\ &= 1450\,\mathrm{N}\ \text{due to pressure}\end{aligned}$$

Force due to atmospheric pressure of shaft end = $10^5 \times \pi/4(0.065)^2$ = 331.83 N. Total cnd thrust = 1781.83 N.

6.6 In the plenum $P_{01} = 1.25 \times 10^5\,\mathrm{N\,m^{-2}}$, $T_{01} = 300\,\mathrm{K}$. At maximum radius in the inlet

$$u_{IT} = 1575 \times 0.15 = 236.25\,\mathrm{m\,s^{-1}}$$

For zero inlet whirl

$$W_1^2 = V_1^2 + 236.25^2 \tag{A.2}$$

$$a = \sqrt{(\gamma R T_1)}, \qquad M_n = \frac{W_1}{a_1} = \frac{W_1}{\sqrt{\gamma R T_1}} = 0.85$$

Therefore

$$W_1 = 0.85\sqrt{(1.4 \times 287 \times T_1)} \tag{A.3}$$

and

$$300 = T_1 + \frac{V_1^2}{2 \times 1.005 \times 10^3} \tag{A.4}$$

Solution of equations (A.2)–(A.4) yields:

$$V_A = V_1 = 165.3\,\mathrm{m\,s^{-1}}, \qquad T_1 = 286.4\,\mathrm{K}$$

and from the gas laws $p_1 = 1.063 \times 10^5\,\mathrm{N\,m^{-2}}$. This gives

$$\rho_1 = 1.29\,\mathrm{kg\,m^{-3}}$$

$$\dot{m} = 1.29 \times 165.3 \times \pi/4(0.3^2 - 0.1^2) = 13.43\,\mathrm{kg\,s^{-1}}$$

Turning now to the outlet diameter of the impeller

$$u_2 = 0.3 \times 1575 = 472.5\,\mathrm{m\,s^{-1}}$$

Using Stanitz,

$$V_{u2} = 394.54\left(= 1 - \frac{0.62\pi}{12}\right)$$

Therefore

$$472.5 \times 394.54 = 1.005 \times 10^3 \Delta T$$

$$\Delta T = 185.5\,^\circ\mathrm{C}$$

and

$$\frac{P_{02}}{P_{01}} = \left(1 + 0.82 \times \frac{185.5}{300}\right)^{3.5} = 4.2/1$$

$$P_{02} = 5.25 \times 10^5\,\mathrm{N\,m^{-2}}$$

7.1 Power output = 0.75 MW, therefore

$$\text{hydraulic power} = \frac{0.75 \times 10^6}{0.955 \times 0.92 \times 0.97}$$

$$= 0.88\,\mathrm{MW}$$

Since head = 2.5 m

$$Q = \frac{0.88 \times 10^6}{2.5\,g \times 10^3} = 36.2\,\mathrm{m^3\,s^{-1}}$$

Assuming hub:tip ratio = 0.4, then $D_H = 2\,\mathrm{m}$ and

$$V_A = \frac{36.2}{\pi/4(5^2 - 2^2)} = 2.195\,\mathrm{m\,s^{-1}}$$

$$u_{tip} = 7.15 \times 5/2 = 17.875\,\mathrm{m\,s^{-1}}$$

$$u_H = 7.15 \times 2/2 = 7.15\,\mathrm{m\,s^{-1}}$$

$$gH_E = 2.5\,g \times 0.92 = 22.56\,\mathrm{J\,kg^{-1}}$$

Therefore, at the tip section,

$$22.56 = 17.875\,V_{u1}(V_{u2} = 0), \qquad V_{u1} = 1.262\,\mathrm{m\,s^{-1}}$$

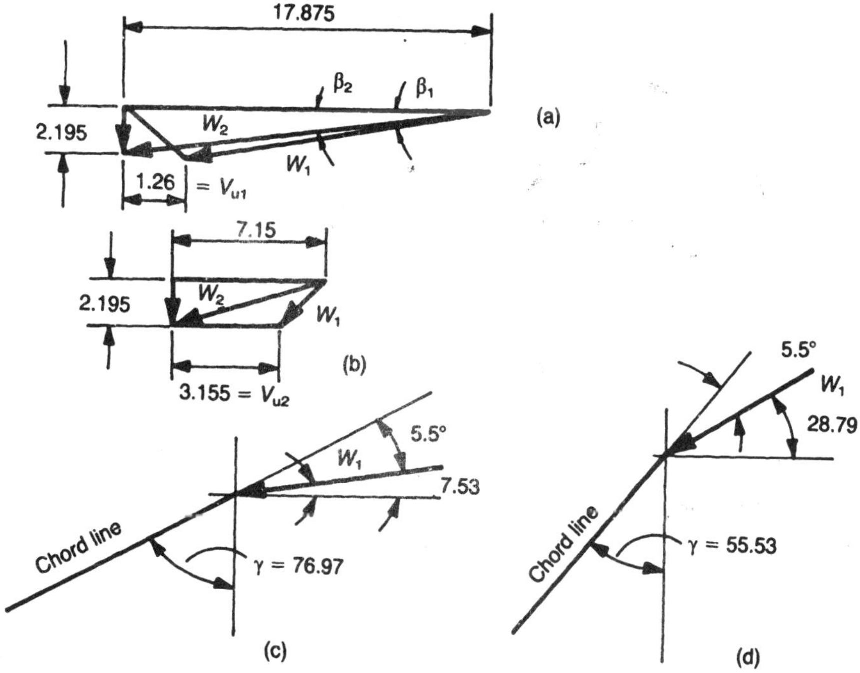


**Figure A.8** Solution 7.1.

From Fig. A.8(a); $\beta_2 = 7°$, $\beta_1 = 7.53°$, $\beta_m = 7.265°$ ($\beta_m = (\beta_1 + \beta_2)/2$), $\alpha_1 = 60.1°$.

Using Clark 'Y' data (Fig. 4.8), $C_L$ (optimum) = 0.97 at $\alpha = 5.5°$, which gives

$$22.56 = \frac{2.195}{2 \times 17.875} \times 0.97 \times \frac{c}{s} \text{ cosec } 7.265$$

$$c/s = 0.149, \qquad s/c = 6.67$$

At the hub section (Fig. A.8(b)) $V_{u1} = 3.195$, $\beta_1 = 28.79°$, $\beta_2 = 17.07°$ and $\beta_m = 22.93°$. Therefore

$$22.56 = \frac{2.195}{2 \times 7.15} \times 0.97 \times \frac{c}{s} \textit{ cosec } 22.93$$

$$c/s = 1.15, \qquad s/c = 0.866$$

Assuming $h/c = 3$ and $c_{hub} = 1.5/3 = 0.5\,\text{m}$ then

$$s = 0.866 \times 0.5 = 0.43\,\text{m}$$

and

$$Z = \frac{\pi \times 2}{0.43} = 14.6 \text{ (i.e. 15 blades)}$$

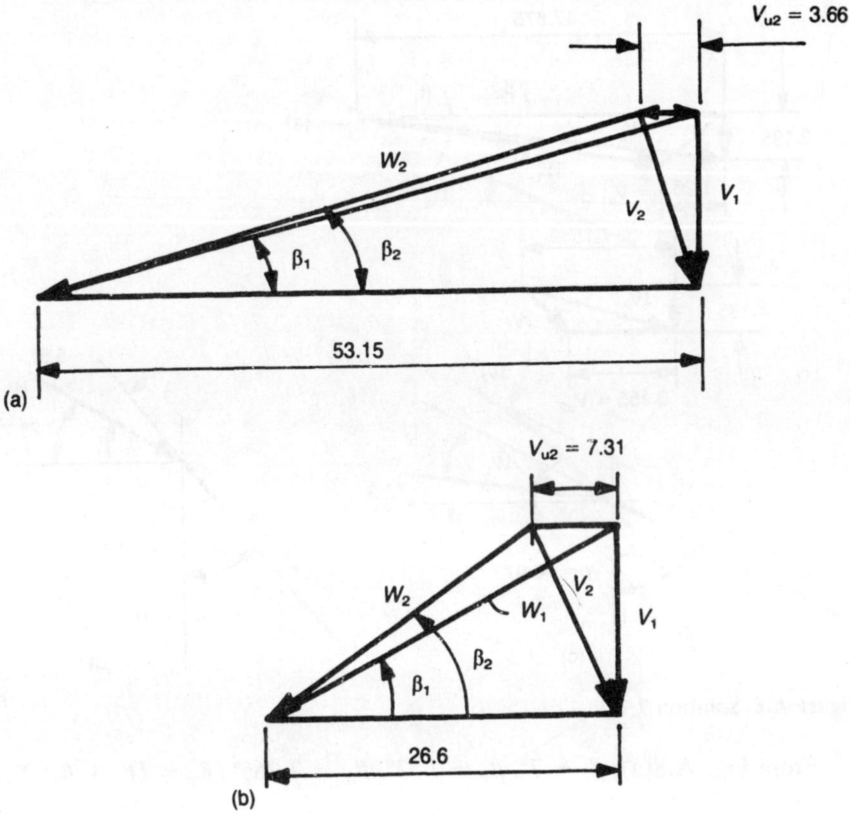


**Figure A.9** Solution 7.2.

to give $s = 0.42$ and $c_{hub} = 0.482$ m.

At the tip section

$$s = \frac{\pi \times 5}{16}, \qquad c = 0.163\,\text{m}$$

Stagger angles are shown in Figs A.8(c) and (d)

7.2 Since $Q = 4\,\text{m}^3\,\text{s}^{-1}$ and $V_A = 15\,\text{m}\,\text{s}^{-1}$, the flow area $= 4/15 = 0.267\,\text{m}^2$. If the hub:tip ratio $= 0.5$ then $D_T = 0.673$. Rounding: $D_T = 700$ mm, $D_H = 350$ mm and $V_A = 13.86\,\text{m}\,\text{s}^{-1}$.

Since $\Delta p = 175\,\text{N}\,\text{m}^{-2}$, $gH = 175/12$ and

$$gH_E = \frac{175}{1.2 \times 0.75} = 194.4\,\text{J}\,\text{kg}^{-1}$$

Since $w = 1450 \times \pi/30$, $u_H = 26.6\,\text{m}\,\text{s}^{-1}$ and $u_T = 53.15\,\text{m}\,\text{s}^{-1}$ (zero inlet whirl is assumed).

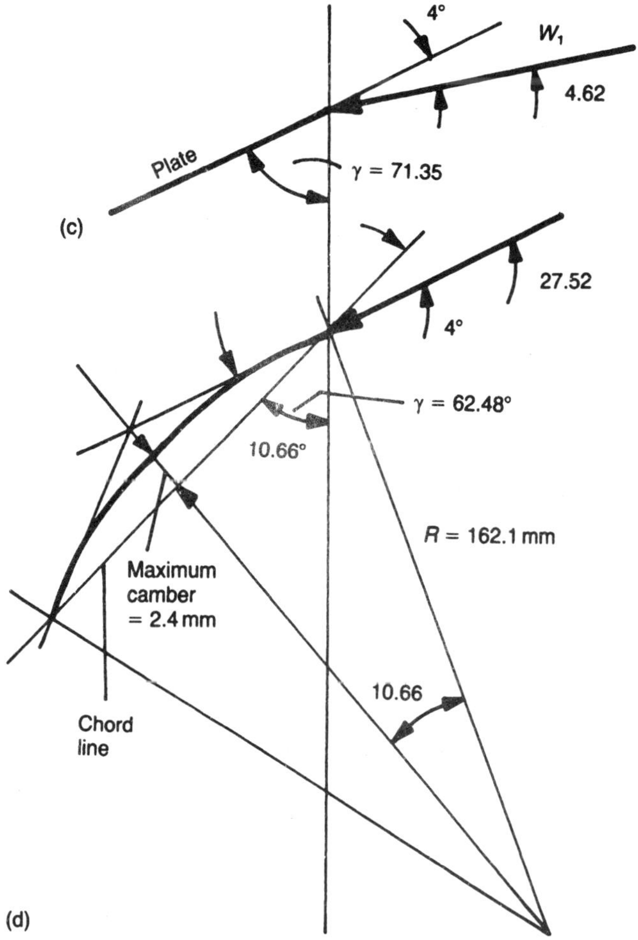


At the tip section $V_{u2} = 3.66$ and (Fig. A.9(a)) $\beta_1 = 14.62°$, $\beta_2 = 15.65°$ and $\beta_m = 15.14°$. Thus

$$\frac{194.4}{53.15^2} = \frac{13.86}{2 \times 53.15^2} \times C_L \times \frac{c}{s} \operatorname{cosec} 15.14$$

$$\left| C_L \frac{c}{s} \right|_{tip} = 0.1378$$

Similarly, for the hub section, $V_{u2} = 7.31$ and (Fig. A.9(b)) $\beta_1 = 27.52°$, $\beta_2 = 35.7°$ and $\beta_m = 31.61°$. This gives

$$\left| C_L \frac{c}{s} \right| = 0.553$$

Since the deflections are very small, flat plate data can be used at the tip, and at the hub a cambered plate. Assuming a camber of 4%,

$$\text{maximum } \frac{C_L}{C_D} \text{ is at } \alpha = 4^\circ \text{ and } C_L = 0.8$$

Therefore

$$0.8 \times \frac{c}{s} = 0.553, \qquad \frac{s}{c} = 1.447.$$

Since $h = 175\,\text{mm}$ if $b/c = 3/1$, $c = 60\,\text{mm}$. This gives $s_{\text{hub}} = 0.087$ and $Z = 12.64 = 13$ blades.

At the tip, $C_L = 0.4$, $s/c = 2.903$ and $C_{\text{tip}} = 0.76\,\text{m}$. For the tip section, Fig. A.9(c) shows $\gamma = 71.35^\circ$ while, for the hub section, Fig. A.9(d) shows $\gamma = 62.48^\circ$ with a camber radius of 162.1 mm.

*Comment*: A constant section may be possible, with a 2% camber at the hub.

7.3

$$u_H = 7.15 \times 2.5/2 = 8.9375\,\text{m}\,\text{s}^{-1}$$

$$V_A = 50 \Big/ \left(\frac{\pi}{4}\right)(5^2 - 2.5^2) = 3.4\,\text{m}\,\text{s}^{-1}$$

$$\text{Hub spacing} = \frac{\pi \times 2.5}{12} = 0.654$$

Therefore

$$|c/s|_{\text{hub}} = 0.765$$

$$\frac{g\,2.5}{0.92 \times 8.9375^2} = \frac{3.4}{2 \times 8.9375} \times 0.96 \times 0.765 \operatorname{cosec} \beta_m$$

$$(C_L = 0.96 \text{ at } C_L/C_D \text{ max @ } 5^\circ\ \alpha)$$

In Fig. A.10(a) $\beta_m = 24.75$, and $\beta_2 = 20.83^\circ$. Now

$$\beta_m = \frac{\beta_1 + \beta_2}{2} = 24.75$$

So that $\beta_1 = 28.67^\circ$, $V_{u1} = 2.72\,\text{m}\,\text{s}^{-1}$ and

$$gH_E = 2.72 \times 8.9375 = 24.3\,\text{J}\,\text{kg}^{-1}$$

Hydraulic power $= 50 \times 10^3 \times 24.3 = 1.22\,\text{MW}$

Shaft power $= 1.22 \times 0.92 = 1.12\,\text{MW}$

$$90 - \gamma = 23.67^\circ = \text{setting angle in Fig. A.10(b)}$$

$$\gamma = 66.33^\circ$$

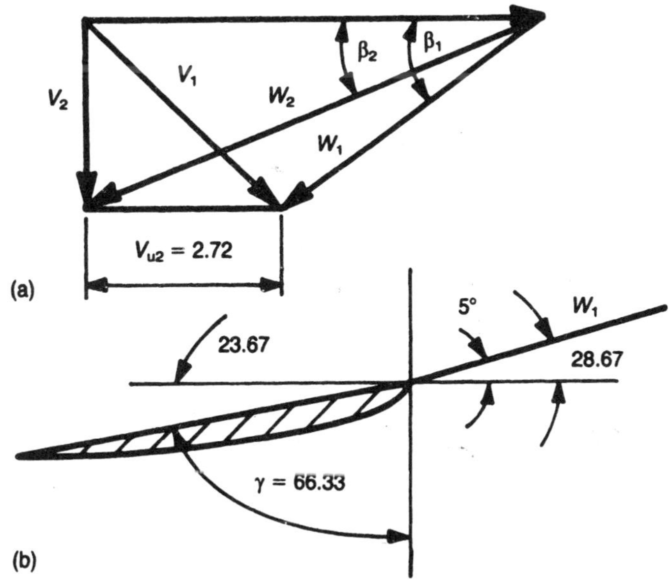


**Figure A.10** Solution 7.3.

7.4 Electrical output = 45 MW
Hydraulic power = g × 11 × 500 × 10³ = 53.955 MW

$$\eta_o = 0.834 \left[ = \frac{45}{53.955} \right]$$

$$\eta_H = \frac{0.834}{0.96 \times 0.97} = 0.896$$

Therefore

$$gH_E = 11\,g \times 0.896 = 96.69\,\mathrm{J\,kg^{-1}}$$
$$= u_1 V_{u4}, \text{ since } V_{u2} = 0$$

$$u_{tip} = 65.2 \times \frac{\pi}{30} \times 4 = 27.31\,\mathrm{m\,s^{-1}}$$

$$u_{hub} = 9.56\,\mathrm{m\,s^{-1}}$$

$$V_{u1tip}\ 3.54\,\mathrm{m\,s^{-1}} \text{ and } V_A = 11.336\,\mathrm{m\,s^{-1}}$$

From the tip velocity diagram (Fig. A.11(a)), $\beta_1 = 25.5°$ and $\beta_2 = 22.54°$. Assuming the free vortex rule applies,

$$V_{u1hub} = 3.54 \times \frac{1}{0.35} = 10.114\,\mathrm{m\,s^{-1}}$$

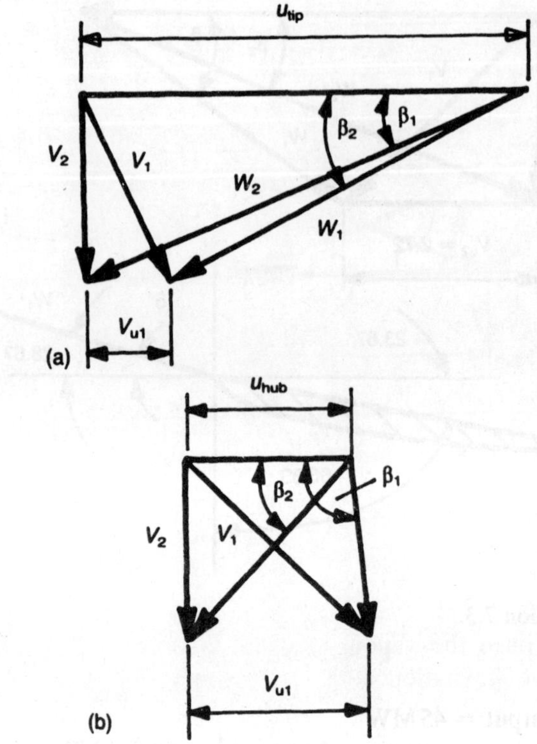


**Figure A.11** Solution 7.4.

From the hub velocity diagram (Fig. A.11(b)), $\beta_1 = 92.8°$ and $\beta_2 = 49.86°$. For the draft tube

$$\text{inlet area} = \pi\, 8^2/4 = 50.27\,\text{m}^2$$
$$\text{outlet area} = 1.5 \times 50.27 = 75.4\,\text{m}^2$$

Therefore the inlet velocity is $9.95\,\text{m s}^{-1}$ and the outlet velocity is $6.63\,\text{m s}^{-1}$. Theoretical draft tube regain $= \tfrac{1}{2}(9.95^2 - 6.63^2) = 27.52\,\text{J kg}^{-1}$; actual regain $= 0.8 \times 27.52 = 22.018\,\text{J kg}^{-1}$; draft tube loss $= 5.502\,\text{J kg}^{-1}$.

Applying the energy equation to the draft tube

$$\frac{p_A}{\rho} = \frac{9.95^2}{2} + 8g = \frac{10^5}{10^3} + \frac{6.63^2}{2} + 0 + 5.502$$

and, taking the draft-tube exit centreline as datum,

$$\frac{p_A}{\rho} = 58.36\,\text{J kg}^{-1}, \qquad p_A = 0.584 \times 10^5\,\text{N m}^{-2}\ \text{(abs)}$$

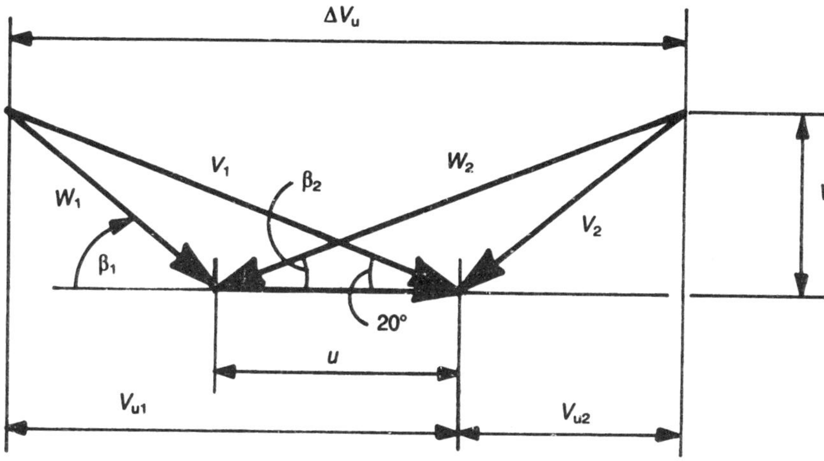


**Figure A.12** Solution 8.1.

*Comment*: Since the vapour pressure is only $2500\,\mathrm{N\,m^{-2}}$ there is no general risk of cavitation.

8.1 From steam tables $v_g = 3.992\,\mathrm{m^3\,kg^{-1}}$.

$$\dot{m} = 20\,000/3600 = 5.56\,\mathrm{kg\,s^{-1}}$$

which gives

$$Q = \frac{20\,000}{3600} \times 3.992 \times 0.93 = 20.64\,\mathrm{m^3\,s^{-1}}$$

Power output $= 0.25 \times 10^6\,\mathrm{J\,s^{-1}}$ and losses = 30%, therefore

$$\text{hydraulic input} = \frac{0.25 \times 10^6}{0.7} = 0.357 \times 10^6\,\mathrm{J\,s^{-1}}$$

$$\frac{0.357 \times 10^6}{5.56} = 0.642 \times 10^5 = u\Delta V_u$$

$$V_A = 0.7u = 0.7 \times \frac{1500\pi}{30} \times \frac{D_m}{2} = 54.98D_m$$

and $u = 78.54D_m$.

From the velocity triangles at mean blade height (Fig. A.12)

$$V_{u1} = V_A \cot 20 = 151.06D_m$$

$$V_{u2} = V_A \cot 20 - u = 151.06D_m - 78.54D_m = 72.54D_m$$

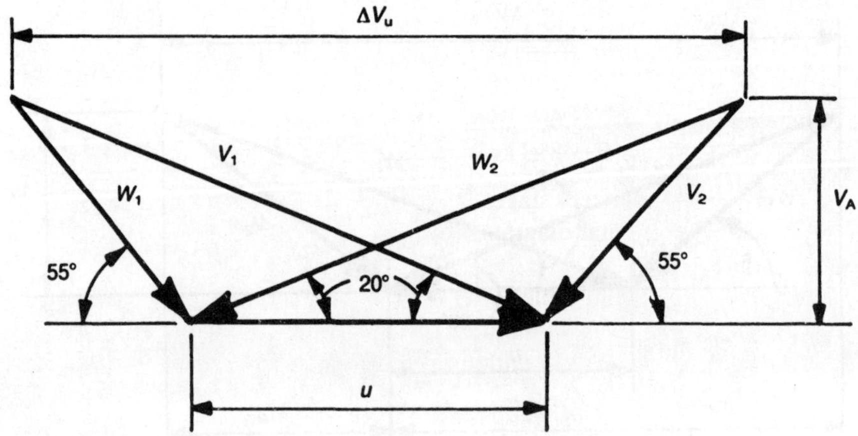


**Figure A.13** Solution 8.2.

Therefore

$$0.642 \times 10^5 = 78.54 D_m^2 (151.06 + 72.52)$$

$$D_m = 1.912\,\text{m}$$

Assuming blade height small compared to $D_m$

$$u_m = \frac{1500\pi}{30} \times \frac{1.912}{2} = 150.17\,\text{m}\,\text{s}^{-1}$$

Therefore

$$V_A = 105.12\,\text{m}\,\text{s}^{-1}$$

$$\frac{20.64}{105.12} = \pi \times 1.912 \times h, \qquad h = 0.033\,\text{m}$$

Therefore the tip diameter is 1.945 m and the hub diameter is 1.879 m.

8.2 Mean diameter = 1.5 m

$$U_m = \frac{1.5}{2} \times \frac{\pi \times 3000}{30} = 235.6\,\text{m}\,\text{s}^{-1}$$

The velocity diagrams can now be drawn for the mean section, and Figure A13 results

$$\therefore V_A = 15.1\,\text{m}\,\text{s}^{-1}$$

$$\Delta V_u = 396.8\,\text{m}\,\text{s}^{-1}$$

From tables, $v_g = 3.42\,\text{m}^3\,\text{kg}^{-1}$ at the outlet

$$\therefore \dot{m} = \frac{115.1 \times \pi \times 1.5 \times 0.035}{3.42} = 5.55\,\mathrm{kg\,s^{-1}}$$

$$\begin{aligned} \text{Diagram work} &= 235.6 \times 396.8 \\ &= 93.48\,\mathrm{kJ\,kg^{-1}} \text{ per stage} \end{aligned}$$

Since 4 stages

$$\begin{aligned} \text{Total diagram work} &= 4 \times 93.48 \\ &= 373.94\,\mathrm{kJ\,kg^{-1}} \\ \text{from tables inlet enthalpy} &= 3793\,\mathrm{kJ\,kg^{-1}} \\ \text{outlet enthalpy} &= 2683\,\mathrm{kJ\,kg^{-1}} \\ \therefore \text{enthalpy change} &= 1110\,\mathrm{kJ\,kg^{-1}} \\ \therefore \eta_{\mathrm{THERMAL}} &= \frac{373.94}{1110} = 33.7\% \end{aligned}$$

8.3 $\text{Total diagram work} = (\dot{m}u\Delta V_u) \times 10$

$$= 16.5\,\mathrm{MW}$$

$$\therefore u\Delta V_u = \frac{16.5 \times 10^6}{10 \times 70} = 23571\,\mathrm{J\,kg^{-1}} \text{ for one stage,}$$

$$u_m = \frac{3000\pi}{30} \times \frac{1.5}{2} = 235.62\,\mathrm{m\,s^{-1}}$$

Figure A14(a) can be drawn,

$$\begin{aligned} \therefore \Delta V_u &= 100.04\,\mathrm{m\,s^{-1}} \\ V_{u2} &= 67.81\,\mathrm{m\,s^{-1}} \\ V_{u1} &= 167.81\,\mathrm{m\,s^{-1}} \\ \beta_1 &= 55.86^\circ \\ \beta_2 &= 30.79^\circ \\ \varepsilon &= 25.09^\circ \\ V_1 &= W_2 = 195.34\,\mathrm{m\,s^{-1}} \\ V_2 &= W_1 = 120.83\,\mathrm{m\,s^{-1}} \end{aligned}$$

$$T_{01} = T_1 + \frac{195.34^2}{2 \times 1.145 \times 10^3} = 1000,$$

$$T_1 = 983.3\,\mathrm{K}$$

$$a = \sqrt{(1.33 \times 287 \times 983.3)} = 612.66\,\mathrm{m\,s^{-1}}$$

$$\text{maximum } M_n = 0.32$$

$$\frac{P_0 - p}{\frac{1}{2}\rho v^2} = \left[1 + \frac{M_n^2}{4} + \frac{(2-k)}{24} M_n^4\right]$$

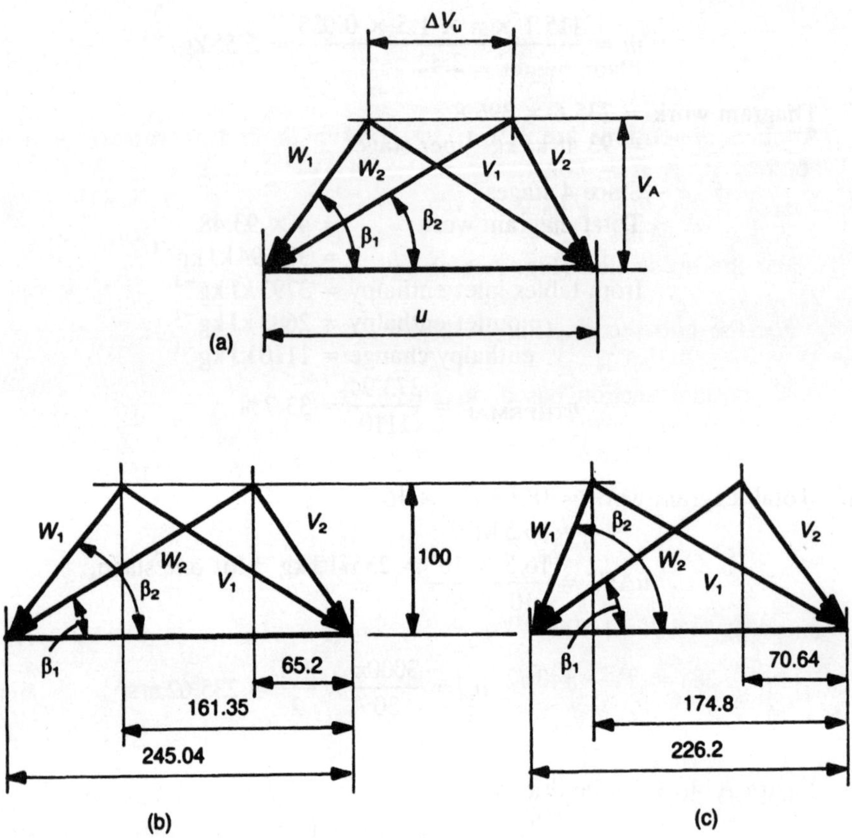


**Figure A.14** Solution 8.3.

Since $M_n$ is low, $\rho$ can be assumed incompressible and $k = 1.33$. Thus

$$\frac{P_0 - p}{\frac{1}{2}\rho v^2} = 1.0256\rho = \frac{p}{287 \times 983.3} = 3.543 \times 10^{-6}p$$

Therefore

$$7.5 \times 10^5 - p = \frac{1.0256}{2} \times 195.34^2 \times 3.543 \times 10^{-6}p$$

$$p = 7.0138 \times 10^5\,\text{N}\,\text{m}^{-2}$$

$$\rho = 2.485\,\text{kg}\,\text{m}^{-3}$$

$$Q = \frac{70}{2.485} = 28.17\,\text{m}^3\,\text{s}^{-1}$$

This gives

$$\text{blade height} = \frac{28.17}{100 \times \pi \times 1.5} = 0.06\,\text{m}$$

Annulus dimensions are 1.56 m and 1.44 m. Since free vortex blading applies, $V_u \times R = \text{const}$, $u_{tip} = 245.04\,\text{m s}^{-1}$, $V_{u1\,tip} = 161.35\,\text{m s}^{-1}$, $V_{u2\,tip} = 65.2\,\text{m s}^{-1}$, $u_{hub} = 226.2\,\text{m s}^{-1}$, $V_{u1\,hub} = 174.8\,\text{m s}^{-1}$ and $V_{u2\,hub} = 70.64\,\text{m s}^{-1}$.

For the tip section (Fig. A.14(b)) $\beta_1 = 50.08°$, $\beta_2 = 29.08°$ and $\varepsilon = 21°$.

For the hub section (Fig. A.14(c)) $\beta_1 = 62.8°$, $\beta_2 = 32.7°$ and $\varepsilon = 31.1°$.

A constant section based on mean section will give acceptable incidence angles.

8.4 For air, $C_p = 1.005 \times 10^3\,\text{J kg}^{-1}$ so that for the first stage

$$1.005 \times 10^3 \times 20 = 0.925 u \Delta V_u$$

$$u \Delta V_u = 21.73\,\text{kJ kg}^{-1}$$

Since $V_A = V_1 = 140\,\text{ms}^{-1}$

$$288 = T_1 + 140^2/2 \times 1.005 \times 10^3, \qquad T_1 = 278.25\,\text{K}$$

$$\frac{P_{01}}{P_1} = \left(\frac{T_{01}}{T_1}\right)^{\gamma/(\gamma-1)}, \qquad p_1 = 0.895 \times 10^5\,\text{N M}^{-2}$$

$$\rho_1 = \frac{0.895 \times 10^5}{287 \times 278.25} = 1.121\,\text{kg m}^{-3}$$

$$a = \sqrt{(1.4 \times 287 \times 278.25)} = 334.4\,\text{m s}^{-1}$$

to give

$$W_1 = 0.95 \times 334.4 = 317.65\,\text{m s}^{-1}$$

Since $V_A = V_1$, then $V_{u1} = 0$ and

$$u = \sqrt{(317.65^2 - 140^2)} = 285.13\,\text{m s}^{-1}$$

$$\Delta V = V_{u2} = 76.21\,\text{m s}^{-1}$$

From Fig. A.15(a) $\beta_{1T} = 26.15°$ and $\beta_{2T} = 33.83°$. Since $u_T = 285.13\,\text{m s}^{-1}$

$$285.13 = \frac{6000\pi}{30} \times \frac{D_T}{2}, \qquad D_T = 0.908\,\text{m}$$

$$D_H = 0.6 \times 0.908 = 0.545\,\text{m}$$

$$\rho = 1.121\,\text{kg m}^{-3}$$

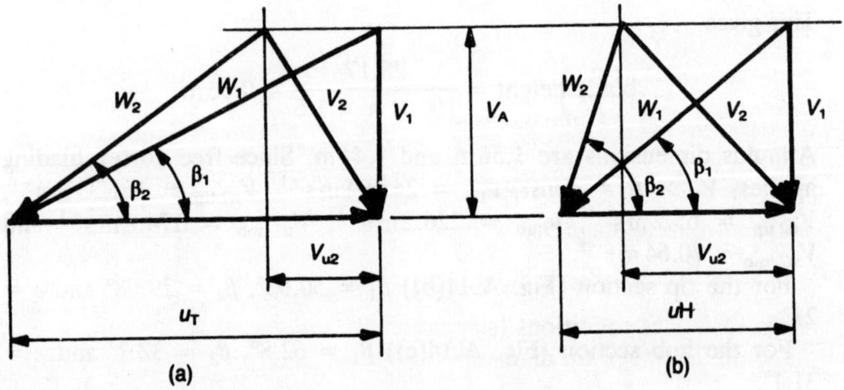


**Figure A.15** Solution 8.4.

and therefore

$$\dot{m} = 1.121 \times 140 \times \frac{\pi}{4}(0.908^2 - 0.545^2) = 65\,\text{kg s}^{-1}$$

$$\left|\frac{P_{02}}{P_{01}}\right|_{\text{stage 1}} = \left(1 + 0.89 \times \frac{20}{288}\right)^{3.5} = 1.233:1$$

If free vortex conditions, $V_{u1H}$ and $V_{u2H}$ are found:

$$V_{u1} = 0,$$

$$V_{u2H} \times 0.545 = 76.21 \times 0.908 = 126.97\,\text{m s}^{-1}$$

$$u_H = 171.14\,\text{m s}^{-1}$$

From Fig. A.15(b) $\beta_{1H} = 39.285°$ and $\beta_{2H} = 72.49°$

8.5 Since the fluid is helium $C_p = 5.193 \times 10^3$, $M = 4.003$ and

$$R = \frac{8.314}{4.003} = 2.077 \times 10^3\,\text{J kg}^{-1}\,\text{K}^{-1}$$

Thus $C_v = 3.116$ and

$$n = \frac{C_p}{C_v} = 1.67$$

$$\left(\frac{n}{n-1}\right) = 2.493$$

$$3.75 = \left(1 + 0.85 \times \frac{\Delta T}{350}\right)^{2.493}$$

Therefore $\Delta T = 287.8\,°\text{C}$ and $\Delta T_{\text{stage}} = 23.23\,°\text{C}$.

From data sheets $\eta_p = 0.87$ (Fig. 1.17) and $\Omega = 0.855$ (Fig. 4.20).

$$\text{stage}\left(\frac{p_2}{p_1}\right) = \left(1 + 0.87 \times \frac{23.33}{T_{01}}\right)^{2.493}$$

For the first stage $P_{02} = 5.71\,\text{bar}$ and $T_{02} = 373.23\,\text{K}$.

$$\left(\frac{P_{03}}{P_{02}}\right)_{\text{stage 2}} = \left(1 + 0.87 \times \frac{23.23}{373.23}\right)^{2.43}$$

$$P_{03} = 6.559\,\text{bar}, \qquad T_{03} = 396.46\,\text{K}$$

and so on up to stage 5:

$$P_{05} = 8.348\,\text{bar}, \qquad T_{05} = 442.92\,\text{K}$$

These are inlet conditions (stagnation) to the 5th stage.

At the 5th stage tip diameter

$$u_{\text{T}} = \frac{3.5}{2} \times \frac{3000\pi}{30} = 549.78\,\text{m}\,\text{s}^{-1}$$

Assuming 50% reaction at the tip,

$$5.193 \times 10^3 \times 23.23 = 0.855 \times \Delta V_{\text{u}} \times 549.78$$

$$\Delta V_{\text{u}} = 256.63\,\text{m}\,\text{s}^{-1}$$

also (Fig. A.16(a)) $V_1 = 189.45\,\text{m}\,\text{s}^{-1}$, $\varepsilon = 22.74°$, $\beta_1 = \alpha_2 = 73.43°$ (to axial) and $\beta_2 = \alpha_1 = 50.69°$ (to axial). Applying the free vortex laws

$$V_{\text{u1}} \times R = \text{const}$$

$$V_{\text{u2}} \times R = \text{const}$$

therefore the annulus dimensions are needed.

At inlet to stage 5, $T_{05} = 442.92\,\text{K}$ and $P_{05} = 8.348\,\text{bar}$ which gives

$$442.92 = T_5 + \frac{189.45^2}{2 \times 5.193 \times 10^3}$$

$$\text{T}_5 = 439.46\,\text{K}$$

$$\frac{P_{05}}{p_5} = \left(\frac{442.92}{439.46}\right)^{2.493}, \qquad p_5 = 8.157\,\text{bar}$$

Therefore

$$\rho = \frac{8.187 \times 10^5}{2.077 \times 10^3 \times 439.46} = 0.897\,\text{kg}\,\text{m}^{-3}$$

$$Q = \frac{350}{0.897} = 390.23\,\text{m}^3\,\text{s}^{-1}$$

$$\frac{390.23}{120} = \frac{\pi}{4}(3.5^2 - d_{\text{h}}^2), \qquad d_{\text{h}} = 2.524\,\text{m}$$

Thus $V_{\text{u1 hub}} = 203.26\,\text{m}\,\text{s}^{-1}$, $V_{\text{u2 hub}} = 559.13\,\text{m}\,\text{s}^{-1}$ and

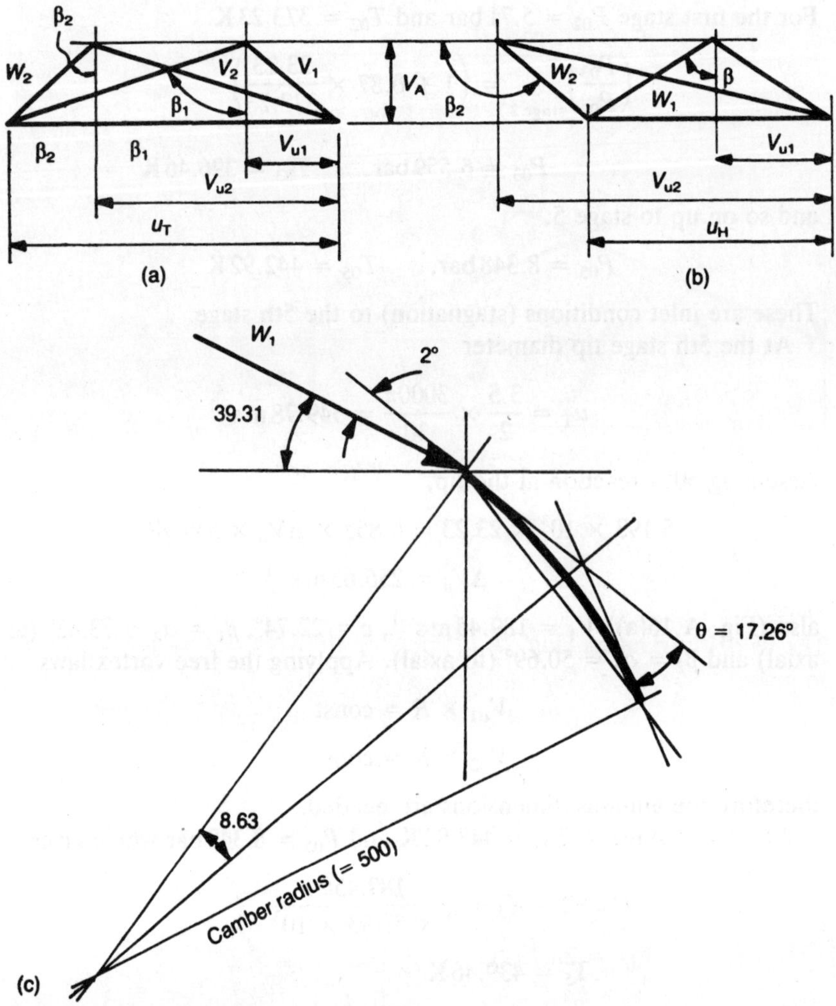


**Figure A.16** Solution 8.5.

$$u_{hub} = \frac{3000\pi}{30} \times \frac{2.524}{2} = 396.47\,\text{m s}^{-1}$$

From Fig. A.16(b) $\alpha_1 = 59.44°$, $\beta_1 = 58.16°$, $\alpha_2 = 77.89°$ and $\beta_2 = 143.58°$.

Returning to the tip section, Fig. A.16(c) shows the camber section, where $\theta/2 = 8.63°$, $\theta = 17.26°$, $\gamma = 48.69 - 8.63 = 40.06°$ to the axial direction, $\beta_1' = 48.69°$ to axial and $\beta_2' = \gamma - 8.63 = 31.43°$ to axial.

9.1 $$u_2 = \frac{3.3}{2} \times 18 = 29.7\,\text{m s}^{-1}$$

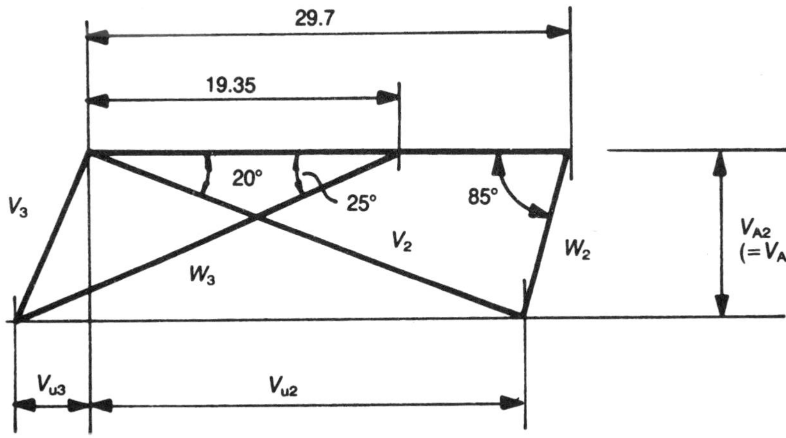


**Figure A.17** Solution 9.1.

$$u_3 = \frac{2.15}{2} \times 18 \text{ (mean dia)} = 19.35\,\text{m}\,\text{s}^{-1}$$

The velocity triangles (Fig. A.17) may be drawn at the mean diameter, to give

$$V_{u2} + W_2 \cos 85 = 29.7, \qquad 29.7 - W_2 \cos 85 = V_{u2}$$

$$V_{u2} \tan 20 = V_{A2} = W_2 \sin 85$$

therefore

$$(29.7 - W_2 \cos 85) \tan 20 = W_2 \sin 85$$

$$V_2 = 30.64\,\text{m}\,\text{s}^{-1},$$

$$V_{u2} = 28.79\,\text{m}\,\text{s}^{-1}, \qquad V_{A2} = 10.48\,\text{m}\,\text{s}^{-1}$$

$$A_2 = 3\,\text{m}^2, \qquad Q = 3 \times 10.48 = 31.44\,\text{m}^3\,\text{s}^{-1}$$

$$A_3 = 3.1\,\text{m}^2, \qquad V_{A3} = 10.14\,\text{m}\,\text{s}^{-1}, \qquad V_4 = 10.73\,\text{m}\,\text{s}^{-1}$$

Thus

$$V_{u3} = W_3 \cos 25 - 19.35$$

$$W_3 = 10.14 \text{ cosec } 25 = 23.998\,\text{m}\,\text{s}^{-1}$$

$$V_{u3} = 4.65, \qquad V_3 = 11.16\,\text{m}\,\text{s}^{-1}$$

Therefore

$$\text{Euler } gH = 28.79 \times 29.7 + 19.35 \times 4.65$$
$$= 855.06 + 89.98 = 945.04\,\text{J}\,\text{kg}^{-1}$$

and

$$\text{runner power} = 945.04 \times 31.44 \times 10^3 = 29.71\,\text{MW}$$

The total 'head' drop is found by applying the energy equations. Using (1) to indicate inlet to guide vanes, (2) as inlet to runner, (3) as inlet to draft tube and (4) as draft tube outlet, then from (1) to (2) over guide vanes

$$gh_1 + \frac{V_1^2}{2} + gZ_1 = gh_2 + \frac{V_2^2}{2} + gZ_2 + 0.05\frac{V_2^2}{2} \qquad \text{(A.5)}$$

From (2) to (3) over runner

$$gh_2 + \frac{V_2^2}{2} + gZ_2 = gh_3 + \frac{V_3^2}{2} + gZ_3 + 0.2\frac{W_3^2}{2} + \text{work done} \qquad \text{(A.6)}$$

From (3) to (4) over draft tube

$$gh_3 + \frac{V_3^2}{2} + gZ_3 = gh_A + \frac{V_4^2}{2} + gZ_4 + 0.5\frac{V_3^2}{2} \qquad \text{(A.7)}$$

From equation (A.7)

$$\begin{aligned} gh_3 &= gh_A - 31.14 + 115.13 - 31.88 \\ &= gh_A + 52.11 \end{aligned}$$

From equation (A.6)

$$gh_2 + \frac{30.64^2}{2} + g \times 4 = (gh_A \times 52.11) + \frac{11.16^2}{2}$$

$$+ 945.04 + g \times 25 + 0.2 \times \frac{23.998^2}{2}$$

$$gh_2 = gh_A + 640.26$$

From equation (A.5)

$$\begin{aligned} gh_1 &= gh_A + 640.26 + \frac{30.64^2}{2} + 0.05 \times \frac{30.64^2}{2} \\ &= gh_A + 1133.135 \end{aligned}$$

Total energy drop = 1133.135 J kg$^{-1}$, therefore $\eta_{\text{hydraulic}} = 0.834$.

9.2 From Fig. A.18(a)

$$\eta_{TS} = \frac{h_{01} - h_{03}}{h_{01} - h_{3ss}}\left(= \frac{C_P}{C_p}\left(\frac{T_{01} - T_s}{T_{01} - T_{3ss}}\right)\right)$$

$$0.86 = \frac{850 - 670}{\Delta T_{is}} \qquad \therefore \Delta T_{is} = 209.3\,\text{K}$$

$\Delta T_{is}$ for nozzle = 104.65 since 50% reaction.

$$\eta_N = \frac{C_P\,(T_{01} - T_2)}{C_p\,(T_{01} - T_{2s})} = 0.97, \qquad T_{01} - T_2 = 0.97 \times 104.65$$

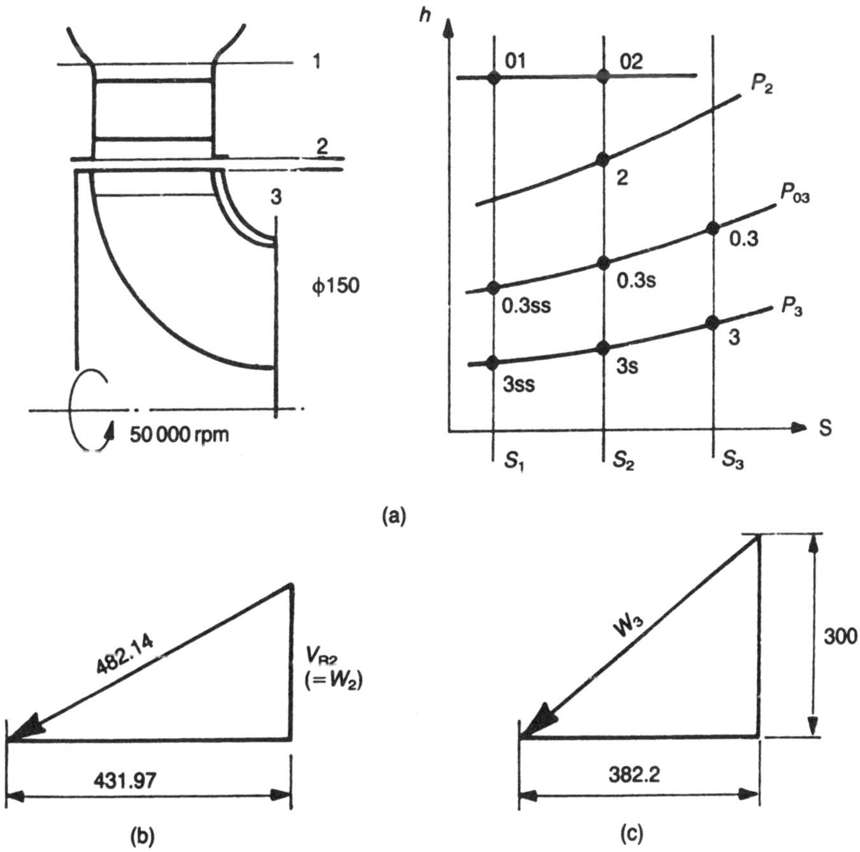


**Figure A.18** Solution 9.2.

$$T_2 = 748.5\,\text{K}$$

Thus nozzle velocity = $\sqrt{(1.145 \times 10^3 \times 101.51 \times 2)} = 482.14\,\text{m s}^{-1}$. The inlet triangle becomes Fig. A.18(b). Since

$$u_T = 50\,000\,\pi/30 \times \frac{0.165}{2} = 431.97\,\text{m s}^{-1}$$

$$\alpha_1 = 26.37°$$

$$W_2 = V_{R2} = 214.15\,\text{m s}^{-1}$$

Peripheral nozzle area = $\pi \times 0.175 \times 0.03 \times 0.85 = 0.014\,\text{m}^2$. Now, $p_2$ is required in order to find $\rho_2$:

$$\frac{P_{02}}{p_2} = \left(\frac{850}{748.5}\right)^4 = 1.67$$

$$p_2 = \frac{3 \times 10^5}{1.67} = 1.8 \times 10^5\,\mathrm{N\,m^{-2}}$$

Thus

$$\rho_2 = \frac{1.8 \times 10^5}{287 \times 748.5} = 0.838\,\mathrm{kg\,m^{-3}}$$

$$\dot{m} = 0.014 \times 214.15 \times 0.838 = 2.52\,\mathrm{kg\,s^{-1}}$$

Assuming $V_{A3} = V_3 = 300$

$$670 = T_3 + \frac{300^2}{2 \times 1.145 \times 10^3}, \qquad T_3 = 630.7\,\mathrm{K}$$

$$\frac{P_{01}}{p_3} = \left(\frac{850}{640.7}\right)^4, \qquad p_3 = 0.968 \times 10^5\,\mathrm{N\,m^{-2}}$$

$$\rho_3 = 0.502\,\mathrm{kg\,m^{-3}}$$

$$\frac{2.52}{0.502} = 300 \times A_3$$

$$A_3 = 0.0167\,\mathrm{m^2}$$

$$D_3 = \sqrt{\frac{4}{\pi} \times 0.0167} = 0.146\,\mathrm{m}$$

Thus

$$u_{3\,\mathrm{tip}} = \frac{50\,000\pi}{30} \times \frac{0.146}{2} = 382.2\,\mathrm{m\,s^{-1}}$$

From Fig. A. 18(c)

$W_3 = 485.9\,\mathrm{m\,s^{-1}}$, $\quad a_3 = \sqrt{(1.33 \times 287 \times 630.7)} = 490.66\,\mathrm{m\,s^{-1}}$

Thus $M_n = 0.99$

9.3 Since zero outlet whirl $u_{out}$ is not needed and

$$u_T = \frac{0.15}{2} \times 50000 \times \frac{\pi}{30} = 392.7\,\mathrm{m\,s^{-1}}$$

The inlet gas triangle (ideal) is shown in Fig. A.19.

$$479.4 = \sqrt{(1.145 \times 10^3 \times \Delta T \times 2)}$$

$$\Delta T = 10^\circ\mathrm{C} = T_{01} - T_2$$

$$\eta_N = \frac{T_{01} - T_2}{T_{01} - T_{2s}} = 0.97, \qquad T_{2s} = 839.69\,\mathrm{K}$$

Since 50% reaction, isentropic turbine drop = 20.62° and

$$\eta_{TS} = 0.86 = \frac{C_p(850 - T_3)}{C_p(20.62)}$$

$$T_3 = 832.27\,\mathrm{K}$$

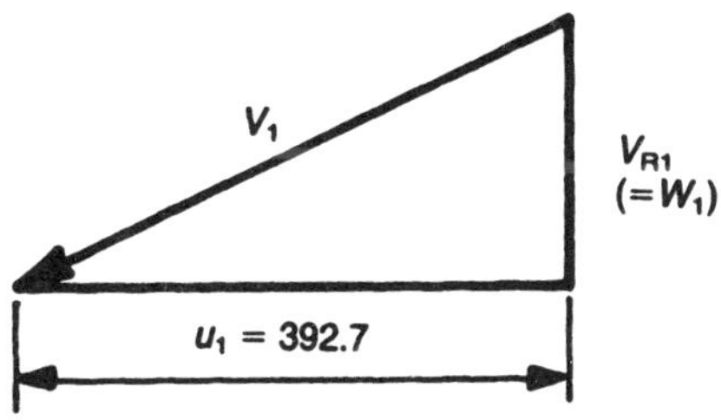


**Figure A.19** Solution 9.3.

and $$T_{3s} = 850 - 20.62 = 829.38\,\text{K}$$

$$\frac{p_{01}}{p_3} = \left(\frac{850}{829.38}\right)^4 = 1.103$$

$$p_3 = 2.72 \times 10^5\,\text{N m}^{-2}$$

For the nozzle guide vanes

$$\frac{p_{01}}{p_2} = \left(\frac{850}{839.69}\right)^4 = 1.05$$

$$p_2 = 2.857 \times 10^5\,\text{N m}^{-2}$$

and

$$\rho_2 = \frac{2.857 \times 10^5}{287 \times 839.69} = 1.186\,\text{kg m}^{-3}$$

Now, $\dot{m} = 0.56\,\text{kg s}^{-1}$, therefore $Q = 0.472\,\text{m}^3\text{s}^{-1}$ and, for the throat

$$A = \frac{0.472}{479} = 9.86 \times 10^{-4}\,\text{m}^2$$

The 'diagram' power is

$$\dot{m} \times uV_{u1} = 0.56 \times 392.7^2 = 83.36\,\text{kW}$$

9.4 At plane 2, $N = 50\,000$, $u_2 = 392.7\,\text{m s}^{-1}$ and $u_{3T} = 274.9\,\text{m s}^{-1}$. From Fig. A.20(b)

$$V_2 = 479.4\,\text{m s}^{-1} = \sqrt{(2 \times 1.145 \times 10^3 \times \Delta T_{ACT})}$$

$$\Delta T_{ACT} = 100.36 \text{ (nozzle)}$$

Turbine $\Delta T = 2 \times 100.36 = 200.72\,°\text{C}$, $T_{01} = 1000\,\text{K}$.

$$\eta_{TS} = 0.83 = \frac{200.72}{1000 - T_{3s}}, \qquad T_{3s} = 758.18\,\text{K}$$

$$\frac{P_{01}}{p_3} = \frac{3.5 \times 10^3}{p_3} = \left(\frac{1000}{758.19}\right)^4, \qquad p_3 = 1.157 \times 10^5\,\text{N m}^{-2}$$

$$\rho_3 = \frac{1.157 \times 10^5}{287 \times 758.19} = 0.532\,\text{kg m}^{-3}$$

Power generated = 120 kW, therefore

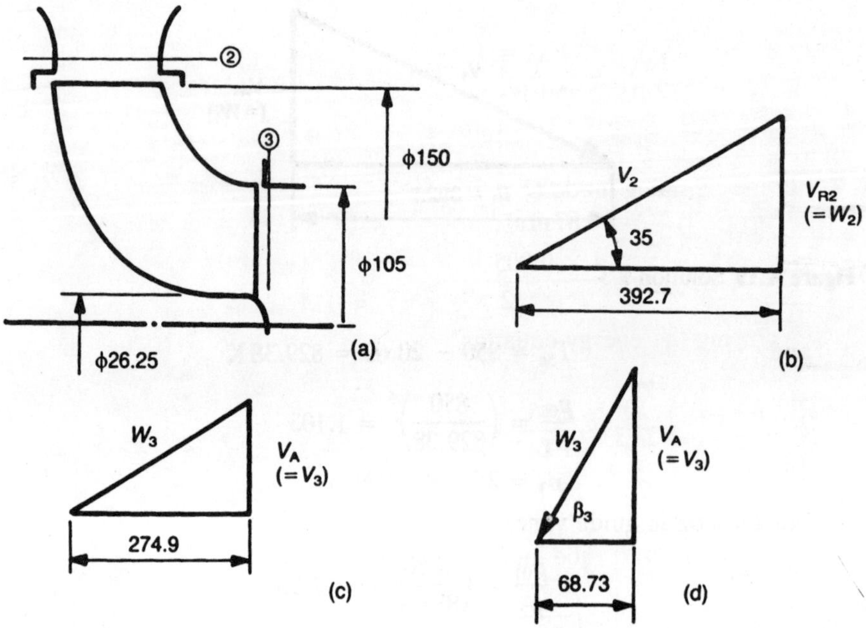


**Figure A.20** Solution 9.4.

$$\frac{120 \times 10^3}{0.8} = 1.145 \times 10^3 \times 200.8 \times \dot{m}$$

$$\dot{m} = 0.653\,\mathrm{kg\,s^{-1}}$$

Area of flow $= \frac{\pi}{4}(1.05^2 - 0.02625^2) = 0.0081\,\mathrm{m^2}$, therefore

$$0.653 = 0.0081 \times 0.532 \times V_{A3}, \qquad V_{A3} = 151.2\,\mathrm{m\,s^{-1}}$$

Thus at the exducer the maximum diameter velocity triangle is given by Fig. A.20(c):

$$\beta_3 = 28.8°$$

$$W_3 = 313.7\,\mathrm{m\,s^{-1}}$$

$$a = \sqrt{(1.33 \times 287 \times 758.19)} = 537.96\,\mathrm{m\,s^{-1}}$$

$$\text{relative } M_n = 0.583$$

At the exducer the hub diameter $u_H = 68.73$ and (Fig. A.20(d)) $\beta_3 = 65.56°$.

*Second part of solution*

If 0% reaction then $\Delta T_{\text{nozzle}}$ = turbine $\Delta T$ and, following the solution through

$$\rho_3 = 0.829$$

$$\dot{m} = 2 \times 0.653 = 1.306\,\mathrm{kg\,s^{-1}}$$

$$V_{A3} = 194.49\,\mathrm{m\,s^{-1}}$$

$a$ is now $579\,\mathrm{m\,s^{-1}}$ and $W_3 = 336.74\,\mathrm{m\,s^{-1}}$, which gives

$$\text{relative } M_n = 0.582$$

9.5 In the pumping mode $Q = 74\,\mathrm{m^3\,s^{-1}}$ in the draft tube, which gives $V_2 = 2.16\,\mathrm{m\,s^{-1}}$, $V_1 = 4.87\,\mathrm{m\,s^{-1}}$ and $V_m = 3.52\,\mathrm{m\,s^{-1}}$. The friction loss is

$$\frac{4 \times 0.005 \times 78 \times 3.52^2}{2 \times 5.5} = 1.76\,\mathrm{J\,kg^{-1}}$$

Applying the energy equation from draft tube exit to its inlet (planes 2 to 1)

$$\frac{10^5}{10^3} + \frac{2.16^2}{2} + 29g = \frac{p_1}{10^3} + \frac{4.87^2}{2} + 0 + 1.76$$

thus $p_1 = 0.373 \times 10^6\,\mathrm{N\,m^{-2}}$.

In turbining mode $Q = 76.5\,\mathrm{m^3\,s^{-1}}$, $V_1 = 5.03\,\mathrm{m\,s^{-1}}$ and $V_2 = 2.23\,\mathrm{m\,s^{-1}}$. Draft tube loss is

$$0.8\left(\frac{5.03^2}{2} - \frac{2.23^2}{2}\right) = 8.13\,\mathrm{J\,kg^{-1}}$$

Thus thc lowest draft tube pressure is in pumping mode and, since the inlet head is 318 m

$$p_{\text{inlet}} = 318g \times 10^3 = 3.1196 \times 10^6\,\mathrm{N\,m^{-2}}$$

Applying the energy equation from draft tube inlet to its exit,

$$\frac{p_1}{10^3} + \frac{5.03^2}{2} + 0 = \frac{10^5}{10^3} + \frac{2.23^2}{2} + 29g + 8.13$$

$$p_1 = 0.383 \times 10^6\,\mathrm{N\,m^{-2}}$$

$$\sigma = \frac{0.373 \times 10^6}{3.1196 \times 10^6} = 0.1197$$

Applying modelling concepts (Chapter 2), if the full size machine has subscript 'p', and model has subscript 'm', then

$$Q_p = 74, \qquad H_p = 318, \qquad N_p = 333, \qquad \eta_p = 88.5\%$$

$$\frac{Q_p}{N_p D_p^3} = \frac{Q_m}{N_m D_m^3}, \qquad \frac{H_p}{N_p^2 D_p^2} = \frac{H_m}{N_m^2 D_m^2}$$

$$\frac{74}{333 \times 10^3} = \frac{Q_m}{3000 \times 1^3}, \qquad Q_m = 0.667\,\mathrm{m^3\,s^{-1}}$$

$$\frac{318}{333^2 \times 10^2} = \frac{H_m}{3000^2 \times 1^2} = 258\,\mathrm{m}$$

Using the Hutton formula (Table 2.1)

$$\frac{1 - \eta_p}{1 - \eta_m} = 0.3 + 0.7\left(\frac{R_{EM}}{R_{EP}}\right)^{0.2} = 0.3 + 0.7\left(\frac{3000}{333} \times \frac{1}{10^2}\right)^{0.2}$$

Since $\eta_p = 88.5$, $\eta_m = 84.3\%$ and therefore

$$\text{model power} = \frac{74 \times 10^3 \times 258g}{0.843} = 2.22\,\text{MW}$$

9.6 Total energy drop over unit is

$$\left(100g + \frac{10^5}{10^3}\right) - \frac{10^5}{10^3} - 0.5g = 99.5g\,\text{J kg}^{-1}$$

Hydraulic power is

$$\frac{50 \times 10^6}{0.9} = 10^3 \times Q \times 99.5g$$

$$Q = 56.92\,\text{g m}^3\,\text{s}^{-1}$$

The radial velocity at entry to runner is

$$\frac{56.92}{\pi \times 6 \times 0.3} = 10.07\,\text{m s}^{-1}$$

The velocity at entry to draft tube is

$$\frac{56.92}{\frac{\pi}{4} \times 4^2} = 4.53\,\text{m s}^{-1}$$

The velocity at exit from draft tube is

$$\frac{56.92}{28.3} = 2.011\,\text{m s}^{-1}$$

Looking at the draft tube (DT)

$$\frac{p_{in}}{\rho} + \frac{4.53^2}{2} + 0 = \frac{p_A}{\rho} + \frac{2.011^2}{2} + 2g + \text{DT loss}$$

$$\text{DT loss} = 0.83\left[\frac{4.52^2}{2} - \frac{2.011^2}{2}\right] = 6.83\,\text{J kg}^{-1}$$

$$P_{in} = 0.1175 \times 10^6\,\text{N m}^{-2}\ \text{(absolute)}$$

Considering now the guide vane path, from Fig. A.21(a):

$$u = 29.3 \times \tfrac{6}{2} = 87.9\,\text{m s}^{-1}$$

$$10.07 = V_1 \sin 25, \qquad V_1 = 23.83\,\text{m s}^{-1}$$

therefore guide vane loss is

$$0.06 \times \frac{23.83^2}{2} = 17.03\,\text{J kg}^{-1}$$

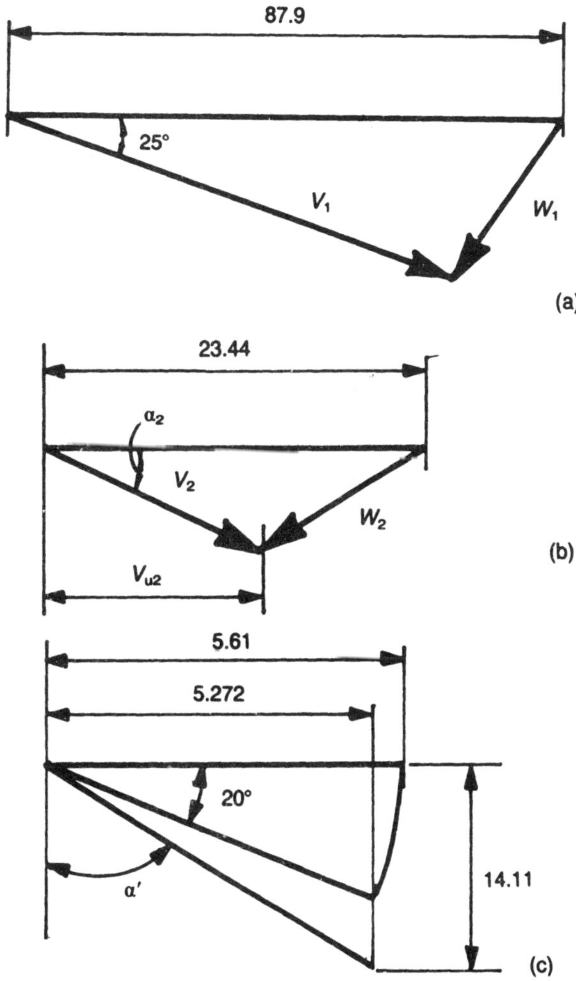


**Figure A.21** Solution 9.5.

Applying the energy equation through the guide vanes

$$100g = \frac{p}{10^3} + \frac{23.83^2}{2} + 0 + 17.03$$

giving the pressure at inlet to the runner of $0.68 \times 10^6\,\mathrm{N\,m^2}$. The inlet velocity triangle is (Fig. A.21(b)) $\alpha_2 = 21.69°$ and $\beta_2 = 31.02°$.

Construction as shown by Fig. 7.7, gives Fig. A.21(c); the nozzle angle to triangle is

$$\alpha' = \tan^{-1}\frac{5.272}{14.11} = 20.49°$$

# References


Abbot, I.M. and Doenhoff, A.E. (1959) *Theory of Wing Sections*, Dover, New York.

Addison, H. (1955) *Centrifugal and Other Rotodynamic Pumps*, 2nd edn, Chapman & Hall, London.

Ahmad, K., Goulas, A. and Baker, R.C. (1981) *Performance Characteristics of Granular Solids Handling Centrifugal Pumps – A literature Review*, BHRA Technical Note 1666: March, British Hydromechanics Research Association.

Ainley, D.G. and Matheison, G.C.R. (1957) *A Method of Performance Estimation for Axial Flow Turbines*, Aero. Res. Council R. and M. 2974.

Anderson, H.H. (1938) Mine Pumps. *Journal of the Mining Society*, Durham.

Anderson, H.H. (1977) *Statistical Records of Pumps and Water Turbines effectiveness*. Proceedings of I. Mech. E. Conf.: Scaling for Performance Prediction in Rotodynamic Machines, Sept., pp. 1–6, Stirling.

Anderson, H.H. (1984) The Area Ratio System. *World Pumps*, p. 201.

Anghern, R., Holler, K. and Barp, B. (1977) A Comparison of theoretical Stress calculations with Experimentally verified stresses in Francis Turbines of High Specific Speed. *Escher-Wyss News*, **50**(1), 25–28.

Balje, O.E. (1981) *Turbomachines*, J Wiley, New York.

Bain, A.G. and Bonnington, S.T. (1970) *The Hydraulic Transport of Solids by Pipe-line*, Pergamon, Oxford.

Barp, B., Schweizer, F. and Flury, E. (1973) Operating Stresses on Kaplan Turbine Blades. *Escher-Wyss News*, **46**(2), 10–15.

Benson, R.S.A. (1970) A Review of Methods of assessing Loss Coefficients in Radial Gas Turbines. *J. Mech. Eng. Science*, **12**, 905–932.

Bosman, C. and Marsh, H. (1974) An Improved method for calculating the flow in Turbomachines including a consistent Loss Model. *J. Mech. Eng. Science*, **16**(1), 25–31.

Bragg, S.L. and Hawthorne, W.R. (1956) Some exact solutions of the flow through Annular Cascade Discs. *J. Aero. Science*, April, 243–249.

Brown, L.E. (1972) Axial flow Compressor and Turbine Loss coefficients: a comparison of several parameters. *Trans. ASME.* **94**. Ser. A, (3), 193–201.

Bunjes, J.H. and Op de Woerd, J.G.H. (1982) *Centrifugal Pump Performance prediction oy Slip and Loss Analysis.* Proceedings of Inst. Mech. Eng. Conf.: Centrifugal Pumps – Hydraulic Design, 16 Nov. 1982, London, pp. 17–24.

Burgess, K.E. and Reizes, J.A. (1976) The effects of Sizing, Specific Gravity, and Concentration on the Performance of Centrifugal Pumps. *Proc. I. Mech. E.*, **190**, 391–398.

Burn, J.D. (1978) Steam Turbines for Industrial Mechanical Drives. *GEC J. Sci. Tech.*, **45**(1), 3–10.

Burton, J.D. (1966) *A Theoretical and Experimental Analysis of the flow in Regenerative Pumps and Turbines.* PhD thesis, Southampton University.

Busemann, A. (1928) Das Forderhohenverhaltnis Radialer Kreiselpumpen mit Logarithmish-spiralgen Schaufeln. *Z. Agnew. Math. Mechm.*, **18**, 372.

Bush, A.R., Fraser, W.H. and Karassik, I.J. (1976) Coping with Pump Progress: The sources and solutions of Centrifugal Pump Pulsations, Surges and Vibrations, *Pump World*, **2**(1) (Worthington Pump Corp Inc.).

Carchedi, F. and Wood, G.R. (1982) Design and Development of a 12:1 pressure ratio Compressor for the Ruston 6MW Gas Turbine. *Trans. ASME. J. Eng. Power*, **104**, 823–831.

Carter, A.D.S. (1948) Three Dimensional Flow Theories for Axial Flow Compressors and Turbines. *Proc. I. Mech. E.*, **159**(41), 255–268.

Carter, A.D.S. (1961) Blade Profiles for Axial Flow Fans Pumps and Compressors. *Proc. I. Mech. E.*, **175**, 775–806.

Chacour, S. and Graybill, J.E. (1977) IRIS: A computerised High Head Pump-Turbine Design System. *Trans. ASME. J. Fluids Eng.*, **99**, 567–579.

Chen, H. and Baines, N.C. (1992) Analytical Optimisation Design of Radial and Mixed flow Turbines. *Proc. I. Mech. E.*, **206**, 177–187.

Chiappe, E.A. (1982) *Pump Performance prediction using Graphical Techniques and Empirical Formulae.* Proceedings of I. Mech. E. Conf.: Centrifugal Pumps-Hydraulic Design, 28 November 1982, London, pp. 37–44.

Csanady, G.J. (1964) *Theory of Turbomachines.* McGraw-Hill, New York.

Craig, H.R.M. and Cox, A. (1970/71) Performance Estimation of Axial Flow Turbines. *Proc. I. Mech. E.*, **185**, 407–424.

Dalgleish, W. and Whitaker, J. (1971) *Scale Effects for Backward Curved Centrifugal Fans.*, Report 940, November, National Engineering Laboratory.

Dubas, M. and Schuch, M. (1987) Static and Dynamic Analysis of Francis Turbine Runners. *Sulzer Tech. Review*, **3**, 32–51.

Duncan, A.B. (1986) *A Review of the Pump Rotor Axial equilibrium problem – Some Case Studies*. Proceedings of I. Mech. E. Seminar: Radial Loads and Axial Thrusts on Centrifugal Pumps, 5 Feb., London, pp. 39–51.

Dunham, J. (1965) *Non-axisymetric flows in Axial Compressors*, Monograph in Mechanical Engineering No. 3, Institution of Mechanical Engineers.

Dunham, J. (1970) A Review of Cascade Data on Secondary Losses in Turbines. *J. Mech. Eng. Sci.*, **12**(1), 48–59.

Eck, B. (1973) *Fans*, Pergamon Press, Oxford.

Eckert, B. and Schnell, H. (1961) *Axial und Radial Kompressoren*, 2nd edn, Springer Verlag, Berlin.

Emmons, H.W., Pearson, C.E. and Grant, H.P. (1955) Compressor Surge and Stall propagation. *Trans. ASME*, **77**, 455–469.

Ferguson, T.B. (1963) One Dimensional Incompressible Flow in a Vaneless Diffuser. *Engineer*, **215**, 562–565.

Ferguson, T.B. (1969) *Radial Vaneless Diffusers*. Proceedings of 3rd Conference on Fluid Mechanics in Turbomachines, 1969, Budapest.

Ferri, A. (1949) *Elements of the Aerodynamics of Supersonic Flows*, McMillan, New York.

Fielding, L. (1981) The effect of Irreversibility on the Capacity of a Turbine Blade Row. *Proc. I. Mech. E.*, **195**(6), 127–137.

Goulas, A. and Truscott, G.F. (1986) *Dynamic Hydraulic Loading on a Centrifugal Pump Impeller*. Proceedings of I. Mech. E. Seminar: Radial Loads and Axial Thrusts on Centrifugal Pumps, 5 Feb., London, pp. 53–64.

Gregory-Smith, D.G. (1982) Secondary Flows and Losses in Axial Flow Turbines. *Trans. ASME J. Eng. Power*, **104**, 819–822.

Grein, H. and Bachmann, K.M.J. (1976) Commissioning problems of a Large Pump-turbine. *Escher-Wyss News*, **49**(2), 15–22.

Grein, H. and Staehle, M. (1978) Fatigue cracking in stay vanes of large Francis Turbines. *Escher-Wyss News*, **51**(1), 33–37.

Grist, E. (1986) The volumetric Performance of Cavitating Centrifugal Pumps; Part 1 Theoretical Analysis and Method of Prediction; Part 2 Predicted and Measured Performance. *Proc. I. Mech. E.*, **200**, papers 58 and 59.

Guelich, J., Jud, W. and Hughes, S.F. (1986) *Review of Parameters influencing Hydraulic forces on Centrifugal Impellers*. I. Mech. E. Seminar: Radial Loads and Axial Thrusts on Centrifugal Pumps, 5 Feb., London, pp. 1–16.

Guy, H.L. (1939) Some Research on Steam Turbine Nozzle Efficiencies. *J. Inst. Civil Eng.*, **13**, 91.

Hart, M., Hall, D.M. and Singh, G. (1991) *Computational Methods for the Aerodynamic Development of Large Steam Turbines*. Proceedings of I. Mech. E. Conf. on Turbomachinery: Latest developments in a changing scene, pp. 11–24.

Hay, N., Metcalfe, R. and Reizes, J.A. (1978) A Simple method for the selection of Axial Fan Blade Profiles. *Proc. I. Mech. E.*, **192**(25), pp. 269–275.

Hayward, A.T.J. (1961) *Aeration in Hydraulic Systems – its assessment and control.* Proceedings of I. Mech. E. Conf.: Hydraulic Power Transmission and Control, Nov., London, paper 17.

Hennsler, H.D. and Bhinder, F.S. (1977) *The Influence of Scaling on the Performance of Small Centrifugal Compressors.* Proceedings of I. Mech. E. Conf.: Scaling for Performance Prediction in Rotodynamic Machines, 6–8 Sept., Stirling, pp. 113–122.

Hesketh, J.A. and Muscroft, J. (1990) Steam Turbine Generators for Sizewell 'B' Nuclear Power Station. *Proc. I. Mech. E.*, **204**, 183–191.

Hesselgreaves, J.E. and McEwen, D. (1976) *A Computer Aided Design Method for Axial Flow Pumps and Fans.* Proceedings of I. Mech. E. Conf.: Design of Pumps and Turbines, Sept., NEL, paper 3.4.

Hiett, G.F. and Johnston, I.H. (1964) Experiments concerning the Aerodynamic Performance of Inward Radial Flow Turbines. *Proc. I. Mech. E.*, **178**, part 3, 28–41.

Hill, J.M. and Lewis, R.I. (1974) Experimental Investigations of Strongly Swept Turbine Cascades with Low Speed Flow. *J. Mech. Eng. Sci.*, **16**(1), 32–40.

Horlock, J.H. (1958) *Axial Flow Compressors*, Butterworth, London.

Horlock, J.H. (1966) *Axial Flow Turbines*, Butterworth, London.

Howell, A.R. (1945) Fluid Dynamics of Axial Compressors. *Proc. I. Mech. E.*, **153**, 441–452.

Hughes, S.J., Salisbury, A.G. and Turton, R.K. (1988) *A Review of CAE techniques for Rotodynamic Pumps.* Proceedings of I. Mech. E. Conf.: Use of Cad/CAM for Fluid Machinery Design and Manufacture, 21 Jan., London, pp. 9–19.

Huppert, N.C. and Benser, W.A. (1953) Some Stall and Surge Phenomena in Axial Flow Compressors. *J. Aer. Sci.*, **120**, Dec., 835–845.

Hutton, S.P. and Furness, R.A. (1974) *Thermodynamic Scale effects in Cavitating Flows and Pumps.* Proceedings of I. Mech. E. Conf.: Cavitation, Sept., Edinburgh.

Jamieson, A.W.H. (1955) Gas Turbine Principles and Practice (ed. R. Cox), Newnes, London, Chap. 9.

Johnston, J.P. and Dean, R.C. (1966) Losses in Vaneless Diffusers of Centrifugal Compressors and Pumps. *Trans. ASME J. Eng. Power*, **88**, 49–62.

Kearton, W.J. (1958) *Steam Turbine Theory and Practice*, 7th edn, Pitman, London.

Klein, J. (1974) *Cavitation Problems on Kaplan Runners.* Proceedings of I. Mech. E. Conf.: Cavitation, Sept., Edinburgh, paper C180/74, pp. 303–308.

Knapp, R.T. and Daily, J.W. (1970) *Cavitation*, Eng. Soc. Monograph, McGraw-Hill, New York.

Knoerschild, E.M. (1961) The Radial Turbine for low Specific Speeds and low Velocity Factors. *Trans. ASME J. Eng. Power,* **83**, 1–8.

Kobayashi, K., Honjo, M., Tashiro, H. and Nagayama, T. (1991) *Verification of flow patterns for three dimensional designed blades.* Proceedings of I. Mech. E. Conf.: Turbomachinery – latest developments in a Changing scene, 19–20 March, London, pp. 25–32.

Konno, D. and Ohno, T. (1986) *Experimental Research on Axial Thrust Loads of Double Suction Centrifugal Pumps.* I. Mech. E. Seminar: Radial Loads and Axial Thrusts in Centrifugal Pumps, 5 Feb., London, pp. 65–77.

Kurokawa, J. and Toyokura, T. (1976) *Axial Thrust, Disc Friction Torque and Leakage Loss of Turbomachinery.* Proceedings of Int. Conf.: Design and Operation of Pumps and Turbines, Sept., NEL, paper 5.2.

Laali, A.R. (1991) *A New approach for Assessment of the Wetness Losses in Steam Turbines.* Proceedings of I. Mech. E. Conf.: Turbomachinery – latest developments in a changing scene, 19–20 March, London, pp. 155–166.

Lack, P.A. (1982) *Flow modulation with good Energy Saving on Centrifugal Fans with Inlet Guide Vanes.* Proceedings of Int. Conf.: Fan Design and Applications. British Hydromechanics Research Association (BHRA), Sept., London, paper K4.

Lakhwani, C. and Marsh, H. (1973) Rotating Stall in an isolated Rotor Row and a Single Stage Compressor. Proceedings of I. Mech. E. Conf.: Heat and Fluid Flow in Steam and Gas Turbine Plant. 3–5 April, Warwick, pp. 149–157.

Lakshminarayana, B. (1982) Fluid Dynamics of Inducers. *Trans. ASME J. Eng. Power,* **104**, 411–427.

Le Bot, Y. (1970) Analyse des Résultats expérimentaux obtenus en fonctionne-ment sain sur un compresseur exial subsonique de L'université de Cambridge prévision des régimes instationnaires. ONERA reports NT 8 1381 (E).

Lush, P.A. (1987a) Design for Minimum Cavitation. *Chartered Mech. Engineer*, Sept., 22–24.

Lush, P.A. (1987b) Materials for Minimum Cavitation. *Chartered Mech. Engineer*, Oct., 31–33.

Markov, N.H. (1958) *Calculation of the Aerodynamic Characteristics of Turbine Blades.* Translation from the Russian by Associated Technical Services, British Library (Lending).

Marscher, W.D. (1986) *The Effects of Fluid Forces at various operating conditions on the Vibrations of Vertical Turbine Pumps.* I. Mech. E. Seminar: Radial Loads and Axial Thrusts on Centrifugal Pumps, 5 Feb., London, pp. 17–38.

Marsh, H. (1966) *A Digital Computer Programme for the throughflow Fluid*

*Mechanics in an abitrary Turbomachine using a Matrix Method*, report R282, National Gas Turbine Establishment.

Matthias, H.B. (1966) *The Design of Pump Suction Bends*. IAHR/VDI Symposium: Pumps in Power Stations. Sept., Braunschweig, paper E3, pp. E21–E30.

McCutcheon, A.R.S. (1978) Aerodynamic Design and Development of a High Pressure Ratio Turbocharger. Proceedings of I. Mech. E. Conf.: Turbocharging, London, paper C73/78.

McKenzie, A.B. (1980) The Design of Axial Compressor Blading based on tests of Low Speed Compressors. *Proc. I. Mech. E.*, **194**, 103–112.

Meier, W. (1966) Separate Hydraulic Machines or Reversible Pump-Turbines for Storage Plants. *Escher-Wyss News*, **39**(3), 31–37.

Meier, W., Muller, J., Grew, H. and Jacques, M. (1971) Pump-Turbines and Storage Pumps. *Escher-Wyss News*, **44**(2), 3–23.

Merry, H. (1976) *Effect of twophase Liquid/Gas flow, on the performance of Centrifugal Pumps*. Proceedings of I. Mech. E. Conf.: Pumps and Compressors for Offshore Oil and Gas, 29 June–1 July, Aberdeen, pp. 61–68.

Milne, A.J. (1986) *A Comparison of Pressure Distribution and Radial Loads on Centrifugal Pumps*. I. Mech. E. Seminar: Radial Thrusts and Axial Loads on Centrifugal Pumps, 5 Feb., London, pp. 73–87.

Miller, D.C. (1977) *The Performance prediction of Scaled Axial Compressors from compressor tests*. Proceedings of I. Mech. E. Conf.: Scaling for Performance Prediction in Rotodynamic Machines, 6–8 Sept., Stirling, pp. 69–78.

Muhlemann, E.H. (1972) Arrangements of Hydraulic Machines for Pumped Storage and comparison of cost, efficiency and starting time. *Escher-Wyss News*, **45**(1), 3–11.

Myles, D.J. and Watson, J.T.R. (1964) *The Design of Axial Flow Fans by Computer part 1: basic Frame Sizes*. Report 145, National Engineering Laboratory.

Myles, D.J., Bain, R.W. and Buxton, G.H.L. (1965) *The Design of Axial Flow Fans by Computer part 2*. Report 181, National Engineering Laboratory.

Myles, D.J. (1969/70) An Analysis of Impeller and Volute Losses in Centrifugal Fans. *Proc. I. Mech. E.*, **184**, 253–278.

Necce, R.E. and Daily, J.M. (1960) Roughness Effects on Frictional Resistance of enclosed Rotating Discs. *Trans. ASME J. Basic Eng.*, **82**, 553–562.

Nechleba, M. (1957) *Hydraulic Turbines*, Artia, Prague.

Nixon, R.A. (1965) *Examination of the problem of Pump Scale Laws*. Symposium on Pump Design and Testing, April, NEL, p. D2.1.

Nixon, R.A. and Cairney, W.D. (1972a) *Scale Effects in Centrifugal Cooling Water Pumps for Thermal Power Stations*. Report 505, National Engineering Laboratory.

Nixon, R.A. and Otway, F.O.J. (1972b) *The use of Models in Determining the performance of Large Circulating Water Pumps.* Proceedings of I. Mech. E. Conf.: Site Testing of Pumps, Oct., London, pp. 113–132.

Odrowaz-Pieniazek, S. (1979) Solids-handling Pumps – a Guide to Selection. *Chem. Eng.*, Feb., 94–101.

Oschner, K. (1988) Stand der Technik bei Inducern fur Kreiselpumpen. *Chemic Technik*, **17**, 174–180.

Osterwalder, J. and Ettig, C. (1977) Determination of Individual Losses and Scale Effects by Model Tests with a Radial Pump. Proceedings of I. Mech. E. Conf.: Scaling for Performance Prediction in Rotodynamic Machines, 6–8 Sept., Stirling, pp. 105–112.

Osterwalder, J. (1978) Efficiency Scale-up for Hydraulic Roughness with due consideration for Surface Roughness. *IAHR J. Hyd. Res.*, **16**, 55–76.

Pampreen, R.C. (1973) Small Turbomachinery Compressor and Fan Aerodynamics. *Trans. ASME J. Eng. Power*, **95**, 251–256.

Parsons, N.C. (1972) The Development of Large Wet Steam Turbines. *Trans. North-East Coast Inst. Eng. Shipbuilding*, **89**, 31–42.

Pearsall, I.S. (1973) Design of Pump Impellers for Optimum Cavitation Performance. *Proc. I. Mech. E.*, **187**, 667–678.

Pearsall, I.S. (1978) *New Developments in Hydraulic Design Techniques.* Proceedings of pumps conference: Inter-Flow 1978.

Perkins, H.J. (1970) *The Analysis of Steady Flow Through Turbomachines.* GEC. Power Engineering MEL. Report no. W/M3c, p. 1641.

Petrie, K. (1964/5) Development of a Small Single and Two Shaft Gas Turbine for Military Applications. *Proc. I. Mech. E.*, **179**, pp. 343–364.

Pfleiderer, C. (1961) *Die Kreiselpumpen*, Springer-Verlag, Berlin.

Pollard, D. (1973) *A Method for the Hydraulic Design of Pumps and Fans using Computer Techniques.* Proceedings of I. Mech. E. Conf.: Computer Aided Design of Pumps and Fans, 12–13 Sept., Newcastle, paper C242/73.

Rachmann, D. (1967) *Physical Characteristics of Self Priming Phenomena Centrifugal Pumps*, SP 911, British Hydromechanics Research Association.

Railly, J.W. (1961) Three Dimensional Design of Multistage Axial Flow Compressors. *J. Mech. Eng. Sci.*, **3**(3), 214–224.

Railly, J.W. and Howard, J.H.G. (1962) Velocity Profile Development in Axial Flow Compressors. *J. Mech. Eng. Sci.*, **4**(2), 166–176.

Riegels, F.W. (1961) *Aerofoil Sections, Result from wind Tunnel Investigations, Theoretical Foundations*, Butterworth, London.

Richardson, J. (1982) *Size, Shape and Similarity.* Proceedings of I. Mech. E. Conf.: Centrifugal Pumps – Hydraulic Design, 16 Nov., London, pp. 29–36.

Rogers, G.F.C. and Mayhew, Y.R. (1967) *Engineering Thermodynamics*, 2nd edn, Longman, London.

Rohlik, H.E. (1968) *Analytical Determination of Radial Inflow Turbine*

*Design Geometry for Maximum Efficiency.* TN D4384, National Aeronautical and Space Agency.

Ruden, P. (1944) *Investigation of Axial Flow Fans* (Translation) TN 1062, National Aeronautical and Space Agency.

Rutschi, K. (1961) The Effects of the Guide Apparatus on the Output and Efficiency of Centrifugal Pumps. *Schweiz. Bauz.*, **79**, 233–240.

Scrivener, C.T.J., Connolly, C.F., Cox, J.C. and Dailey, G.M. (1991) *Use of CFD in the Design of a Modern Multi-stage Aero-engine LP Turbine.* Proceedings of I. Mech. E. Conf.: Turbomachinery – latest Developments in a changing Scene, 19–20 March, pp. 1–10.

Shapiro, H.A. (1953) *The Dynamics and Thermodynamics of Compressible Fluid Flow*, Ronald Press, New York, Vols 1 and 2.

Sherstyuk, A.N. and Kosmin, V.M. (1966) Meridional Profiling of Vaneless Diffusers. *Thermal Engineering* 13(2), 64–69.

Sherstyuk, A.N. and Sekolov, A.I. (1969) The Effects of the slope of Vaneless Diffuser Walls on the Characteristics of a Mixed Flow Compressor. *Thermal Engineering*, **16**(8), 116–121.

Smith, A. (1975) Experimental Development of Wet Steam Turbines, *Parsons Tech. Review*, Autumn, 1–11.

Smith, A.G. and Fletcher, P.G. (1954) *Observations on the Surging of various Low Speed Fans and Compressors*, memorandum M219, National Gas Turbine Establishment.

Soderberg, C.R. (1949) Unpublished notes, Gas Turbine Laboratory, MIT.

Stahl, H.A. and Stepannof, A.J. (1956) Thermodynamic Aspects of Cavitation in Centrifugal Pumps. *Trans. ASME*, **78**, 1691–1693.

Stahler, A.F. (1965) Slip Factor of a Radial Bladed Centrifugal Compressor. *Trans. ASME J. Eng. Power*, **87**, 181–192.

Stanitz, J.D. (1952) Some Theoretical Aerodynamic Investigations of Impellers in Radial and Mixed Flow Centrifugal Compressors. *Trans. ASME*, **74**, 743–797.

Stenning, A.M. (1953) *Design of Turbines for High Efficiency Low Power Output Applications*, DACL report 79, MIT.

Steppanof, A.J. (1957a) *Pumps and Blowers: Two Phase Flow*, John Wiley, New York.

Steppanof, A.J. (1957b) *Centrifugal and Axial Flow Pumps*, John Wiley, New York.

Stirling, T.E. (1982) *Analysis of the Design of two pumps using NEL methods*. Proceedings of I. Mech. E. Conf.: Centrifugal Pumps – Hydraulic Design, 16 Nov., London, pp. 55–73.

Stirling, T.E. and Wilson, G. (1983) *A Theoretically Based CAD method for Mixed Flow Pumps.* 8th Technical Conference British Pump Manufacturers Association , 29–31 March, Cambridge, pp. 185–204.

Stodola, A. (1945) *Steam and Gas Turbines*, 6th edn, Peter Smith, New York.

Susada, S., Kitamura, N. and Otaku, C. (1974) Experimental Studies on Centrifugal Pump with Inducer for Water Jetted Propulsion. Proceedings of I. Mech. E. Conf.: Cavitation, 3–5 Sept., Edinburgh, paper C165–C174.

Suter, P. and Traupel, W. (1959) *Untersuchungen uber den Ventilationsverlust von Turbinenradern. Mittielungen aus den Inst. fur Thermische Turbomaschinen* no. 4, Translation BSRA 917, British Library (Lending).

Sutton, M. (1968) *Pump Scale Laws as Affected by Individual Component Losses.* Proceedings of I. Mech. E. Conf.: Model Testing of Hydraulic Machinery and Associated Structures, 18–19 April, London, paper 10, pp. 76–82.

Tanuma, T. and Sakamoto, T. (1991) *The Removal of Water from Steam Turbine Stationary Blades by Suction Slots.* Proceedings of I. Mech. E. Conf.: Turbomachinery – Latest Developments in a Changing Scene, 19–20 March, paper C423/022, pp. 179–190.

Thorne, E.W. (1979) *Design by the Area Ratio Method.* Proceedings of 6th Tech. Conf. of BPMA Pumps 1979, 28–30 March, Canterbury, pp. 89–102.

Thorne, E.W. (1982) *Analysis of an End Suction Pump by the Area Ratio Method.* Proceedings of I. Mech. E. Conf.: Centrifugal Pumps – Hydraulic Design. 16 Nov., London, pp. 25–28.

Tillner, W., Fritsch, H., Kruft, R., and Lehmann, W., Louis, H., Masendorf, G. (1990) *The Avoidance of Cavitation Damage* (ed. W.J. Bartz) (in German), English translation Maxwell, A.J. (ed. Turton, R.K.) 1993, MEP, London.

Tomita, Y., Yamasaki, S. and Sasahara, J. (1973) The Scale Effect and Design Method of the Regenerative Pump with Non-Radial Vanes. *Bull. JSME*, **16**(98), August, pp. 1176–1183.

Turton, R.K. (1966) *The Effect of the Discharge Diffuser Arrangement on the Performance of a Centrifugal Pump.* Proceedings of International Association for Hydraulic Research/Verein Deutsche Ingen. (IAHR/VDI) Conf.: Pumps in Power Stations, Sept., paper G3.

Turton, R.K. and Goss, M. (1982) *The Fluctuating Radial and Axial Thrusts experienced by Centrifugal Pumps.* IAHR Symposium: Operating Problems of Power Stations and Power Plants, 13–17 Sept., Amsterdam, pp. 22.1–22.12.

Turton, R.K. and Goss, M. (1983) *A Study of the Fluctuating and Steady Forces on Conventional Centrifugal Pumps.* 8th BPMA technical Conf.: Pumps the Heart of the Matter, 29–31 March, Cambridge, paper 18.

Turton, R.K. (1984) The use of Inducers as a Way of achieving Low NPSH values for a Centrifugal Pump. *World Pumps*, March, 77–82.

Turton, R.K. (1994) *Rotodynamic Pump Design*, Cambridge University Press.

Varley, F.A. (1961) Effects of Impeller Design and Roughness on the

Performance of Centrifugal Pumps. *Proc. I. Mech. E.*, **175**, 955–989.

Wallace, F.J. (1958) Theoretical Assessment of the Performance Characteristics of Inward Radial Flow Turbines. *Proc. I. Mech. E.*, **172**, 931–950.

Wallis, R.A. (1961) *Axial Flow Fans*, Newnes, London.

Watabe, K. (1958) On Fluid Friction of Rotating Rough Disc in Rough Vessel. *JSME*, **1**(1), 69–74.

Weinig, F. (1935) *Die Stromung um die Schaufeln von Turbomaschinen*, Joh. Ambr. Barth, Leipzig.

Weisner, F.J. (1967) A Review of Slip Factors for Centrifugal Impellers. *Trans. ASME J. Eng. Power*, **894**(4), 558.

Whitfield, A. (1974) Slip Factor of a Centrifugal Compressor and its variation with Flow Rate. *Proc. I. Mech. E.*, **188**, 415–421.

Whitfield, A. and Sutton, A.J. (1989) The Effect of Vaneless Diffuser Geometry on the Surge Margin of Turbocharger Compressors. *Proc. I. Mech. E.*, **203**(D), 91–98.

Whitfield, A., Doyle, M.D. and Firth, M.R. (1993) Design and Performance of a High Pressure Ratio Turbocompressor. Parts 1 and 2. *Proc. I. Mech. E.*, **207**, 115–131.

Wilkinson, D.H. (1968) *A Numerical Solution of the Analysis and Design Problems for the Flow past one or more Aerofoils or Cascades*, R&M 3545, Aeronautical Research Council.

Willis, D.J. and Truscott, G.F. (1976) *A Survey of Solids Handling Pumps and Systems Part 2: Main Survey*, TN 1363, British Hydromechanics Research Association.

Wilson, G. (1963) *The Development of a Mixed Flow Pump with a Stable Characteristic*, Report 110, National Engineering Laboratory.

Wilson, D.G. (1987) New Guidelines for the preliminary design performance prediction of Axial Flow Turbines. *Proc. I. Mech. E.*, **201**(A4), 279–290.

Wilson, J.H. and Goulburn, J.R. (1976) *An Experimental Examination of Impeller passage Friction and Disc Friction Losses associated with High Speed Machinery*. Proceedings of Int. Conf.: Design and Operation of Pumps and Turbines. NEL, Sept., paper 5.1.

Wisclicenus, G. (1965) *Fluid Mechanics of Turbomachines*, Dover Press, New York, Vols 1 and 2.

Wolde, T.T., Moelke, W.H. and Mendte, K.W. (1974) *Experiences with Acoustical Methods for the Detection of Cavitation in Pumps*. Proceedings of I. Mech. E. Conf.: Cavitation, 3–5 Sept., Edinburgh, paper C187/74.

Wood, H.J. (1963) Current Technology of Radial Inflow Turbines for Compressible Fluids. *Trans. ASME J. Eng. Power*, **85**, 72–83.

Woods-Ballard, W.R. (1982) Comparative Energy Savings in Fan Systems. Proceedings of Int. Conf.: Fan Design and Applications, Sept., BHRA, paper K1.

Worster, R.C. (1963) The Flow in Volutes and its effect on Centrifugal Pump Performance. *Proc. I. Mech. E.*, **177**, 843–875.

Worster, R.C. and Copley, D.M. (1961) *Pressure Measurements at the Blade Tips of a Centrifugal Pump Impeller*, RR 7010, British Hydromechanics Research Association.

Wu, C.H. (1952) *A General Theory of Three Dimensional Flow in Subsonic and Supersonic Turbomachines of Axial Radial and Mixed Flow Types*, TN 2604, National Aeronautical and Space Agency.

Yamasaki, S. and Tomita, Y. (1971) Research on the Performance of the Regenerative Pump with non-radial Vanes. (1st report: performance and inner flow) *Bull. JSME*, **14**(77), 1178–1186.

Zweifel, O. (1945) *The Spacing of Turbomachine Blading especially with Large Angular Sections*, Brown Boveri Review, December, pp. 436–444.

# Bibliography


Dixon, S.L. (1975) *Fluid Mechanics, Thermodynamics of Turbomachinery* 2nd edn, Pergamon Press, Oxford.

Ferguson, T.B. (1963) *The Centrifugal Compressor Stage*, Butterworth, London.

Japiske, D. (1976) Review-progress in numerical turbomachinery analysis, *Trans. ASME J. Fluid Eng.*, **98**, 592–606.

Karassik, I.J. (1981) *Centrifugal Pump Clinic*, Dekker, New York.

Karassik, I.J., Krusch, W.C., Frazer, W.H. and Messina, J.P. (1976) *Pump Handbook*, McGraw-Hill, New York.

Lazarkiewicz, H. and Troskolanski, A.T. (1965) *Impeller Pumps*, Pergamon Press, Oxford.

Neumann, B. (1991) *The Interaction between Geometry and Performance of a Centrifugal Pump*, Mechanical Engineering Publications, London.

Pearsall, I.S. (1974) Cavitation. *Chartered Mechanical Engineer*, **21**, 79–85.

Sayers, A.T. (1990) *Hydraulic and Compressible Flow Turbomachines*, McGraw-Hill, London.

Shepherd, D.G. (1956) *Principles of Turbomachines*, MacMillan, New York.

Turton, R.K. (1993) *An Introductory Guide to Pumps and Pumping Systems*, Mechanical Engineering Publications, London.

Woods of Colchester (1952) *Woods Guide to Practical Fan Engineering*, Colchester.

Young, F.R. (1989) *Cavitation*, McGraw-Hill London.

# Index